Generalized Functions and Their Applications

Generalized Functions and Their Applications

Edited by

R. S. Pathak
Banaras Hindu University
Varanasi, India

PLENUM PRESS • NEW YORK AND LONDON

Library of Congress Cataloging-in-Publication Data

Generalized functions and their applications / edited by R.S. Pathak.
p. cm.
"Proceedings of an International Symposium on Generalized Functions and Their Applications, held December 23-26, 1991, in Varanasi, India"--CIP copr.
Includes bibliographical references and index.
ISBN 0-306-44404-6
1. Theory of distributions (Functional analysis)--Congresses.
I. Pathak, R. S.
QA324.G45 1993
515'.782--dc20 92-42065
CIP

Proceedings of an International Symposium on
Generalized Functions and Their Applications,
held December 23–26, 1991, in Varanasi, India

ISBN 0-306-44404-6

A Division of Plenum Publishing Corporation
233 Spring Street, New York, N.Y. 10013

Printed in the United States of America

PREFACE

The International Symposium on Generalized Functions and Their Applications was organized by the Department of Mathematics, Banaras Hindu University, and held December 23-26, 1991, on the occasion of the Platinum Jubilee Celebration of the university. More than a hundred mathematicians from ten countries participated in the deliberations of the symposium. Thirty lectures were delivered on a variety of topics within the area.

The contributions to the proceedings of the symposium are, with a few exceptions, expanded versions of the lectures delivered by the invited speakers. The survey papers by Komatsu and Hoskins and Sousa Pinto provide an up-to-date account of the theory of hyperfunctions, ultradistributions and microfunctions, and the nonstandard theory of new generalized functions, respectively; those by Stankovic and Kanwal deal with structures and asymptotics. Choquet-Bruhat's work studies generalized functions on manifold and gives applications to shocks and discrete models. The other contributions relate to contemporary problems and achievements in theory and applications, especially in the theory of partial differential equations, differential geometry, mechanics, mathematical physics, and systems science.

The proceedings give a very clear impression of the present state of the art in this field and contain many challenges, ideas, and open problems. The volume is very helpful for a broad spectrum of readers: graduate students to mathematical researchers.

The patron of the symposium was Professor C.S. Jha (Vice-Chancellor, Banaras Hindu University), under whose guidance it was successfully organized by a committee consisting of K. Lal, S.N. Lal, V.V. Menon, R.N. Pandey, R.R. Sharma, and R.S. Pathak (Chairman). On behalf of the committee I thank all the mathematicians who took part in the symposium, some of whom had to travel long distances to come to Varanasi, braving many hardships during the journey.

The committee thanks the university administration, their colleagues, the office staff, and the research students of the Department of Mathematics and the Department of Applied Mathematics for their help and co-operation in organizing the symposium.

The committee wishes to record its appreciation of the following funding agencies for their generous financial support: Council of Scientific and Industrial Research (New Delhi), Indian National Science Academy (New Delhi), National Board for Higher Mathematics (Department of Atomic Energy, Government of India), the University Grants commission (Banaras Hindu University), and the Faculty of Science, Institute of Technology (Banaras Hindu University).

I am personally grateful to the contributors to the proceedings, many of whom undertook to provide their articles in camera-ready form. Plenum Publishing Corporation deserves special thanks for their continued co-operation in the process of publishing the proceedings.

I dedicate this volume with love and gratitude to my alma mater, the Banaras Hindu University, on the occasion of its Platinum Jubilee.

Varanasi, September 1992 R.S. Pathak

CONTENTS

REPRODUCING KERNELS ON THE UNIT SPHERE

F. Brackx, R. Delanghe, F. Sommen [1], N. Van Acker

University of Ghent
Department of Mathematical Analysis
Galglaan 2
B-9000 Ghent/Belgium

ABSTRACT

Reproducing kernels are constructed for Hilbertmodules of those Cliffordalgebra-valued nullsolutions of certain differential operators in the unit ball of the Euclidean space $\mathbf{R}^m$, which moreover satisfy specific boundary conditions on the unit sphere. These nullsolutions are complexified to the Lie ball in $\mathbf{C}^m$, thus forming a closed submodule of the Hilbertmodule L_2^+ which consists of holomorphic functions in the Lie ball with L_2-boundary behaviour on the Lie sphere. Projecting the so-called Cauchy-Hua kernel of L_2^+ on this closed submodule, followed by restricting the reproducing property to the Euclidean unit ball, leads to the reproducing kernel of the initial Hilbertmodule.

1 CLIFFORD ALGEBRA AND CLIFFORD ANALYSIS

Generalizing in a rather natural way the notion of holomorphicity in the complex plane leads to the theory of the so-called monogenic functions defined in the Euclidean space $\mathbf{R}^m$ and taking values in the Clifford-algebra $\mathbf{R}_m$; this theory is nowadays known as Clifford-analysis.

The associative, but non-commutative, Clifford algebra $\mathbf{R}_m$ is the ring extension of the real number field $\mathbf{R}^m$ with symbols

$$(e_1, \ldots, e_m)$$

satisfying the multiplication rules

$$e_j^2 = -1, \quad j = 1, \ldots, m$$

and

$$e_j e_k + e_k e_j = 0, \quad j \neq k, \quad j, k = 1, \ldots, m.$$

[1]Research Associate, NFWO, Belgium

Generalized Functions and Their Applications, Edited by
R.S. Pathak, Plenum Press, New York, 1993

The basis of the Clifford algebra $\mathbf{R}_m$ is explicitly given by

$$(e_A = e_{h_1} \cdots e_{h_r},\ 1 \leq h_1 < \ldots < h_r \leq m)$$

where

$$A = \{h_1, \ldots, h_r\} \subseteq \{1, \ldots, m\} = M,$$

and where $e_\emptyset = 1$ denotes the identity. A general clifford number thus has the form $a = \sum_{A \subseteq M} e_A a_A$. A linear combination of the form $\underline{x} = \sum_{i=1}^{m} e_i x_i, x_i \in \mathbf{R}$, is called a vector; a real-linear combination of the $e_i e_j,\quad i \neq j$ is called a bivector, etc; $e_1 \cdots e_m$ is called the pseudoscalar.

The geometric product of two vectors $\underline{x}$ and $\underline{y}$ splits up into a scalar part and a bivector part: $\underline{x}\,\underline{y} = \underline{x} \cdot \underline{y} + \underline{x} \wedge \underline{y}$, where $\underline{x} \cdot \underline{y}$ is the commutative inner product and $\underline{x} \wedge \underline{y}$ is the anti-commutative bivector-valued outer product. The anti-involution on the Clifford-algebra is determined by its restriction to the vectors: $\bar{1} = 1,\quad \bar{e}_j = -e_j,\quad j = 1, \ldots, m$. Consequently for a vector $\underline{x}$ we have that $\bar{\underline{x}} = -\underline{x}$ and $\underline{x}\,\bar{\underline{x}} = \bar{\underline{x}}\,\underline{x} = |\underline{x}|^2$. For an arbitrary clifford number a the product $a\bar{a}$ is not necessarily scalar-valued. The real part of this product is defined as (the square of) the euclidean norm and denoted by $|a|^2$. For a detailed study of Clifford-algebra, we refer to [4].

The functions under consideration are defined in $\mathbf{R}^m$ and take values in the Clifford-algebra $\mathbf{R}_m$. The Dirac-operator or vector-derivative is given by $\partial_{\underline{x}} = \sum_{j=1}^{m} e_j \partial_{x_j}$. The Dirac-operator factorizes the m-dimensional laplacian: $\partial_{\underline{x}}{}^2 = -\Delta_{\underline{x}}$. A function $f : \Omega \to \mathbf{R}_m$, Ω open in $\mathbf{R}^m$, is said to be (left) monogenic in Ω if it is in $C_1(\Omega)$ and satisfies $\partial_{\underline{x}} f = 0$ in Ω. A polynomial which is homogeneous of degree k and monogenic is called an inner monogenic of degree k. The module of these inner monogenics of degree k is denoted by $\mathcal{M}_{+,k}$. For a detailed study of this higher dimensional function theory we refer to [1] and [2].

2 LIE BALL AND LIE SPHERE

The Lie-ball in $\mathbf{C}^m$ (see [7]) is given by

$$LB(0,1) = \{\underline{z} \in \mathbf{C}^m : \quad L(\underline{z}) < 1\}$$

where $L(\underline{z})$ is the so-called Lie-norm of $\underline{z} = (z_1, \ldots, z_m)$, given by:

$$L(\underline{z})^2 = |\underline{z}|^2 + \sqrt{|\underline{z}|^4 - |\underline{z}^2|^2}.$$

It is clear that the Lie ball is a subset of the complex unit ball in $\mathbf{C}^m$. The boundary of the Lie ball ($L(\underline{z}) = 1$) has those points $\underline{z} = (\underline{x}, \underline{y})$ in common with the unit sphere in $\mathbf{C}^m$, for which $\underline{x}$ and $\underline{y}$ are linearly dependent. This special part of the complete boundary $\partial LB(0,1)$, which is also a part of the complex unit sphere, is the Lie sphere. This compact, real-analytic manifold can be represented as follows (see [8] and [5]):

$$LS^{m-1} = \{e^{i\theta}\underline{\omega} : \quad \theta \in \mathbf{R}, \quad \underline{\omega} \in S^{m-1}\}.$$

Moreover it can be proved that LS^{m-1} is the Shilov-boundary of the Lie ball, i.e. each function which is holomorphic in the Lie ball and continuous up to the boundary attains in modulus its maximum on the Lie sphere. The importance of the Lie ball is illustrated by the theorem of Siciak (see [9]) which states that:

(i) If a series of homogeneous polynomials $\sum_k R_k(\underline{x})$ converges normally in $\mathring{B}_m(1)$, then its complexification converges normally in $LB(0,1)$.

(ii) The Lie ball is the largest domain for which the previous result (i) holds in general, since the Lie ball is a harmonic hull for the unit ball $\mathring{B}_m(1)$.

3 HILBERTMODULES ON THE LIE SPHERE

We consider the Hilbertmodule $L_2(LS^{m-1})$ of clifford-valued L_2-functions on LS^{m-1}, provided with the inner product :

$$(f,g) = \frac{1}{2\pi A_{m-1}} \int_0^{2\pi} \int_{S^{m-1}} \overline{f(e^{i\theta}\underline{\omega})} g(e^{i\theta}\underline{\omega})\, d\theta\, dS(\underline{\omega}).$$

By $L_2^+(LS^{m-1})$ we denote the space of holomorphic functions f in $LB(1)$ for which the radial boundary value $f(e^{i\theta}\underline{\omega})$ belongs to $L_2(LS^{m-1})$. In [5] and [8] it is proved that $L_2^+(LS^{m-1})$ provided with the above inner product is a Hilbertmodule with reproducing kernel, the so-called Cauchy-Hua kernel:

$$K_+(\underline{z}, e^{i\theta}\underline{\omega}) = \sum_{k,l=0}^{\infty} (e^{i\theta}\underline{\omega})^l \overline{C_{m,k}(\underline{z}, e^{-i\theta}\underline{\omega})}\underline{z}^l = \frac{1}{(-(\underline{\omega} - e^{-i\theta}\underline{z})^2)^{m/2}}$$

where $C_{m,k}(\underline{z}, e^{i\theta}\underline{\omega})$ is the complex extension of

$$C_{m,k}(e^{i\psi}\underline{\omega}', e^{i\theta}\underline{\omega}) = e^{ik(\psi+\theta)}[C_k^{m/2}(t) + \underline{\omega}'\,\underline{\omega} C_{k-1}^{m/2}(t)],$$

$C_\mu^\lambda(t)$ denotes a Gegenbauer-polynomial in $t = \langle \underline{\omega}, \underline{\omega}' \rangle$. This decomposition of the Cauchy-Hua kernel provides an example of a decomposition into so-called spherical monogenics of degree (k,l) on LS^{m-1}. By a spherical monogenic of degree (k,l) on LS^{m-1} is meant a function of the form

$$(e^{i\theta}\underline{\omega})^l P_k(e^{i\theta}\underline{\omega}), \quad l \in \mathbf{Z}, \quad P_k \in \mathcal{M}_{+,k}.$$

They are common eigenfunctions of the operators $\Gamma = -\sum_{i<j} e_{ij}(z_i\partial_{z_j} - z_j\partial_{z_i})$ and $E = -i\partial_\theta$.

4 CONSTRUCTION OF REPRODUCING KERNELS ON S^{m-1}: THE METHOD

Consider the clifford differential operator $P(\underline{x}, \partial_{\underline{x}})$ with polynomial coefficients and the module $\mathcal{M}_P$ satisfying the following conditions:

(i) each function in $\mathcal{M}_P$ is a nullsolution of P in the unit ball $\mathring{B}_m(1)$

(ii) each function in $\mathcal{M}_P$ has radial L_2-boundary values on S^{m-1}

(iii) $\mathcal{M}_P$ is a Hilbertmodule with reproducing kernel for the inner product

$$(f,g) = \frac{1}{A_{m-1}} \int_{S^{m-1}} \overline{f(\underline{\omega})} g(\underline{\omega})\, dS(\underline{\omega})$$

where in the integrand the L_2-boundary values of f and g appear

(iv) each function in $\mathcal{M}_P$ satisfies an auxiliary condition—to be specified later on—which is related to the nature of the differential operator P.

The aim of the construction is to obtain the reproducing kernel $k(\underline{x}, \underline{\omega}) \in \mathcal{M}_P$ for which

$$f(\underline{x}) = (k(\underline{x}, \underline{\omega}), f(\underline{\omega})) = \frac{1}{A_{m-1}} \int_{S^{m-1}} \overline{k(\underline{x}, \underline{\omega})} f(\underline{\omega})\, dS(\underline{\omega})$$

for all $f \in \mathcal{M}_P$. The consecutive steps in this construction procedure are now described in detail; the above mentioned condition (iv) will follow from this discussion.

Step 1 Take a function f in $\mathcal{M}_P$; then it is at least continuous in $\mathring{B}_m(1)$ and by restricting it to spheres with radius $r, r < 1$, it can be decomposed into the following $L_2(\mathring{B}_m(1))$-convergent series (see [6]):

$$f(r\underline{\omega}) = \sum_{k=0}^{\infty} [P_k f(r,\underline{\omega}) + \underline{\omega} P'_k f(r,\underline{\omega})]$$

where $P_k f$ and $P'_k f$ are inner spherical monogenics of degree k, still depending upon r. In the module of inner spherical monogenics these $P_k f$,$P'_k f$ may be expanded with respect to the basis $\{P_k^{(j)}, j = 1, \ldots, K(m,k)\}$:

$$P_k f(r,\underline{\omega}) = \sum_j P_k^{(j)}(\underline{\omega}) A_k^{(j)} f(r) \quad \text{and} \quad \underline{\omega} P'_k f(r,\underline{\omega}) = \underline{\omega} \sum_j P_k^{(j)}(\underline{\omega}) B_k^{(j)} f(r)$$

with clifford C_∞-functions $A_k^{(j)} f(r)$ and $B_k^{(j)} f(r)$. This leads to

$$f(r\underline{\omega}) = \sum_{k,j} \sum_{A \subseteq M} [A_{k,A}^{(j)} f(r) + \underline{\omega} B_{k,A}^{(j)} f(r)] P_k^{(j)}(\underline{\omega}) e_A,$$

with scalar C_∞-functions $A_{k,A}^{(j)} f(r)$ and $B_{k,A}^{(j)} f(r)$, the series still being $L_2(\mathring{B}_m(1))$-convergent. In general this expansion cannot be rewritten as

$$f(r\underline{\omega}) = \sum_{k=0}^{\infty} [\alpha_k(r^2) P_k(\underline{x}) + \underline{x} \beta_k(r^2) \tilde{P}_k(\underline{x})] \tag{1}$$

which moreover should be normally convergent in $\mathring{B}_m(1)$. This will only be the case for specific differential operators $P(\underline{x}, \partial_{\underline{x}})$ which we term as "of Frobenius type" and are defined as follows:

Definition 4.1 *A differential operator $P(\underline{x}, \partial_{\underline{x}})$ is said to be of Frobenius type iff each nullsolution f in $\mathring{B}_m(1)$ can be expanded into a normally convergent series in $\mathring{B}_m(1)$ of the form* (1) *where moreover, for each $k \in \mathbf{N}$, the functions α_k and β_k have normally convergent power series expansions in $r \in]-1,1[$ of the form*

$$\alpha_k(r^2) = \sum_{s=0}^{\infty} a_s r^{2s}, \quad \beta_k(r^2) = \sum_{s=0}^{\infty} b_s r^{2s}.$$

Let us give an example of such a Frobenius type differential operator. Consider the operator $\partial_{\underline{x}} - \lambda$, $\lambda \in \mathbf{C}_0$. Its nullsolutions in $\mathring{B}_m(1)$ can be expanded as follows (see [3]):

$$f(r\underline{\omega}) = (\lambda r)^{1-m/2} \sum_{k=0}^{\infty} [J_{k+m/2-1}(\lambda r) - \underline{\omega} J_{k+m/2}(\lambda r)] P_k(\underline{\omega})$$

where $J_\nu(z)$ denotes the Besselfunction of the first kind of order ν, and where this series is normally convergent in $\mathring{B}_m(1)$. It follows that here

$$\begin{aligned} r^k \alpha_k(r^2) &= (\lambda r)^{1-m/2} J_{k+m/2-1}(\lambda r) \\ r^{k+1} \beta_k(r^2) &= -(\lambda r)^{1-m/2} J_{k+m/2}(\lambda r); \end{aligned}$$

taking into account the series decomposition for Besselfunctions, this indeed proves that the operator $(\partial_{\underline{x}} - \lambda)$ is a differential operator of the Frobenius type.

Step 2 Nullsolutions of differential operators of the Frobenius type admit a complexification to the Lie ball, which is done quite easily. It indeed suffices to put $r^2 = z, z \in \mathbf{C}$, to make both functions $\alpha_k(z)$ and $\beta_k(z)$ holomorphic in the unit disk $\{|z| < 1\}$ in the complex plane. Hence:

Lemma 4.1 *If $P(\underline{x}, \partial_{\underline{x}})$ is a differential operator of the Frobenius type, then the functions $\alpha_k(\sum_{j=1}^m x_j^2)$ and $\beta_k(\sum_{j=1}^m x_j^2)$, considered as functions of $\underline{x} \in \mathbf{R}^m$, can be extended holomorphically to the Lie ball.*

Proof.
As we already remarked the Lie ball is a subset of the complex unit ball, the last one being a subset of $\{\underline{z} \in \mathbf{C}^m : |\underline{z}^2| < 1\}$. □

The complexification of the nullsolution

$$f(\underline{x}) = \sum_{k=0}^{\infty} [\alpha_k(r^2) P_k(\underline{x}) + \underline{x}\beta_k(r^2)\tilde{P}_k(\underline{x})]$$

in $\mathring{B}_m(1)$ of the Frobenius type differential operator $P(\underline{x}, \partial_{\underline{x}})$ is now given by

$$f(\underline{z}) = \sum_{k=0}^{\infty} [\alpha_k(\underline{z}^2) P_k(\underline{z}) + \underline{z}\beta_k(\underline{z}^2)\tilde{P}_k(\underline{z})].$$

We want this complexification to be in $L_2^+(LS^{m-1})$, i.e. holomorphic in the Lie ball with L_2-boundary values on LS^{m-1}. To guarantee this, we need some supplementary conditions.

Theorem 4.1 *Let $P(\underline{x}, \partial_{\underline{x}})$ be a differential operator of the Frobenius type the nullsolutions in $\mathring{B}_m(1)$ of which are given by the normally convergent series*

$$f(\underline{x}) = \sum_{k=0}^{\infty} [\alpha_k(r^2) P_k(\underline{x}) + \underline{x}\beta_k(r^2)\tilde{P}_k(\underline{x})].$$

If

(i)

$$\sup_{|z|\leq 1} |\alpha_k(z)| = c_1^* \quad \textit{and} \quad \sup_{|z|\leq 1} |\beta_k(z)| = c_2^* \quad \textit{for each } k \in \mathbf{N}$$

(ii)

$$\sum_{k=0}^{\infty} ||P_k||^2_{L_2(S^{m-1})} < \infty \quad \textit{and} \quad \sum_{k=0}^{\infty} ||\tilde{P}_k||^2_{L_2(S^{m-1})} < \infty$$

then the complexified series of $f(\underline{x})$ will be in $L_2^+(LS^{m-1})$.

Proof.
It is clear that $f(\underline{x})$ in $\mathring{B}_m(1)$ can be rewritten in the form of a double-series, which is still normally convergent in $\mathring{B}_m(1)$:

$$f(\underline{x}) = \sum_{k=0}^{\infty}\sum_{s=0}^{\infty} (a_s r^{2s} P_k(\underline{x}) + \underline{x} b_s r^{2s} \tilde{P}_k(\underline{x})).$$

Using the theorem of Siciak, it follows that

$$f(\underline{z}) = \sum_{k=0}^{\infty}\sum_{s=0}^{\infty} (a_s \underline{z}^{2s} P_k(\underline{z}) + \underline{z} b_s \underline{z}^{2s} \tilde{P}_k(\underline{z}))$$

converges normally in $LB(1)$. Hence $f(\underline{z})$ represents a holomorphic function in the Lie ball. The L_2-boundary behaviour on LS^{m-1} of $f(\underline{z})$ is equivalent with

$$\lim_{r<1, r\to 1} \sum_{k=0}^{\infty} \int_0^{2\pi} [|\alpha_k(r^2 e^{2i\theta})|^2 ||P_k||_2^2 + |\beta_k(r^2 e^{2i\theta})|^2 ||\tilde{P}_k||_2^2] \, d\theta < \infty.$$

The left hand side is smaller than

$$2\pi \sum_{k=0}^{\infty} (c_1^* ||P_k||_2^2 + c_2^* ||\tilde{P}_k||_2^2)$$

which clearly converges in view of the hypothesis made. □

It is interesting noticing that if a nullsolution $f(\underline{x})$ of the Frobenius type differential operator $P(\underline{x}, \partial_{\underline{x}})$ satisfies the conditions of theorem 4.1, then automatically $f(\underline{x})$ shows L_2-boundary behaviour on S^{m-1}. Indeed, a necessary and sufficient condition thereto is

$$\lim_{r<1, r\to 1} \sum_{k=0}^{\infty} [|\alpha_k(r^2)|^2 ||P_k||_2^2 + |\beta_k(r^2)|^2 ||\tilde{P}_k||_2^2] < \infty$$

which clearly is fulfilled. At this stage of the construction process we are able to redefine the initial Hilbertmodule $\mathcal{M}_P$ more precisely:

(M1) $\mathcal{M}_P \subset ker P|_{\mathring{B}_m(1)}$, P being of Frobenius type

(M2) each function in $\mathcal{M}_P$ satisfies the conditions of theorem 4.1

(M3) $\mathcal{M}_P$ is a Hilbertmodule with reproducing kernel for the above mentioned inner product.

Step 3 Take $f \in \mathcal{M}_P$ then its complexification $f(\underline{z})$ obtained as explained in step 2 will belong to $L_2^+(LS^{m-1})$ and moreover will satisfy $P(\underline{z}, \partial_{\underline{z}})f = 0$ in $LB(1)$, $P(\underline{z}, \partial_{\underline{z}})$ denoting the complexification of the original operator $P(\underline{x}, \partial_{\underline{x}})$. If H denotes the module of the complexifications of all functions in $\mathcal{M}_P$, then it is clear that H is a submodule of $L_2^+(LS^{m-1})$. Hence the Hilbertmodule H will inherit a reproducing kernel K_H from $L_2^+(LS^{m-1})$ by projecting its reproducing Cauchy-Hua kernel K_+ on H:

$$K_H(\underline{z}, e^{i\theta}\underline{\omega}) = \Pi_{H, e^{i\theta}\underline{\omega}} K_+(\underline{z}, e^{i\theta}\underline{\omega}).$$

It should be noticed that it is sufficient to project K_+ only in the variable $e^{i\theta}\underline{\omega}$, since we also have for each $f \in H$:

$$\int_0^{\pi} \int_{S^{m-1}} \overline{[K_+(\underline{z}, e^{i\theta}\underline{\omega}) - \overline{\Pi_{H,\underline{z}} \overline{K_+(\underline{z}, e^{i\theta}\underline{\omega})}}]} f(e^{i\theta}\underline{\omega}) \, d\theta \, dS(\underline{\omega}) = 0$$

or still

$$K_H(\underline{z}, e^{i\theta}\underline{\omega}) = \overline{\Pi_{H,\underline{z}} \overline{K_+(\underline{z}, e^{i\theta}\underline{\omega})}}.$$

Step 4 The reproducing property obtained for $f(\underline{z})$ in H:

$$\begin{aligned} f(\underline{z}) &= (K_+(\underline{z}, e^{i\theta}\underline{\omega}), f(e^{i\theta}\underline{\omega}))_{L_2(LS^{m-1})} \\ &= \frac{1}{\pi A_{m-1}} \int_0^{\pi} d\theta \int_{S^{m-1}} \overline{K_+(\underline{z}, e^{i\theta}\underline{\omega})} f(e^{i\theta}\underline{\omega}) \, dS(\underline{\omega}) \end{aligned}$$

is now restricted to $\underline{x} \in \mathring{B}_m(1)$. By integrating w.r.t. the variable θ, the relation thus obtained can be transformed into a formula of the desired form:

$$f(\underline{x}) = \frac{1}{A_{m-1}} \int_{S^{m-1}} \overline{k(\underline{x}, \underline{\omega})} f(\underline{\omega}) \, dS(\underline{\omega}).$$

5 CONSTRUCTION OF REPRODUCING KERNELS ON S^{m-1}: EXAMPLES

(i) The operator $\partial_{\underline{x}} - \lambda$, $\lambda \in \mathbf{C}_0$

We already showed that the operator $\partial_{\underline{x}} - \lambda$ is of Frobenius type. Now we consider the module of nullsolutions of $(\partial_{\underline{x}} - \lambda)$ in $\mathring{B}_m(1)$ which show L_2-boundary behaviour on S^{m-1}:

$$H_2^\lambda(\mathring{B}_m(1)) = \left\{f : \quad (\partial_{\underline{x}} - \lambda)f = 0 \text{ in } \mathring{B}_m(1) \text{ and } f(\underline{\omega}) \in L_2(S^{m-1})\right\}$$

provided with the inner product

$$(f,g) = \frac{1}{A_{m-1}} \int_{S^{m-1}} \overline{f(\underline{\omega})} g(\underline{\omega})\, dS(\underline{\omega}).$$

It is known that $H_2^\lambda(\mathring{B}_m(1))$ is a Hilbertmodule with reproducing kernel(see [3]). In order to make $H_2^\lambda(\mathring{B}_m(1))$ an $\mathcal{M}_P$-module with $P = \partial_{\underline{x}} - \lambda$, condition (M2) still has to be proved. As the eigenfunctions of the Dirac-operator can be written as

$$f(r\underline{\omega}) = (\lambda r)^{1-m/2} \sum_{k=0}^{\infty} [J_{k+m/2-1}(\lambda r) - \underline{\omega} J_{k+m/2}(\lambda r)] P_k(\underline{\omega}),$$

this series being normally convergent in $\mathring{B}_m(1)$, the assumed $L_2(S^{m-1})$-boundary behaviour explicitly means that

$$\sum_{k=0}^{\infty} [|J_{k+m/2-1}(\lambda)|^2 + |J_{k+m/2}(\lambda)|^2] ||P_k||^2_{L_2(S^{m-1})} < \infty.$$

As far as the first condition of theorem 4.1 is concerned, we have that

$$|\alpha_k(r^2)| = |\lambda|^k \frac{|J_{k+m/2-1}(\lambda r)|}{|\lambda r|^{k+m/2-1}}.$$

Using the following estimate (see [10]): for $\mu \in \mathbf{R}$, $\mu > -1/2$, and $z = x + iy \in \mathbf{C}$ is

$$|J_\mu(z)| \leq |\tfrac{1}{2} z|^\mu \frac{e^{|y|}}{\Gamma(\mu+1)},$$

it is found that

$$|\alpha_k(r^2)| \leq \frac{e^{|r \operatorname{Im} \lambda|} |\lambda|^k}{2^{k+m/2-1} \Gamma(k+m/2)}$$

where the right hand side tends to zero as $k \to \infty$ since $\Gamma(k+m/2) \approx k!$ for large k. The function $\beta_k(r^2)$ can be treated in a similar way. To verify the second condition of theorem 4.1 we now can use the asymptotic behaviour of the Besselfunctions (see [12]):

$$|J_\mu(x+iy)| \sim \frac{1}{\sqrt{2\pi}} \frac{e^{|y|}}{(x^2+y^2)^{1/4}}$$

for $\lambda \in \mathbf{C}$ with sufficiently large $\operatorname{Im} \lambda$, which shows that for these values of λ, the Besselfunctions satisfy $|J_\mu(\lambda)| > 1$. Hence it follows that

$$\sum_{k=0}^{\infty} ||P_k||^2_{L_2(S^{m-1})} \leq \sum_{k=0}^{\infty} [|J_{k+m/2-1}(\lambda)|^2 + |J_{k+m/2}(\lambda)|^2] ||P_k||^2_{L_2(S^{m-1})}$$

this last series being convergent due to the imposed $L_2(S^{m-1})$- boundary behaviour. According to 'step 4' of the general procedure we now get

$$f(r\underline{\omega}') = \frac{1}{\pi A_{m-1}} \int_0^\pi d\theta \int_{S^{m-1}} \overline{K_+(r\underline{\omega}', e^{i\theta}\underline{\omega})} f(e^{i\theta}\underline{\omega})\, dS(\underline{\omega}).$$

Making use of the expansion of f and the orthogonality on S^{m-1} of the inner and outer spherical monogenics, we end up with

$$\begin{aligned} f(r\underline{\omega}') &= \frac{1}{2\pi A_{m-1}} \sum_{k,l=0}^{\infty} r^{k+2l} \int_0^{2\pi} e^{-i(k+2l)} (\lambda e^{i\theta})^{1-m/2} J_{k+m/2-1}(\lambda e^{i\theta})\, d\theta \\ &\quad \int_{S^{m-1}} (C_k^{m/2} + \underline{\omega}'\,\underline{\omega} C_{k-1}^{m/2}) P_k\, dS(\underline{\omega}) \\ &\quad - \frac{1}{2\pi A_{m-1}} \sum_{k,l=0}^{\infty} r^{k+2l+1} \underline{\omega}' \int_0^{2\pi} e^{-i(k+2l+1)} (\lambda e^{i\theta})^{1-m/2} J_{k+m/2}(\lambda e^{i\theta})\, d\theta \\ &\quad \int_{S^{m-1}} (C_k^{m/2} + \underline{\omega}'\,\underline{\omega} C_{k-1}^{m/2}) P_k\, dS(\underline{\omega}). \end{aligned}$$

The integrals w.r.t. θ can now be calculated by means of the classical series decomposition of the Besselfunctions, resulting in

$$f(r\underline{\omega}') = \frac{1}{A_{m-1}} \sum_{k=0}^{\infty} (\lambda r)^{1-m/2} [J_{k+m/2-1}(\lambda r) - \underline{\omega}' J_{k+m/2}(\lambda r)] \int_{S^{m-1}} (C_k^{m/2} + \underline{\omega}'\,\underline{\omega} C_{k-1}^{m/2}) P_k(\underline{\omega})\, dS(\underline{\omega}).$$

Hence, putting

$$K_\lambda(r\underline{\omega}', \underline{\omega}) = r^{1-m/2} \sum_{k=0}^{\infty} \frac{[J_{k+m/2-1}(\lambda) - \underline{\omega} J_{k+m/2}(\lambda)](C_k^{m/2} + \underline{\omega}\,\underline{\omega}' C_{k-1}^{m/2})[\overline{J_{k+m/2-1}(\lambda r)} + \underline{\omega}' \overline{J_{k+m/2}(\lambda r)}]}{|J_{k+m/2-1}(\lambda)|^2 + |J_{k+m/2}(\lambda)|^2}$$

we finally get

$$f(r\underline{\omega}') = \frac{1}{A_{m-1}} \int_{S^{m-1}} \overline{K_\lambda(r\underline{\omega}', \underline{\omega})} f(\underline{\omega})\, dS(\underline{\omega}).$$

The function K_λ belongs to H_2^λ, it is anti-symmetric and it has the reproducing property; thus it is the reproducing kernel of the Hilbertmodule H_2^λ. Notice that if we formally make λ tend to zero then we obtain the classical Szegö-kernel $K_0(r\underline{\omega}', \underline{\omega})$.

(ii) The operator $\Delta_{\underline{x}} + \lambda^2$, $\lambda \in \mathbf{C}_0$ with $J_{k+m/2-1}(\lambda) \neq 0$

The nullsolutions of this operator in $\mathring{B}_m(1)$ can be written as follows:

$$f(r\underline{\omega}) = (\lambda r)^{1-m/2} \sum_{k=0}^{\infty} J_{k+m/2-1}(\lambda r) S_k f(\underline{\omega}),$$

where $S_k f$ is a spherical harmonic function of degree k. As the module $\mathcal{H}_k$ of spherical harmonics of degree k has an orthogonal decomposition into inner and outer spherical monogenics (see [1]), we also have

$$f(r\underline{\omega}) = (\lambda r)^{1-m/2} \sum_{k=0}^{\infty} [J_{k+m/2-1}(\lambda r) P_k f(\underline{\omega}) + \underline{\omega} J_{k+m/2-1}(\lambda r) \tilde{P}_k f(\underline{\omega})]$$

with P_kf, $\tilde{P}_kf$ inner spherical monogenics of degree k. So in this case the functions α_k and β_k are given by

$$r^k\alpha_k(r^2) = r^{k+1}\beta_k(r^2) = (\lambda r)^{1-m/2}J_{k+m/2-1}(\lambda r).$$

Hence it is clear that the operator $(\Delta_{\underline{x}} + \lambda^2)$ is a differential operator of the Frobenius type. Now we consider the module of nullsolutions of $(\Delta_{\underline{x}} + \lambda^2)$ in $\mathring{B}_m(1)$ which also show L_2-boundary behaviour on S^{m-1}:

$$h_2^\lambda(\mathring{B}_m(1)) = \left\{f : \quad (\Delta_{\underline{x}} + \lambda^2)f = 0 \text{ in } \mathring{B}_m(1) \text{ and } f_-) \in L_2(S^{m-1})\right\}.$$

Provided with the same inner product as in the first example, h_2^λ is clearly a Hilbert-module. In order to make it an $\mathcal{M}_P$- module with $P = \Delta_{\underline{x}} + \lambda^2$, condition (M2) still has to be proved. From the first example we know that the first condition of theorem 4.1 is satisfied, while the second one is verified in a similar way:

$$\sum_{k=0}^{\infty}[||P_kf||^2_{L_2(S^{m-1})} + ||\tilde{P}_kf||^2_{L_2(S^{m-1})}] \leq \sum_{k=0}^{\infty}|J_{k+m/2-1}(\lambda)|^2[||P_kf||^2_{L_2(S^{m-1})} + ||\tilde{P}_kf||^2_{L_2(S^{m-1})}]$$

this last series being convergent if $\operatorname{Im}\lambda$ is sufficiently large; this is precisely the assumed L_2-boundary behaviour on S^{m-1}. For this operator it is more convenient to consider the decomposition of the Cauchy-Hua kernel into spherical harmonic functions $e^{ik\theta}S_{k-2l}(\underline{\omega})$ on the Lie sphere (see [8])(instead of its decomposition into spherical monogenics on LS^{m-1}):

$$\begin{aligned} K_+(r\underline{\omega}', \underline{\omega}) &= \sum_{n=0}^{\infty}\sum_{l=0}^{[n/2]} r^n\tilde{N}(m, n-2l)P_{n-2l}(m;t)e^{in\theta} \\ &= \sum_{n=0}^{\infty} r^n e^{in\theta}C_n^{m/2}(t) \end{aligned}$$

where $P_k(m;t)$ is the so-called m-dimensional Legendre polynomial of degree k in the variable $t = \langle\underline{\omega}, \underline{\omega}'\rangle$, and where $\tilde{N}(m,k)$ is the dimension of $\mathcal{H}_k$. Reproducing the function at $\underline{x} = r\underline{\omega}'$ then leads to

$$f(r\underline{\omega}') = \frac{1}{\pi A_{m-1}}\int_0^\pi\int_{S^{m-1}}\sum_{n=0}^{\infty}\sum_{l=0}^{[n/2]} r^n\tilde{N}(m, n-2l)P_{n-2l}(m;t)e^{-in\theta}f(e^{i\theta}\underline{\omega})\,d\theta\,dS(\underline{\omega}).$$

Taking into account the explicit decomposition for $f(e^{i\theta}\underline{\omega})$, and simplifying the result using the orthogonality on S^{m-1} of spherical harmonics, we obtain

$$\begin{aligned} f(r\underline{\omega}') = \frac{1}{2\pi A_{m-1}}\sum_{n=0}^{\infty}\sum_{l=0}^{[n/2]} r^n\tilde{N}(m, n-2l)\int_{S^{m-1}} P_{n-2l}(m;t)S_{n-2l}f(\underline{\omega})\,dS(\underline{\omega}) \\ \int_0^{2\pi}\lambda^{1-m/2}e^{i\theta(1-m/2-n)}J_{n-2l+m/2-1}(\lambda e^{i\theta})\,d\theta. \end{aligned}$$

Again the integrals w.r.t. the variable θ can be calculated. If we then put $n - 2l = k$ and alternate summations we obtain

$$f(r\underline{\omega}') = \frac{1}{A_{m-1}}\sum_{k=0}^{\infty}\tilde{N}(m,k)(\lambda r)^{1-m/2}J_{k+m/2-1}(\lambda r)\int_{S^{m-1}} P_k(m;t)S_kf(\underline{\omega})\,dS(\underline{\omega}).$$

The function

$$P_{\lambda^2}(r\underline{\omega}', \underline{\omega}) = r^{1-m/2}\sum_{k=0}^{\infty}\frac{J_{k+m/2-1}(\lambda)\tilde{N}(m,k)P_k(m;\langle\underline{\omega}, \underline{\omega}'\rangle)\overline{J_{k+m/2-1}(\lambda r)}}{|J_{k+m/2-1}(\lambda)|^2}$$

belongs to h_2^λ, it is anti-symmetric and it has the reproducing property. Thus it is the reproducing kernel of the Hilbertmodule $h_2^\lambda(\mathring{B}_m(1))$ with the restrictions $\lambda \in \mathbf{C}_0$, $J_{k+m/2-1}(\lambda) \neq 0$, $\mathrm{Im}\,\lambda$ sufficiently large. Notice that if we formally let λ tend to zero in P_{λ^2}, the classical Poisson-kernel is obtained, which is indeed the reproducing kernel of $h_2^\lambda(\mathring{B}_m(1))$ (see [11]).

REFERENCES

[1] F. Brackx, R. Delanghe and F. Sommen. "Clifford Analysis", Pitman, London (1982).

[2] R. Delanghe, F. Sommen and V. Souček. "Clifford Algebra and Spinor-valued functions: a function theory for the Dirac-operator", Kluwer Academic Publishers, Dordrecht (1992).

[3] Xu Zhenyuan. "Boundary value problems and function theory for spin-invariant differential operators", Ph.D. thesis, Gent (1989).

[4] I.R. Porteous. "Topological Geometry", 2nd ed., Cambridge University Press, Cambridge (1981).

[5] F. Sommen, Spherical monogenics on the Lie sphere, *Journal of Functional Analysis* 92, nr. 2, pp. 372–402(1990).

[6] F. Sommen, Spherical Monogenic Functions and Analytic Functionals on the Unit Sphere, *Tokyo Journal of Mathematics* 4, nr. 2, pp. 427–456(1981).

[7] V. Avanissian, Sur les fonctions harmoniques d'ordre quelconque et leur prolongement analytique dans $\mathbf{C}^n$, *Springer Lecture Notes in Math.* 919, pp. 192–281(1981).

[8] M. Morimoto, Analytic functionals on the Lie sphere, *Tokyo J. Math.* 3, nr. 1, pp. 1-35(1980).

[9] J. Sičiak, Holomorphic continuation of harmonic functions, *Ann. Polon. Math.* 29, pp. 67–73(1974).

[10] A. Erdélyi, W. Magnus, F. Oberhettinger and F. Tricomi. "Higher transcendental functions", Vol II, McGraw-Hill, New-York (1953).

[11] S.G. Krantz. "Function Theory of Several Complex variables", John Wiley & Sons, New-York (1982).

[12] A. Angot. "Compléments de Mathématiques", Masson & Cie, Paris (1965).

ULTRADISTRIBUTIONAL BOUNDARY VALUES OF HOLOMORPHIC FUNCTIONS

R. D. Carmichael,[1] R. S. Pathak,[2] and S. Pilipović[3]

[1] Department of Mathematics and Computer Science
Wake Forest University, Winston-Salem, NC 27109, U.S.A.

[2] Department of Mathematics
Banaras Hindu University, Varanasi 221005, India

[3] Institute of Mathematics
University of Novi Sad, 21000 Novi Sad, Yugoslavia

ABSTRACT

Ultradistributions of Beurling type $\mathcal{D}'((M_p), L^s)$ and of Roumieu type $\mathcal{D}'(\{M_p\}, L^s)$ are defined where M_p, $p = 0, 1, 2, 3, \ldots$, is a sequence of positive real numbers which is used to define ultradistributions. A norm growth of the Cauchy integral of elements in $\mathcal{D}'(*, L^s)$, where $*$ is either (M_p) or $\{M_p\}$, is obtained involving the associated function $M^*(\rho)$ corresponding to the sequences M_p; the Cauchy integral is a holomorphic function of $z \in \mathcal{R}^n + iC \subset \mathcal{C}^n$ where C is a regular cone in $\mathcal{R}^n$. We define holomorphic functions in tubes $\mathcal{R}^n + iC$ which are motivated by this norm growth and show that these functions are representable by Fourier-Laplace transforms; certain of these functions obtain ultradistributional boundary values. Problems for future research are indicated.

1 INTRODUCTION

The ultradistributions of class M_p of Beurling type $\mathcal{D}'((M_p), L^s)$ and of Roumieu type $\mathcal{D}'(\{M_p\}, L^s)$ are defined in [8]; these ultradistribution spaces generalize the Schwartz distributions $\mathcal{D}'_{L^r}$ [12, pp. 199-200]. Cauchy and Poisson integrals of elements in $\mathcal{D}'((M_p), L^s)$ and $\mathcal{D}'(\{M_p\}, L^s)$ are constructed and studied in [8]. The Cauchy integral is shown to be a holomorphic function in the tube $T^C = \mathcal{R}^n + iC$, where C is a regular cone in $\mathcal{R}^n$, and is shown to satisfy a pointwise growth estimate and to obtain an ultradistributional boundary value. Similarly the Poisson integral is shown to obtain such a boundary value.

Generalized Functions and Their Applications, Edited by
R.S. Pathak, Plenum Press, New York, 1993

In this paper we prove that the Cauchy integral defined in [8] satisfies a growth condition in the L^s norm which involves the associated function $M^*(\rho)$ of Komatsu [10, p. 31]. This growth condition motivates the study of holomorphic functions in tubes which satisfy this type of norm growth in L^s, and we show that such functions have Fourier-Laplace integral representations and that certain of these functions obtain ultradistributional boundary values.

2 NOTATION AND DEFINITIONS

The notation of this paper is exactly that of [8] and [1]; see [8, p. 87] and [1, p. 1042]. The definitions of a cone C in $\mathcal{R}^n$ (with vertex at the origin $\overline{0} = (0,0,...,0)$ in $\mathcal{R}^n$), projection of a cone denoted pr(C), compact subcone C' of C, and dual cone $C^* = \{t \in \mathcal{R}^n : \langle t, y\rangle \geq 0$ for all $y \in C\}$ are given in [1, p. 1042]. A regular cone is an open convex cone in $\mathcal{R}^n$ such that $\overline{C}$ does not contain any entire straight line.

$\hat{\varphi}(x) = \mathcal{F}[\varphi(t); x]$ and $\mathcal{F}^{-1}[\varphi(t); x]$ will denote the Fourier and inverse Fourier transform of $L^r, 1 \leq r \leq 2$, functions φ and are as defined in [1, p. 1042]. The reader is assumed to be familiar with the properties of these transforms.

In this paper M_p, $p = 0, 1, 2, ...$, is a sequence of positive real numbers which satisfy some of the conditions (M.1), (M.2), (M.3), (M.2)$'$, or (M.3)$'$ given in [8, p. 87].

Examples of the sequences M_p are given in [8, p. 87] and [10, p. 28]; [10] will serve as the basic reference for these sequences. For a sequence M_p the associated functions $M(\rho)$ and $M^*(\rho)$ are defined in [10, p. 48] and [10, p. 31], respectively, by

$$M(\rho) = \sup_p \log(\rho^p M_0/M_p) \ , \quad 0 < \rho < \infty, \tag{2.1}$$

and

$$M^*(\rho) = \sup_p \log(\rho^p p! M_0/M_p) \ , \quad 0 < \rho < \infty. \tag{2.2}$$

3 ULTRADISTRIBUTIONS AND CAUCHY INTEGRAL

Let M_p, $p = 0, 1, 2, ...$, be a sequence of positive real numbers. $\mathcal{D}((M_p), L^s)$ (respectively, $\mathcal{D}(\{M_p\}, L^s)$), $1 \leq s \leq \infty$, is the set of all complex valued infinitely differentiable functions $\varphi(t)$ such that there is a constant $N > 0$ for which

$$||D_t^\alpha \varphi(t)||_{L^s} \leq N h^{|\alpha|} M_{|\alpha|} \ , \quad |\alpha| = 0, 1, 2, ..., \tag{3.1}$$

for all $h > 0$ (respectively, for some $h > 0$) where the left side of (3.1) is the L^s norm. $\mathcal{D}'((M_p), L^s)$ and $\mathcal{D}'(\{M_p\}, L^s)$ are the spaces of continuous linear forms, ultradistributions, on $\mathcal{D}((M_p), L^s)$ and $\mathcal{D}(\{M_p\}, L^s)$, respectively. The convergence of sequences in

$\mathcal{D}((M_p), L^s)$ and in $\mathcal{D}(\{M_p\}, L^s)$ and properties of these spaces are stated in [8, section 3] as are characterization results for $\mathcal{D}'((M_p), L^s)$ and $\mathcal{D}'(\{M_p\}, L^s)$.

Let C be a regular cone in $\mathcal{R}^n$. The Cauchy kernel function corresponding to the tube $T^C = \mathcal{R}^n + iC$ is

$$K(z-t) = \int_{C^*} \exp(2\pi i \langle z-t, \eta\rangle)\, d\eta \ , \quad z \in T^C \ , \quad t \in \mathcal{R}^n. \tag{3.2}$$

In [8, pp. 92-94] we have proved the following result.

Theorem 3.1. Let the sequence of positive real numbers M_p, $p = 0, 1, 2, \ldots$, satisfy (M.1) and (M.3)'. We have $K(z-t) \in \mathcal{D}(*, L^s)$, $2 \le s \le \infty$, as a function of $t \in \mathcal{R}^n$ for $z \in T^C$ where $*$ is either (M_p) or $\{M_p\}$.

Let $U \in \mathcal{D}'(*, L^s)$, $2 \le s \le \infty$, where $*$ is either (M_p) or $\{M_p\}$ and the sequence M_p satisfies (M.1) and (M.3)'. Because of Theorem 3.1 we form

$$C(U, z) \;=\; \langle U_t, K(z-t)\rangle \ , \quad z \in T^C, \tag{3.3}$$

which is the Cauchy integral of U and is a well defined function of $z \in T^C = \mathcal{R}^n + iC$. Properties of this Cauchy integral are obtained in [8, section 5].

A Poisson integral of $U \in \mathcal{D}'(*, L^s)$, $2 \le s \le \infty$, is defined in [8, section 6] and boundary value properties are obtained for this integral there.

4 NORM GROWTH OF THE CAUCHY INTEGRAL

Throughout this section C is a regular cone in $\mathcal{R}^n$, and the sequence M_p, $p = 0, 1, 2, \ldots$, satisfies (M.1) and (M.3)'.In [8, Theorem 5.1, p. 95] a pointwise growth estimate for the Cauchy integral of elements $U \in \mathcal{D}'(*, L^s)$, $2 \le s < \infty$, was obtained. We now prove a norm growth estimate for this Cauchy integral in this section. Before doing so we state the following needed lemma which is proved in [9, Lemma 3.1].

Lemma 4.1. Let C be a regular cone in $\mathcal{R}^n$. For any n-tuple α of nonnegative integers we have

$$\left(t^\alpha I_{C^*}(t) e^{-2\pi\langle y,t\rangle}\right) \in L^r \tag{4.1}$$

for all r, $1 \le r \le \infty$, for $y \in C$ where $I_{C^*}(t)$ is the characteristic function of the dual cone C^* of C.

Using Lemma 4.1, we now obtain a norm growth estimate on the Cauchy integral defined in (3.3); this norm growth will motivate the growth on holomorphic functions which we consider later in this paper.

Theorem 4.1. Let C be a regular cone in $\mathcal{R}^n$, and let the sequence M_p, $p = 0, 1, 2, \ldots$, satisfy (M.1) and (M.3)'. If $U \in \mathcal{D}'((M_p), L^s)$, $2 \le s < \infty$, then for each compact subcone $C' \subset C$ there exists a constant $T = T(C') > 0$ depending on $C' \subset C$ such that

$$\|C(U;z)\|_{L^s} = \left(\int_{\mathcal{R}^n} |C(U;x+iy)|^s \, dx\right)^{1/s} \tag{4.2}$$
$$\leq \begin{cases} K(U)\exp(M^*(T/|y|)), & y \in C' \subset C, \text{ if } s = 2, \\ K(U,C',s,r,n)\, |y|^{-n(s-r)/rs} \exp(M^*(T/|y|)), & \\ \qquad y \in C' \subset C, \text{ if } 2 < s < \infty, \end{cases}$$

where $K(U)$ is a constant depending on U if $s = 2$ and $K(U,C',s,r,n)$ is a constant depending on U, C', s, r, and n if $2 < s < \infty$. If $U \in \mathcal{D}'(\{M_p\}, L^s)$, $2 \leq s < \infty$, then for each compact subcone $C' \subset C$ and arbitrary constant $T > 0$, which may or may not depend on $C' \subset C$, the inequality (4.2) holds where $K(U)$ is a constant depending on U if $s = 2$ and $K(U,C',s,r,n)$ is a constant depending on U, C', s, r, and n if $2 < s < \infty$.

Proof. We prove (4.2) for the case that $U \in \mathcal{D}'(\{M_p\}, L^s)$, $2 \leq s < \infty$. (The proof of (4.2) for the case $U \in \mathcal{D}'((M_p), L^s)$, $2 \leq s < \infty$, is given in [9].) Using the representation theorem [8, Theorem 3.2, p. 90] and Fubini's theorem we obtain

$$\begin{aligned} C(U;z) &= \sum_{|\alpha|=0}^{\infty} \langle D_t^\alpha g_\alpha(t), K(z-t)\rangle \\ &= \sum_{|\alpha|=0}^{\infty} (-1)^{|\alpha|} \langle g_\alpha(t), D_t^\alpha K(z-t)\rangle \\ &= \sum_{|\alpha|=0}^{\infty} \int_{\mathcal{R}^n} g_\alpha(t) \int_{\mathcal{R}^n} I_{C^*}(\eta)\eta^\alpha e^{2\pi i\langle z-t,\eta\rangle}\, d\eta\, dt \\ &= \sum_{|\alpha|=0}^{\infty} \int_{\mathcal{R}^n} I_{C^*}(\eta)\eta^\alpha e^{2\pi i\langle z,\eta\rangle} \mathcal{F}^{-1}[g_\alpha(t);\eta]\, d\eta \end{aligned} \tag{4.3}$$

where each $g_\alpha(t) \in L^r$, $1/r + 1/s = 1$, and $I_{C^*}(\eta)$ is the characteristic function of C^*. Since each $g_\alpha(t) \in L^r$, $1 < r \leq 2$, for each s, $2 \leq s < \infty$, $1/r + 1/s = 1$, then $\mathcal{F}^{-1}[g_\alpha(t);\eta] \in L^s$, $2 \leq s < \infty$. If $s = 2$ then $r = 2$, and by Lemma 4.1 each of the integrands in the last term in (4.3) is in $L^1 \cap L^2$. For the case $s = 2$ we have using Parseval's equality that

$$\begin{aligned} &\left\|\int_{\mathcal{R}^n} I_{C^*}(\eta)\eta^\alpha e^{2\pi i\langle z,\eta\rangle}\mathcal{F}^{-1}[g_\alpha(t);\eta]d\eta\right\|_{L^2} \\ &\quad = \left\|\mathcal{F}\left[\eta^\alpha I_{C^*}(\eta)e^{-2\pi\langle y,\eta\rangle}\mathcal{F}^{-1}[g_\alpha(t);\eta];x\right]\right\|_{L^2} \\ &\quad = \left\|\eta^\alpha I_{C^*}(\eta)e^{-2\pi\langle y,\eta\rangle}\mathcal{F}^{-1}[g_\alpha(t);\eta]\right\|_{L^2} \\ &\quad \leq \left(\sup_{\eta\in\mathcal{R}^n}\left|\eta^\alpha I_{C^*}(\eta)e^{-2\pi\langle y,\eta\rangle}\right|^2 \int_{\mathcal{R}^n}\left|\mathcal{F}^{-1}[g_\alpha(t);\eta]\right|^2 d\eta\right)^{1/2}. \end{aligned} \tag{4.4}$$

Now let $s > 2$ and note that by Hölder's inequality

$$\int_{\mathcal{R}^n} \left| I_{C^*}(\eta)\eta^\alpha e^{-2\pi\langle y,\eta\rangle} \mathcal{F}^{-1}[g_\alpha(t);\eta] \right|^r d\eta \tag{4.5}$$

$$\leq \left(\int_{\mathcal{R}^n} \left(\left| \mathcal{F}^{-1}[g_\alpha(t);\eta] \right|^r \right)^{s/r} d\eta \right)^{r/s} \left\| \left| I_{C^*}(\eta)\eta^\alpha e^{-2\pi\langle y,\eta\rangle} \right|^r \right\|_{L^{s/(s-r)}}$$

where both terms on the right of (4.5) are finite since $\mathcal{F}^{-1}[g_\alpha(t);\eta] \in L^s$ and because of Lemma 4.1. Thus by (4.5) and Lemma 4.1, each of the integrands in the last term in (4.3) is in $L^1 \cap L^r$ for the case $s > 2$, $1/r + 1/s = 1$. By the Parseval inequality and (4.5) we then have for the case $s > 2$ that

$$\left\| \int_{\mathcal{R}^n} I_{C^*}(\eta)\eta^\alpha e^{2\pi i\langle z,\eta\rangle} \mathcal{F}^{-1}[g_\alpha(t);\eta]\, d\eta \right\|_{L^s}$$

$$= \left\| \mathcal{F}\left[\eta^\alpha I_{C^*}(\eta) e^{-2\pi\langle y,\eta\rangle} \mathcal{F}^{-1}[g_\alpha(t);\eta]; x \right] \right\|_{L^s}$$

$$\leq \left\| \eta^\alpha I_{C^*}(\eta) e^{-2\pi\langle y,\eta\rangle} \mathcal{F}^{-1}[g_\alpha(t);\eta] \right\|_{L^r} \tag{4.6}$$

$$\leq \left(\int_{\mathcal{R}^n} \left(\left| \mathcal{F}^{-1}[g_\alpha(t);\eta] \right|^r \right)^{s/r} d\eta \right)^{1/s}$$

$$\bullet \left(\int_{\mathcal{R}^n} \left| I_{C^*}(\eta)\eta^\alpha e^{-2\pi\langle y,\eta\rangle} \right|^{rs/(s-r)} d\eta \right)^{(s-r)/sr}.$$

From [7, Lemma 4.3.2, p. 155], given a compact subcone $C' \subset C$ there is a $\delta = \delta(C') > 0$ such that

$$\langle y,t\rangle \geq \delta|y||t| \ ,\ y \in C' \ ,\ t \in C^*. \tag{4.7}$$

Using (4.7) and the estimates [9, (3.4) and (3.5)] we have

$$\sup_{\eta\in\mathcal{R}^n} \left| \eta^\alpha I_{C^*}(\eta) e^{-2\pi\langle y,\eta\rangle} \right| \leq \sup_{\eta\in C^*} \left(|\eta|^{|\alpha|} \exp(-2\pi\delta|y||\eta|) \right)$$

$$\leq \begin{cases} 1 & ,\ |\alpha| = 0, \\ \left(\dfrac{|\alpha|}{2\pi\delta|y|} \right)^{|\alpha|} & ,\ |\alpha| = 1,2,..., \end{cases} \tag{4.8}$$

for $y \in C' \subset C$ and $\delta = \delta(C') > 0$.

For the case $s = 2$, we use (4.3), (4.4), (4.8), and the Parseval equality to obtain for $y \in C' \subset C$ that

$$
\begin{aligned}
\|C(U;z)\|_{L^2} &\le \sum_{|\alpha|=0}^{\infty} \left\| \int_{\mathcal{R}^n} I_{C^*}(\eta)\, \eta^\alpha e^{2\pi i\langle z,\eta\rangle} \mathcal{F}^{-1}[g_\alpha(t);\eta]\, d\eta \right\|_{L^2} \qquad (4.9)\\
&\le \sum_{|\alpha|=0}^{\infty} \left(\sup_{\eta\in\mathcal{R}^n} \left|\eta^\alpha I_{C^*}(\eta)\, e^{-2\pi\langle y,\eta\rangle}\right|^2 \int_{\mathcal{R}^n} \left|\mathcal{F}^{-1}[g_\alpha(t);\eta]\,\right|^2 d\eta \right)^{1/2}\\
&\le \sum_{|\alpha|=0}^{\infty} \left(\frac{|\alpha|}{2\pi\delta|y|} \right)^{|\alpha|} \|g_\alpha(t)\|_{L^2}
\end{aligned}
$$

where we are using the convention that $|\alpha|^{|\alpha|} = 1$ if $\alpha = \bar{0}$. For $s > 2$ and $C' \subset C$ let us write $\delta = \delta(C') > 0$ in (4.7) as $\delta = \delta_1 + \delta_2$ where $\delta_1 = \delta_1(C') > 0$ and $\delta_2 = \delta_2(C') > 0$. Now using (4.3), (4.6), Parseval's inequality, (4.7) with $\delta = \delta_1 + \delta_2$, (4.8), and [13, Theorem 32, p. 39] we have for $y \in C' \subset C$

$$
\begin{aligned}
\|C(U;z)\|_{L^s} &\le \sum_{|\alpha|=0}^{\infty} \left\| \int_{\mathcal{R}^n} I_{C^*}(\eta)\eta^\alpha e^{2\pi i\langle z,\eta\rangle} \mathcal{F}^{-1}[g_\alpha(t);\eta] d\eta \right\|_{L^s}\\
&\le \sum_{|\alpha|=0}^{\infty} \left\|\mathcal{F}^{-1}[g_\alpha(t);\eta]\,\right\|_{L^s} \left\| I_{C^*}(\eta)\eta^\alpha e^{-2\pi\langle y,\eta\rangle} \right\|_{L^{rs/(s-r)}}\\
&\le \sum_{|\alpha|=0}^{\infty} \|g_\alpha(t)\|_{L^r} \left(\sup_{\eta\in\mathcal{R}^n} |I_{C^*}(\eta)\, \eta^\alpha \exp(-2\pi\delta_1|y||\eta|)| \right) \qquad (4.10)\\
&\qquad \bullet \left(\int_{C^*} \exp(-2\pi\delta_2|y||\eta| rs/(s-r))\, d\eta \right)^{(s-r)/rs}\\
&\le \Omega_n \left(\int_0^\infty w^{n-1} \exp(-2\pi\delta_2|y| w rs/(s-r))\, dw \right)^{(s-r)/rs}\\
&\qquad \bullet \sum_{|\alpha|=0}^{\infty} \|g_\alpha(t)\|_{L^r} \left(\frac{|\alpha|}{2\pi\delta_1|y|} \right)^{|\alpha|}\\
&= \Omega_n \left((n-1)!(2\pi\delta_2|y|rs/(s-r))^{-n} \right)^{(s-r)/rs}\\
&\qquad \bullet \sum_{|\alpha|=0}^{\infty} \|g_\alpha(t)\|_{L^r} \left(\frac{|\alpha|}{2\pi\delta_1|y|} \right)^{|\alpha|}
\end{aligned}
$$

where Ω_n is the surface area of the unit sphere in $\mathcal{R}^n$.

The condition [8, (3.12)] on the sequence $\{g_\alpha(t)\}$ in [8, Theorem 3.2, p. 90], and hence on the sequence $\{g_\alpha(t)\}$ in this theorem, holds for all $k > 0$ and is obviously equivalent to the condition

$$
\sup_\alpha \left(k^{|\alpha|} M_{|\alpha|} \|g_\alpha\|_{L^r} \right) < \infty \qquad (4.11)
$$

for all $k > 0$. For the case $s = 2$ in this theorem and for all $k > 0$ we return to (4.9) and use (4.11) to obtain

$$||C(U;z)||_{L^2} \leq \sup_\alpha \left(k^{|\alpha|} M_{|\alpha|} ||g_\alpha||_{L^2}\right) \tag{4.12}$$

$$\bullet \sum_{|\alpha|=0}^{\infty} \left(\frac{1}{2}\right)^{|\alpha|} (k\pi\delta)^{-|\alpha|} |y|^{-|\alpha|} |\alpha|^{|\alpha|} \left(M_{|\alpha|}\right)^{-1} .$$

Recall from [8, p. 97] that

$$|\alpha|^{|\alpha|} \leq e^{|\alpha|}|\alpha| \ , \ |\alpha| = 1,2,3,..., \tag{4.13}$$

and recall our convention that $|\alpha|^{|\alpha|} = 1$ if $|\alpha| = 0$; using this in (4.12) and putting $T = e/k\pi\delta$ we have from (4.12)

$$||C(U;z)||_{L^2} \leq \sup_\alpha \left(k^{|\alpha|} M_{|\alpha|} ||g_\alpha||_{L^2}\right)$$

$$\bullet \sum_{|\alpha|=0}^{\infty} \left(\frac{1}{2}\right)^{|\alpha|} \left(\frac{T}{|y|}\right)^{|\alpha|} \frac{|\alpha|!}{M_{|\alpha|}} \tag{4.14}$$

$$\leq \sup_\alpha \left(k^{|\alpha|} M_{|\alpha|} ||g_\alpha||_{L^2}\right) \left(\sum_{|\alpha|=0}^{\infty} \left(\frac{1}{2}\right)^{|\alpha|}\right) (M_0)^{-1} \exp(M^*(T/|y|))$$

for $y \in C' \subset C$ where we have used the calculation [8, (5.12)]. (4.14) holds for all $k > 0$; thus $T = e/k\pi\delta > 0$ is arbitrary and may or may not depend on $C' \subset C$ given the choice of the arbitrary $k > 0$. (4.14) thus proves (4.2) for the case that $U \in \mathcal{D}'(\{M_p\}, L^2)$ where

$$K(U) = (M_0)^{-1} \sup_\alpha \left(k^{|\alpha|} M_{|\alpha|} ||g_\alpha||_{L^2}\right) \left(\sum_{|\alpha|=0}^{\infty} \left(\frac{1}{2}\right)^{|\alpha|}\right) .$$

For $s > 2$ we return to (4.10) and proceed using (4.11) and (4.13) to obtain (4.2) for the case $U \in \mathcal{D}'(\{M_p\}, L^s)$, $2 < s < \infty$, similarly as we did for the case $U \in \mathcal{D}'(\{M_p\}, L^2)$, where

$$K(U, C', s, r, n) = \Omega_n \left((n-1)!(2\pi\delta_2 rs/(s-r))^{-n}\right)^{(s-r)/rs} (1/M_0)$$

$$\bullet \sup_\alpha \left(k^{|\alpha|} M_{|\alpha|} ||g_\alpha||_{L^r}\right) \left(\sum_{|\alpha|=0}^{\infty} \left(\frac{1}{2}\right)^{|\alpha|}\right)$$

and $T = e/k\pi\delta_1 > 0$ is arbitrary and may or may not depend on $C' \subset C$ since $k > 0$ is arbitrary.

Under the assumption $U \in \mathcal{D}'((M_p), L^s)$, $2 \leq s < \infty$, the stated conclusions in (4.2) are obtained using the characterization result [8, Theorem 3.1, p. 90] together with [8, (3.11), p. 90] and proceeding with analysis similar to that in the case for $U \in \mathcal{D}'(\{M_p\}, L^s)$ given previously in this proof; see the proof of [9, Theorem 3.1]. For $U \in \mathcal{D}'((M_p), L^s)$, the constant $T = T(C') > 0$ in (4.2) depends on the compact subcone $C' \subset C$. As in the above proof, T is chosen to be $T = e/k\pi\delta$ for the case $U \in \mathcal{D}'((M_p), L^2)$ and $T = e/k\pi\delta_1$ for the cases $U \in \mathcal{D}'((M_p), L^s)$, $2 < s < \infty$, where now $k > 0$ is fixed from [8, (3.9) and (3.11), p. 90]; thus the constant T is fixed and depends on $C' \subset C$ because of the δ for the case $s = 2$ and the δ_1 for the cases $2 < s < \infty$. The proof of Theorem 4.1 is complete.

Pilipović ([11, Proposition 5.1, (12), p. 212] and [11, (12′), p. 214]) has obtained a growth like (4.2) for a Cauchy integral but only for the case $s = 2$ and for the cone C being quadrants in $\mathcal{R}^n$. Theorem 4.1 generalizes these results both for the generality of the cone C and for $2 \leq s < \infty$.

5 HOLOMORPHIC FUNCTIONS

We assume that the sequence M_p, $p = 0, 1, 2, \ldots$, satisfies (M.1) and (M.3)′ throughout this section, which ensures that $M^*(\rho)$, $\rho > 0$, is finite ([8, Lemma 2.1, p. 88] and [10, lines 7-9, p. 94]).

Let B be a proper open subset of $\mathcal{R}^n$. Let $d(y)$ denote the distance from $y \in B$ to the complement of B in $\mathcal{R}^n$. Let $f(z)$ be defined in $T^B = \mathcal{R}^n + iB$ and satisfy

$$||f(x+iy)||_{L^r} = \left(\int_{\mathcal{R}^n} |f(x+iy)|^r \, dx\right)^{1/r} \tag{5.1}$$

$$\leq K \, (1 + (d(y))^{-m})^q \, \exp(M^*(T/|y|)) \, , \quad y \in B,$$

where $K > 0$, $T > 0$, $m \geq 0$, and $q \geq 0$ are all independent of $y \in B$ and M^* is the associated function of the sequence M_p, $p = 0, 1, 2, \ldots$, defined in (2.2). Holomorphic functions in tubes $T^B = \mathcal{R}^n + iB$ which satisfy (5.1) are generalizations of the Hardy $H^p(T^B)$ spaces [14, pp. 90-91].

If $B = C$, an open connected cone in $\mathcal{R}^n$, we have from the formula

$$d(y) = \inf_{t \in pr(C^*)} \langle t, y \rangle \ , \quad y \in C \ ,$$

given in [4, (2.1), p. 82] that $d(y) \leq |y|$, $y \in C$. From this inequality we have for $B = C$ that the term $(1 + (d(y))^{-m})^q$ in (5.1) is a generalizing factor in this growth as opposed to a similar term with $d(y)$ replaced by $|y|$. Further the right side of the growth (5.1) allows for divergence to ∞ as y approaches any point on the boundary of C and not just as y approaches $\bar{0}$ as would be the case if $|y|$ were in place of $d(y)$ in (5.1). Because of the norm growth obtained on the Cauchy integral in Theorem 4.1 and because of the generalizing nature of $d(y)$ in (5.1) as opposed to $|y|$ we are led to consider holomorphic functions in tubes which satisfy (5.1). We show that for the cases

$1 < r \leq 2$, holomorphic functions in tubes T^B which satisfy (5.1) can be represented by Fourier-Laplace integrals. We need several lemmas which we now state; proofs are contained in [6].

Lemma 5.1. Let B be a proper open connected subset of $\mathcal{R}^n$. Let $1 \leq s < \infty$. Let $g(t)$ be a measurable function on $\mathcal{R}^n$ which satisfies

$$||e^{-2\pi\langle y,t\rangle}\, g(t)||_{L^s} \leq K\,(1+(d(y))^{-m})^q\, \exp(M^*(T/|y|)) \ , \ y \in B, \tag{5.2}$$

where $K > 0$, $T > 0$, $m \geq 0$, and $q \geq 0$ are independent of $y \in B$. Then

$$f(z) = \int_{\mathcal{R}^n} g(t)\, e^{2\pi i\langle z,t\rangle}\, dt \ , \ z \in T^B, \tag{5.3}$$

is a holomorphic function of $z \in T^B = \mathcal{R}^n + iB$.

In the following lemma supp(g) denotes the support of the function g.

Lemma 5.2. Let C be an open connected cone in $\mathcal{R}^n$. Let $1 \leq s < \infty$. Let $g(t)$ be a measurable function on $\mathcal{R}^n$ such that (5.2) holds for $y \in C$. Then $\text{supp}(g) \subseteq C^*$ almost everywhere.

Lemma 5.3. Let B denote an open connected subset of $\mathcal{R}^n$ which does not contain $\bar{0}$. Let $1 < r \leq 2$. Let $f(z)$ be holomorphic in T^B and satisfy (5.1). Then for all y and y' in B we have

$$e^{2\pi\langle y,t\rangle} h_y(t) = e^{2\pi\langle y',t\rangle} h_{y'}(t)$$

for almost every $t \in \mathcal{R}^n$ where

$$h_y(t) = \mathcal{F}^{-1}[f(x+iy);t] \ , \ y \in B,$$

is the $L^s, 1/r + 1/s = 1$, inverse Fourier transform of $f(x+iy)$, $y \in B$.

We now show that holomorphic functions in tubes which satisfy (5.1) for $1 < r \leq 2$ have Fourier-Laplace integral representations.

Theorem 5.1. Let B denote an open connected subset of $\mathcal{R}^n$ which does not contain $\bar{0} \in \mathcal{R}^n$. Let $1 < r \leq 2$. Let $f(z)$ be holomorphic in $T^B = \mathcal{R}^n + iB$ and satisfy (5.1). Then there exists a measurable function $g(t), t \in \mathcal{R}^n$, such that (5.2) holds, $1/r + 1/s = 1$, where $K > 0$, $T > 0$, $m \geq 0$, and $q \geq 0$ are independent of $y \in B$, and

$$f(z) = \int_{\mathcal{R}^n} g(t)\, e^{2\pi i\langle z,t\rangle}\, dt \ , \ z \in T^B. \tag{5.4}$$

Proof. Put

$$g(t) = e^{2\pi\langle y,t\rangle} h_y(t) \ , \ y \in B, \tag{5.5}$$

where $h_y(t) = \mathcal{F}^{-1}[f(x+iy);t]$, $y \in B$, is the $L^s, 1/r+1/s = 1$, inverse Fourier transform of $f(x+iy), y \in B$; by the Plancherel Fourier transform theory $h_y(t)$ is an element of L^s since by (5.1) $f(x+iy) \in L^r$ as a function of $x \in \mathcal{R}^n$ for $y \in B$. By Lemma 5.3 $g(t)$ is independent of $y \in B$. From (5.5) we have

$$e^{-2\pi\langle y,t\rangle} g(t) = \mathcal{F}^{-1}[f(x+iy);t] \ , \ \ y \in B. \tag{5.6}$$

Since $f(x+iy) \in L^r$, $1 < r \leq 2$, as a function of $x \in \mathcal{R}^n$ for $y \in B$, then $\left(e^{-2\pi\langle y,t\rangle} g(t)\right) \in L^s$, $1/r + 1/s = 1$, $y \in B$, by the Plancherel Fourier transform theory, and

$$||e^{-2\pi\langle y,t\rangle}\ g(t)||_{L^s} \leq ||f(x+iy)||_{L^r} \leq K\left(1 + (d(y))^{-m}\right)^q \exp(M^*(T/|y|)),$$

$y \in B$, by the Parseval inequality and (5.1); thus (5.2) is obtained. From (5.6) we have

$$f(x+iy) = \mathcal{F}\left[e^{-2\pi\langle y,t\rangle} g(t); x\right] , \ z = x+iy \in T^B, \tag{5.7}$$

by the Plancherel Fourier transform theory with the Fourier transform being in L^r. But by (5.2) and Lemma 5.1 the integral on the right side of (5.4), which is the L^1 transform of $\left(e^{-2\pi\langle y,t\rangle} g(t)\right), y \in B$, is a holomorphic function of $z \in T^B$. (Recall from the proof of Lemma 5.1 given in [6; Lemma 2.1] that $\left(e^{-2\pi\langle y,t\rangle} g(t)\right) \in L^1, y \in B$, since (5.2) is satisfied here.) From the Plancherel Fourier transform theory we know that the L^1 and L^r, $1 < r \leq 2$, Fourier transforms of the same function are equal when both of these transforms exist; hence the desired equality (5.4) follows from (5.7). The proof is complete.

The proof of the following corollary is obtained by combining Theorem 5.1 and Lemma 5.2.

Corollary 5.1. Let C be an open connected cone in $\mathcal{R}^n$. Let $1 < r \leq 2$. Let $f(z)$ be holomorphic in T^C and satisfy (5.1) for $y \in C$. Then there exists a measurable function $g(t)$, $t \in \mathcal{R}^n$, such that (5.2) holds for $y \in C$, $\text{supp}(g) \subseteq C^*$ almost everywhere, and (5.4) holds for $z \in T^C$.

The following result is a dual theorem to Theorem 5.1.

Theorem 5.2. Let B be a proper open connected subset of $\mathcal{R}^n$. Let $1 < r \leq 2$. Let $g(t)$ be a measurable function on $\mathcal{R}^n$ which satisfies

$$||e^{-2\pi\langle y,t\rangle}\ g(t)||_{L^r} \leq K\left(1 + (d(y))^{-m}\right)^q \exp(M^*(T/|y|)) \ , \ \ y \in B, \tag{5.8}$$

where $K > 0$, $T > 0$, $m \geq 0$, and $q \geq 0$ are independent of $y \in B$. Then

$$f(z) = \int_{\mathcal{R}^n} g(t)\ e^{2\pi i\langle z,t\rangle}\ dt \ , \ \ z \in T^B, \tag{5.9}$$

is a holomorphic function of $z \in T^B$ and satisfies

$$||f(x+iy)||_{L^s} \leq K\left(1 + (d(y))^{-m}\right)^q \exp(M^*(T/|y|)) \ , \ \ y \in B. \tag{5.10}$$

Proof. By the proof of Lemma 5.1 given in [6; Lemma 2.1], $f(z)$ is holomorphic in T^B and $\left(e^{-2\pi\langle y,t\rangle}g(t)\right) \in L^1$, $y \in B$. By (5.8), $\left(e^{-2\pi\langle y,t\rangle}g(t)\right) \in L^r$, $1 < r \leq 2$, as a function of $t \in \mathcal{R}^n$ for $y \in B$. Thus the right side of (5.9) can be viewed as both the L^1 and the L^r Fourier transform of $\left(e^{-2\pi\langle y,t\rangle}g(t)\right)$, $y \in B$, and by the Parseval inequality

$$\|f(x+iy)\|_{L^s} = \left\|\mathcal{F}\left[e^{-2\pi\langle y,t\rangle}g(t);x\right]\right\|_{L^s} \leq \left\|e^{-2\pi\langle y,t\rangle}g(t)\right\|_{L^r} \tag{5.11}$$

for $y \in B$ and $1/r + 1/s = 1$. (5.10) now follows from (5.11) and (5.8).

For the case that B is an open connected subset of R^n which does not contain $\bar{0} \in \mathcal{R}^n$ and $r = 2$, Theorem 5.2 is a converse result to Theorem 5.1.

6 ULTRADISTRIBUTIONAL BOUNDARY VALUES

Let C be a regular cone in $\mathcal{R}^n$, and let the sequence M_p, $p = 0, 1, 2, \ldots$, of positive real numbers satisfy (M.1), (M.2), and (M.3)$'$ and be such that $(M_p/p!)$ satisfies (M.1).

We consider functions $f(z)$ which are holomorphic in $T^C = \mathcal{R}^n + iC$ and which satisfy

$$\|f(x+iy)\|_{L^r} \leq K \exp(M^*(T/|y|)) \quad , \quad y \in C, \tag{6.1}$$

for $1 < r \leq 2$, where $K > 0$ and $T > 0$ are constants which are independent of $y \in C$ and M^* is the associated function of the sequence M_p, $p = 0, 1, 2, \ldots$, defined in (2.2); thus the norm growth which we are considering in this section is that in (5.1) with $m = 0$ or $q = 0$ there. In [9] we have proved that holomorphic functions $f(z)$ in T^C which satisfy (6.1) with $1 < r \leq 2$ have ultradistributional boundary values; we state the result here and ask the reader to see [9] for details.

Theorem 6.1. Let $f(z)$ be holomorphic in T^C and satisfy (6.1) with $1 < r \leq 2$. Then there exists an element $U \in \mathcal{D}'((M_p), L^1)$ such that

$$\lim_{\substack{y \to \bar{0} \\ y \in C}} f(x+iy) = U \tag{6.2}$$

in $\mathcal{D}'((M_p), L^1)$.

The proof of Theorem 6.1 uses facts obtained in the results of section 5.

The following result is a dual theorem to Theorem 6.1, and boundary value results are obtained in $\mathcal{D}'((M_p), L^r)$, $1 < r \leq 2$. The proof of the following result, which is contained in [9], uses properties of the spaces $\mathcal{F}\ \mathcal{D}((M_p), L^r)$ and $\mathcal{F}'\ \mathcal{D}((M_p), L^r)$, $1 < r \leq 2$, which are Fourier transform spaces of $\mathcal{D}((M_p), L^r)$ and $\mathcal{D}'((M_p), L^r)$ and which are constructed in [9].

Theorem 6.2. Let $1 < r \leq 2$. Let $g(t)$ be a measurable function on $\mathcal{R}^n$ such that

$$||e^{-2\pi\langle y,t\rangle} g(t)||_{L^r} \leq K \exp(M^*(T/|y|)) \ , \ y \in C, \tag{6.3}$$

where $K > 0$ and $T > 0$ are constants which are independent of $y \in C$. Then

$$f(z) = \int_{\mathcal{R}^n} g(t)\, e^{2\pi i\langle z,t\rangle}\, dt \ , \ z \in T^C, \tag{6.4}$$

is holomorphic in T^C, satisfies (6.1) with L^r replaced by L^s, $1/r + 1/s = 1$, and there is an element $U \in \mathcal{D}'((M_p), L^r)$ such that

$$\lim_{\substack{y \to \bar{0} \\ y \in C}} f(x+iy) = U \tag{6.5}$$

in $\mathcal{D}'((M_p), L^r)$.

7 RESULTS FOR $2 < r < \infty$

In this section we desire to extend results like we have obtained in sections 5 and 6 to the cases $2 < r < \infty$. The reader should read [5, section 2] to become familiar with the terms quadrant cone and polygonal cone given there. We use the information given in [5, section 2] here. Let $\mu = (\mu_1, \mu_2, \ldots, \mu_n)$ be any of the 2^n n-tuples whose components are 0 or 1; $C_\mu = \{y \in \mathcal{R}^n : (-1)^{\mu_j} y_j > 0, \ j = 1, \ldots, n\}$ is called an octant in $\mathcal{R}^n$.

Using analysis similar to that in [5] we obtain Fourier-Laplace integral representations of holomorphic functions in tubes T^C which satisfy (5.1) for $2 < r < \infty$ for C being a quadrant, a quadrant cone, and a polygonal cone. We begin with the following result.

Theorem 7.1. Let C be an open convex cone which is contained in or is any of the 2^n quadrants C_μ in $\mathcal{R}^n$. Let $f(z)$ be holomorphic in T^C and satisfy (5.1) for $2 < r < \infty$. Then there exists a measurable function $g(t)$, $t \in \mathcal{R}^n$, with $\text{supp}(g) \subseteq C^*$ almost everywhere such that

$$||e^{-2\pi\langle y,t\rangle} g(t)||_{L^2} \leq M\,(1 + (\text{d}(y))^{-m})^q \exp(M^*(T/|y|)) \ , \ y \in C \ , \tag{7.1}$$

for constants $M > 0$, $T > 0$, $m \geq 0$, and $q \geq 0$ which are independent of $y \in C$, and

$$f(z) = X(z) \int_{\mathcal{R}^n} g(t)\, e^{2\pi i\langle z,t\rangle}\, dt \ , \ z \in T^C, \tag{7.2}$$

where $X(z)$ is a polynomial in $z \in T^C$.

Proof. For C being contained in or being the quadrant C_μ, put

$$X(z) = \Pi_{j=1}^n \,(1 - i(-1)^{\mu_j} z_j)^{n+2} \ , \ z = x + iy \in T^C. \tag{7.3}$$

We have

$$|1/X(x+iy)| \leq \Pi_{j=1}^{n}\left(1+x_j^2\right)^{-1-n/2} \quad , \quad z \in T^C. \tag{7.4}$$

Put

$$F(z) = f(z)/X(z) \quad , \quad z \in T^C, \tag{7.5}$$

which is holomorphic in T^C. We have

$$\int_{\mathcal{R}^n} |F(x+iy)|^2 \,\mathrm{d}x = \int_{\mathcal{R}^n} |f(x+iy)/X(x+iy)|^2 \,\mathrm{d}x$$

$$\leq \left\| |f(x+iy)|^2 \right\|_{L^{r/2}} \left\| |1/X(x+iy)|^2 \right\|_{L^{r/(r-2)}} \tag{7.6}$$

$$= \left(\int_{\mathcal{R}^n} |f(x+iy)|^r \mathrm{d}x\right)^{2/r} \left(\int_{\mathcal{R}^n} |1/X(x+iy)|^{2r/(r-2)} \,\mathrm{d}x\right)^{(r-2)/r}$$

$$\leq \left(K(1+(\mathrm{d}(y))^{-m})^q \exp(M^*(T/|y|))\right)^2$$

$$\bullet \left(\int_{\mathcal{R}^n} \Pi_{j=1}^{n} (1+x_j^2)^{-2r/(r-2)-nr/(r-2)} \mathrm{d}x\right)^{(r-2)/r}$$

$$\leq (M(1+(\mathrm{d}(y))^{-m})^q \exp(M^*(T/|y|)))^2$$

where

$$M = K\left(\int_{\mathcal{R}^n} \Pi_{j=1}^{n} (1+x_j^2)^{-2r/(r-2)-nr/(r-2)} \mathrm{d}x\right)^{(r-2)/2r}. \tag{7.7}$$

Thus from (7.6) and the fact that $F(z)$ is holomorphic in T^C we can apply Corollary 5.1 to obtain a function $g(t)$, $t \in \mathcal{R}^n$, with $\mathrm{supp}(g) \subseteq C^*$ almost everywhere such that

$$\|e^{-2\pi\langle y,t\rangle} g(t)\|_{L^2} \leq M\,(1+(\mathrm{d}(y))^{-m})^q \exp(M^*(T/|y|)) \quad , \quad y \in C\,, \tag{7.8}$$

which is (7.1), and

$$F(z) = \int_{\mathcal{R}^n} g(t)\, e^{2\pi i\langle z,t\rangle} \,\mathrm{d}t \quad , \quad z \in T^C. \tag{7.9}$$

(7.2) is now obtained from (7.5) and (7.9) with $X(z)$ being given in (7.3). The proof is complete.

We ask if the representation (7.2) can be rewritten in the form $f(z) = \langle V, \exp(2\pi i\langle z,t\rangle)\rangle$, $z \in T^C$, for some ultradistribution V? If so, $g(t)$ will have to possess sufficient properties to allow for $\langle V, \exp(2\pi i\langle z,t\rangle)\rangle$ to be well defined.

If $m = 0$ or $q = 0$ in (6.1) the Fourier-Laplace integral in (7.2) obtains an ultradistributional boundary value as $y = Im(z) \to \bar{0}$, $y \in C$, by Theorem 6.2 since $g(t)$

satisfies (7.8) with $m = 0$ or $q = 0$. Can this fact be used along with (7.2) to prove that $f(z)$ also obtains an ultradistributional boundary value?

Recall the concept of quadrant cone given in [5, p. 86]. We extend Theorem 7.1 to the case that C is in or is a quadrant cone.

Theorem 7.2. Let C be an open convex cone that is contained in or is a quadrant cone in $\mathcal{R}^n$. Let $f(z)$ be holomorphic in T^C and satisfy (5.1) for $2 < r < \infty$. There exists a measurable function $g(t)$, $t \in \mathcal{R}^n$, and a nonsingular linear transformation L such that $\text{supp}(g) \subseteq (L^{-1}(C))^*$ almost everywhere,

$$\left\| e^{-2\pi\langle L^{-1}(y),t\rangle} g(t) \right\|_{L^2} \leq M(k + (\text{d}(L^{-1}(y)))^{-m})^q \exp(M^*(R/|L^{-1}(y)|)) \quad , \quad y \in C, \tag{7.10}$$

for constants $M > 0$, $R > 0$, $k > 0$, $m \geq 0$, and $q \geq 0$ which are independent of $y \in C$, and

$$f(z) = X(L^{-1}(x) + iL^{-1}(y)) \int_{\mathcal{R}^n} g(t) e^{2\pi i \langle L^{-1}(x)+iL^{-1}(y),t\rangle} \text{d}t \quad , \quad x + iy \in \mathcal{R}^n + iC, \tag{7.11}$$

where $X(u + iv)$ is a polynomial, $u + iv = L^{-1}(x) + iL^{-1}(y)$.

Proof. Let Γ denote the quadrant cone that C is contained in or is. There exists a nonsingular linear transformation L (with domain and range being $\mathcal{R}^n$) which maps the first quadrant $C_{\bar{0}}$ onto Γ in a one to one manner such that the boundary of $C_{\bar{0}}$ is mapped to the boundary of Γ. (See [5, p. 86].) Further, if C is properly contained in Γ then $L^{-1}(C)$ is an open convex cone which is contained in $C_{\bar{0}}$, and L maps $L^{-1}(C)$ one to one and onto C with the boundary of $L^{-1}(C)$ being mapped to the boundary of C. ($L^{-1}(C) = C_{\bar{0}}$ if $C = \Gamma$.) For $u + iv \in \mathcal{R}^n + iL^{-1}(C)$ put

$$G(u + iv) = f(L(u) + iL(v)) = f(x + iy), \quad u + iv \in \mathcal{R}^n + iL^{-1}(C) \quad , \quad x + iy \in \mathcal{R}^n + iC. \tag{7.12}$$

$G(u + iv)$ is holomorphic in $\mathcal{R}^n + iL^{-1}(C)$. We have using (5.1) that

$$\int_{\mathcal{R}^n} |G(u + iv)|^r \text{d}u = \frac{1}{|det(L)|} \int_{\mathcal{R}^n} |f(x + iL(v))|^r \text{d}x \tag{7.13}$$

$$\leq \frac{1}{|det(L)|} (K\ (1 + (\text{d}(L(v)))^{-m})^q\ \exp(M^*(T/|L(v)|)))^r$$

for $y = L(v) \in C$. Recalling that the boundary of $L^{-1}(C)$ is mapped to the boundary of C by L we have for $v \in L^{-1}(C)$ that

$$\begin{aligned} \mathrm{d}(L(v)) &= \inf_{y' \in \partial C} |L(v) - y'| \qquad (7.14) \\ &= \inf_{v' \in \partial L^{-1}(C)} |L(v) - L(v')| = \inf_{v' \in \partial L^{-1}(C)} |L(v - v')| \end{aligned}$$

where ∂C denotes the boundary of C. Corresponding to the nonsingular linear transformation L there exist constants $a > 0$ and $b > 0$ (see [5, p. 93]) such that

$$a|w| \leq |L(w)| \leq b|w| \quad , \quad w \in \mathcal{R}^n, \qquad (7.15)$$

with a and b being independent of $w \in \mathcal{R}^n$. Using (7.15) in (7.14) we have

$$\begin{aligned} \mathrm{d}(L(v)) &= \inf_{v' \in \partial L^{-1}(C)} |L(v - v')| \qquad (7.16) \\ &\geq \inf_{v' \in \partial L^{-1}(C)} a|v - v'| = a \ d(v) \ , \ v \in L^{-1}(C). \end{aligned}$$

Using (7.16) and (7.15) in (7.13) we obtain

$$\begin{aligned} \int_{\mathcal{R}^n} |G(u + iv)|^r du &\leq \frac{1}{|det(L)|}(K(1 + (a \ \mathrm{d}(v))^{-m})^q \exp(M^*(T/a|v|)))^r \qquad (7.17) \\ &= \frac{1}{|det(L)|}((K/a^{mq})(a^m + (\mathrm{d}(v))^{-m})^q \exp(M^*(T/a|v|)))^r \end{aligned}$$

for $v \in L^{-1}(C)$ since M^* is an increasing function. From (7.17) and the holomorphicity of $G(u + iv)$ in $\mathcal{R}^n + iL^{-1}(C)$ we have by Theorem 7.1 the existence of a measurable function $g(t)$, $t \in \mathcal{R}^n$, with $\mathrm{supp}(g) \subseteq (L^{-1}(C))^*$ almost everywhere such that

$$||e^{-2\pi\langle v,t\rangle} g(t)||_{L^2} \leq M(k + (\mathrm{d}(v))^{-m})^q \exp(M^*(R/|v|)) \ , \ v \in L^{-1}(C), \qquad (7.18)$$

for some $M > 0$, $k > 0$, $m \geq 0$, $q \geq 0$, and $R > 0$; and

$$G(u + iv) = X(u + iv) \int_{\mathcal{R}^n} g(t) e^{2\pi i\langle u+iv,t\rangle} dt, \ u + iv \in \mathcal{R}^n + iL^{-1}(C), \qquad (7.19)$$

where $X(u + iv)$ is a polynomial. Thus from (7.12) and (7.19) we have

$$f(x + iy) = X(L^{-1}(x) + iL^{-1}(y)) \int_{\mathcal{R}^n} g(t) e^{2\pi i\langle L^{-1}(x)+iL^{-1}(y),t\rangle} dt, \ x + iy \in T^C,$$

which is (7.11); and (7.18) is (7.10). The proof is complete.

If $m = 0$ or $q = 0$ in (5.1), can we obtain an ultradistributional boundary value for $f(z)$ in Theorem 7.2 using (7.12)? We could if we knew an ultradistributional boundary value existed in Theorem 7.1 for $m = 0$ or $q = 0$ there.

Now recall the concept of a polygonal cone given in [5, p. 86] and the fact of the intersection property for the quadrant cones C_j, $j = 1, ..., m$, whose union is the polygonal cone. Let us extend Theorems 7.1 and 7.2 to the case that C can be a polygonal cone.

For C being a polygonal cone, $C = \cup_{j=1}^{m} C_j$ where the C_j are quadrant cones which have the intersection property noted in [5, p. 86]. Let $f(z)$ be holomorphic in T^C and satisfy (5.1) for $2 < r < \infty$. Now $y \in C$ implies $y \in C_j$ for some $j = 1, \ldots, m$; and note that the distance from y to the boundary of C is greater than or equal to the distance from y to the boundary of C_j. Thus $f(z)$ is holomorphic in $\mathcal{R}^n + iC_j$ and satisfies (5.1) for $y \in C_j$ for each $j = 1, \ldots, m$. Thus by Theorem 7.2 for each quadrant cone C_j there is a nonsingular linear transformation L_j mapping $C_{\bar{0}}$ one to one and onto C_j and a function $g_j(t)$ with $\text{supp}(g_j) \subseteq (L_j^{-1}(C_j))^* = C_{\bar{0}}^*$ almost everywhere and a polynomial X_j such that

$$\left\| e^{-2\pi\langle L_j^{-1}(y),t\rangle} g_j(t) \right\|_{L^2} \leq M(k + (\mathrm{d}(L_j^{-1}(y)))^{-m})^q \exp(M^*(R/|L_j^{-1}(y)|)) \quad , \quad y \in C_j, \tag{7.20}$$

and

$$f(x+iy) = X_j(L_j^{-1}(x) + iL_j^{-1}(y)) \int_{\mathcal{R}^n} g_j(t) e^{2\pi i\langle L_j^{-1}(x)+iL_j^{-1}(y),t\rangle} dt, \quad x + iy \in \mathcal{R}^n + iC_j. \tag{7.21}$$

We have proved the following result.

Theorem 7.3. Let C be a polygonal cone in $\mathcal{R}^n$. Let $f(z)$ be holomorphic in T^C and satisfy (5.1) for $2 < r < \infty$. Then there exist quadrant cones $C_j, j = 1, \ldots, m$; nonsingular linear transformations L_j mapping $C_{\bar{0}}$ one to one and onto C_j; functions g_j having $\text{supp}(g_j) \subseteq \bar{C}_{\bar{0}}$ almost everywhere and satisfying (7.20); and polynomials X_j such that (7.21) holds where $C = \cup_{j=1}^{m} C_j$.

A similar result to Theorem 7.3 can be proved for C being a regular cone.

Can an ultradistributional boundary value be obtained for $f(z)$ in Theorem 7.3 and for the corresponding result for C being a regular cone?

8 FUTURE RESEARCH

In the future we shall attempt to answer the questions posed in section 7 concerning ultradistributional boundary values for the functions $f(z)$ considered there. We further desire to obtain Cauchy and Poisson integral representations for the functions $f(z)$ in sections 6 and 7 where possible and to develop the theory of these functions to the greatest extent possible.

Recall the quadrants C_μ defined in section 7. Tillmann [15] has characterized the analytic functions in tubes $\mathcal{R}^n + iC_\mu$ which have distributional boundary values in the Schwartz space $\mathcal{D}'_{L^r}$ [12, pp. 199-200] and in so doing has shown that the Cauchy kernel function corresponding to the tubes $\mathcal{R}^n + iC$ is an element of the test spaces $\mathcal{D}_{L^s}, 1 < s < \infty$. In [2, Lemma 2.1] Carmichael has shown that in the quadrant setting the Cauchy kernel is also in $\dot{B} \cap \mathcal{D}_{L^\infty}$. (See [12, p. 199] for the definitions of $\dot{B}$ and $\mathcal{D}_{L^s}$.) We know that the general Cauchy kernel defined in (3.2) corresponding to arbitrary regular cones C is an element of $\dot{B} \cap \mathcal{D}_{L^q}$ for all q, $1 \le p \le 2$, $1/p + 1/q = 1$ [3, Lemma 3.1, p. 601]; and we have the result Theorem 3.1 for the ultradifferentiable spaces $\mathcal{D}(*, L^s)$, $2 \le s \le \infty$. We ask if $K(z-t)$ as defined in (3.2) is an element of $\mathcal{D}_{L^s}$, $1 < s < 2$, or $\mathcal{D}(*, L^s)$, $1 < s < 2$, as a function of $t \in \mathcal{R}^n$ for $z \in T^C$ for arbitary regular cones C. If this were the case, much analysis concerning the Cauchy integral of distributions and ultradistributions could be extended to the cases $\mathcal{D}'_{L^s}$ and $\mathcal{D}'(*, L^s)$, $1 < s < 2$.

These problems will be considered in future research.

9 ACKNOWLEDGEMENT

The work of S. Pilipović was supported by the U.S. - Yugoslav Joint Fund for Scientific and Technological Cooperation in cooperation with the National Science Foundation under Grant JF 838.

REFERENCES

1. R. D. Carmichael, Analytic functions related to the distributions of exponential growth, SIAM J. Math. Anal. 10, 1041-1068 (1979).

2. R. D. Carmichael, Distributional boundary values in $\mathcal{D}'_{L^p}$. IV, Rend. Sem. Mat. Univ. Padova 63, 203-214 (1980).

3. R. D. Carmichael, Holomorphic functions in tubes which have distributional boundary values and which are H^p functions, SIAM J. Math. Anal. 14, 596-621 (1983).

4. R. D. Carmichael, Generalization of H^p functions in tubes. I, Complex Variables Theory Appl. 2, 79-101 (1983).

5. R. D. Carmichael, Boundary values of generalizations of H^p functions in tubes, Complex Variables Theory Appl. 8, 83-101 (1987).

6. R. D. Carmichael, Extensions of H^p functions, Progr. Math. (Varanasi), to appear (1992).

7. R. D. Carmichael and D. Mitrović, "Distributions and Analytic Functions", Longman Scientific and Technical, Harlow, Essex, England (1989).

8. R. D. Carmichael, R. S. Pathak, and S. Pilipović, Cauchy and Poisson integrals of ultradistributions, Complex Variables Theory Appl. 14, 85-108 (1990).

9. R. D. Carmichael, R. S. Pathak, and S. Pilipović, Holomorphic functions in tubes associated with ultradistributions, Complex Variables Theory Appl., to appear (1992).

10. H. Komatsu, Ultradistributions, I: Structure theorems and a characterization, J. Fac. Sci. Univ. Tokyo Sect. IA Math. 20, 25-105 (1973).

11. S. Pilipović, Boundary value representation for a class of Beurling ultradistributions, Portugaliae Math. 45, 201-219 (1988).

12. L. Schwartz, "Théorie des distributions", Hermann, Paris (1966).

13. L. Schwartz, "Mathematics for the Physical Sciences", Addison-Wesley, Reading, Massachusetts (1966).

14. E. M. Stein and G. Weiss, "Introduction to Fourier Analysis on Euclidean Spaces", Princeton University Press, Princeton, New Jersey (1971).

15. H. G. Tillmann, Distributionen als Randverteilungen analytischer Funktionen. II, Math. Z. 76, 5-21 (1961).

CAUCHY REPRESENTATION OF DISTRIBUTIONS AND APPLICATIONS TO PROBABILITY II

M. Aslam Chaudhry

Department of Mathematical Sciences
King Fahd University of Petroleum and Minerals
Dhahran 31261, Saudi Arabia

ABSTRACT

The Cauchy representation of distributions in the space $(D_{L_p}(R^n))'$ is proved. It is shown that every probability density defines a generalized function on the space $D_{L_p}(R^n)$ $(1 < p < \infty)$ of test functions. Application of these results in probability theory are discussed.

INTRODUCTION

If $f(t)$ $(-\infty < t < \infty)$ is piecewise continuous, bounded function then the function Bremmerman, [2]

$$f^*(z) = \frac{|y|}{\pi} \int_{-\infty}^{\infty} \frac{f(t)}{|t-z|^2} dt \tag{1}$$

is harmonic in $\Delta = \{z : \operatorname{Im} z \neq 0\}$ and

$$\lim_{\epsilon \to 0+} f^*(t + i\epsilon) = f(t) \quad \text{for all } t \in R. \tag{2}$$

The problem of finding a harmonic function assuming given boundary conditions is known as the Dirichlet problem. It is a classical result that the integral (2) solves the Dirichlet problem in Δ.

The distributional analog of this classical result was proved by Bremmerman [2] for the distributions in $(E(R))'$ as follows. If $T \in (E(R))'$ and

$$\hat{T}(z) = \langle T_t, \frac{1}{t-z} \rangle \tag{3}$$

and

Keywords. Cauchy representation, Hilbert transforms, probability density function.
Subject Classification: 46F, 60B11.

Generalized Functions and Their Applications, Edited by
R.S. Pathak, Plenum Press, New York, 1993

$$\overset{*}{T}(z) = sgn(y) \quad (\hat{T}(z) - \hat{T}(\bar{z})) \\ = \frac{|y|}{\pi}\langle T_t, \frac{1}{|t-z|^2}\rangle, \tag{4}$$

then, $\overset{*}{T}(z)$ is harmonic and

$$\lim_{\epsilon\to 0}\int_{-\infty}^{\infty} \overset{*}{T}(x+i\epsilon)\varphi(x)dx = \langle T, \varphi\rangle \tag{5}$$

for all $\varphi \in E$ *such that* ϕ *and each of its derivative are bounded.* The problem was solved for the spaces $(D_{L_p}(R))'$ $(1 < p < \infty)$ by Pandey and Chaudhry [7].

The following representation theorem was proved by Bremmerman [2].

"Let $T \in (E)'$. Let $\varphi \in E$ be a bounded function with each of its derivatives also bounded. Then

$$\lim_{\epsilon\to 0}\int_{-\infty}^{\infty}[\hat{T}(x+i\epsilon) - \hat{T}(x-i\epsilon)]\varphi(x)dx = \langle T, \varphi\rangle \quad ." \tag{6}$$

It should be noted that (6) does not hold for all $\varphi \in E$. Moreover, $T \in (E)'$ does not imply that the distribution generated by $\hat{T}(x+i\epsilon) - \hat{T}(x-i\epsilon)$ belongs to $(E)'$. For example, we have Bremmerman, [2]

$$\hat{\delta}(z) = \frac{1}{2\pi i}\langle \delta_t, \frac{1}{t-z}\rangle = \frac{-1}{2\pi i z}$$

and

$$\hat{\delta}(x+i\epsilon) - \hat{\delta}(x-i\epsilon) = \frac{\epsilon}{\pi(x^2+\pi^2)} \notin (E)'.$$

However, we have the following result due to Carmichael and Mitrović [4]

If $T \in (E(R))'$, the limits

$$\lim_{\epsilon\to 0}\hat{T}(x \pm i\epsilon) = T^{\pm} \in (D(R))' \tag{7}$$

with

$$T^+ - T^- = T \tag{8}$$

and

$$T^+ + T^- = -\frac{1}{\pi i}(T * vp\frac{1}{x}) \tag{9}$$

The relations (8) and (9) are the distributional form of the classical Plemelj relations [4]. These relations are also proved by Mitrovic [5] to be true for the spaces $O'_{\alpha}(R)$.

The representation (3) if it exists is called Cauchy representation of the distribution. It should be noted that the Cauchy representation is not valid for every Schawrtz distribution $T \in (D(R))'$ because $(t-z)^{-1} \notin D(R)$. However, for every $T \in (D(R))'$ its analytic representation exists i.e., there exists a function $f(z)$, analytic in the z-plane except on K, where K is the support of T, such that, Bremmerman [2]

$$\lim_{\epsilon\to 0}\int_{-\infty}^{\infty}[f(x+i\epsilon) - f(x-i\epsilon)]\varphi(x)dx = \langle T, \varphi\rangle \quad \text{for all } \varphi \in D. \tag{10}$$

The existence of the *analytic representation* in (10) is proved by the Mittag-Leffter method which is cumbersome. The Cauchy representation is far more convenient. The space $(E(R))'$ for which the Cauchy representation exists is too small and the Schwartz space $(D(R))'$ of distributions, for which the Cauchy representation does not exist is too large. In order to extend the class of distributions which are representable by the Cauchy integral, Bremmerman [2] introduced the distribution spaces O'_α.

The general structure formula for the spaces O'_α seems unknown. Therefore, it is difficult to find the exact values of the limits $\lim_{y\to 0} \hat{T}(x \pm iy)$ for $T \in O'_\alpha$.

In this paper, we have proved the existence of the Cauchy representation of the distributions in the intermediate spaces $(D_{L_p}(R^n))' \supset E'(R^n)$ $(1 < p < \infty)$ which is an extension of our previous work [1]. It is also proved that every probability density defines a generalized function on the space $D_{L_p}(R^n)$ $(1 < p < \infty)$ of test functions. Since the structure formula for the spaces $(D_{L_p}(R^n))'$ is known, the proof of the result becomes easier. Applications of these results in probability theory are discussed.

THE TESTING FUNCTION SPACE $D_{L_p}(R^n)$ $(1 < p < \infty)$

An infinitely differentiable function φ defined over R^n is said to belong to the space $D_{L_p}(R^n)$ if $D^\alpha\varphi(x)$ belongs to $L_p(R^n)$ for each $1 = 0, 1, 2, 3\ldots.$ We define a sequence of seminorms on $D_{L_p}(R^n)$ as follows. For $\varphi \in D_{L_p}(R^n)$ we define

$$\gamma_{|\alpha|}(\varphi) = \left[\int_{-\infty}^{\infty} |\psi^{|\alpha|}(x)|^p dx\right]^{1/p} \qquad |\alpha| = 0, 1, 2, 3, \ldots. \tag{11}$$

Since γ_0 is a norm, the sequence of seminorms $\{\gamma_{|\alpha|}\}_{|\alpha|=0}^{\infty}$ is separating [10]. A sequence $\{\varphi_\mu\}_{\mu=1}^{\infty}$ in $D_{L_p}(R^n)$ converges to ϕ in $D_{L_p}(R^n)$ if for each $|\alpha| = 0, 1, 2, 3, \ldots.$

$$\gamma_{|\alpha|}(\varphi_\mu - \varphi) \to 0 \qquad \text{as } \mu \to \infty.$$

The space $D_{L_p}(R^n)$ is multinormed and complete [9, 10]. Moreover, the Schwartz space $D(R^n)$ of test functions is dense in $D_{L_p}(R^n)$ $(1 < p < \infty)$. We denote by $(D_{L_p}(R^n))'$ the space of all continuous linear functionals on $D_{L_p}(R^n)$. It should be noted that our definition of the space $(D_{L_p}(R^n))'$ is different from that of Schwartz where he defines $(D_{L_p}(R^n))'$ as the space of continuous linear functions on $D_{L_q}(R^n)$, $\frac{1}{p} + \frac{1}{q} = 1$. The structure formula for $T \in (D_{L_p}(R^n))'$ can now be stated as follows [4, 9].

Theorem 1 *Every $T \in (D_{L_p}(R^n))$. $(1 < p < \infty)$ is equal to a finite linear combination of the distributional derivatives of functions in $Lq(R^n)$, $\frac{1}{p} + \frac{1}{q} = 1$.*

Theorem 2 *Let $T = \sum_{|\beta|\le r} D^\beta f_\beta$, $f_\beta \in L_q(R^n)$ be an element of $(D_{L_p}(R^n))'$. Then, the Cauchy representation*

$$C(T; z) = \frac{1}{(2\pi\, i)^n}\langle T_t, \prod_{j=1}^{n} \frac{1}{(t_j - z_j)}\rangle \tag{12}$$

of T exists. Moreover,

$$\lim_{y\to 0} C(T; x \pm iy) \quad \text{exists in } (D_{L_p}(R^n))' \quad \text{and}$$

$$\lim_{y\to 0} C(T; x \pm iy) = \pm\frac{1}{2^n} \sum_{|\beta|\le r} (-1)^{|\beta|} D^\beta F_\beta$$

where

$$F_\beta = \sum (\frac{1}{\pi i})^\ell H_{j1} H_{j2} \cdots H_{jn}[f_\beta] \tag{13}$$

and $H_j : L_q(R^n) \to L_q(R^n)$ *is the operator defined by [8]*

$$(H_j\varphi)(t_1, t_2, \ldots, t_{j-1}, x_j, t_{j+1}, \cdots t_n))$$
$$= \frac{1}{\pi} pv \int_R \frac{\varphi(t_1 t_2, \ldots, t_{j-1}, y_j, t_{j+1}, \ldots, t_n) dy_j}{(x_j - y_j)} \quad (1 \leq j \leq n). \tag{14}$$

<u>Proof</u> By definition,

$$\begin{aligned} C(T; z) &= \frac{1}{(2\pi i)^n} \langle T_t, \prod_{j=1}^{n} \frac{1}{(t_j - z_j)} \rangle \\ &= \frac{1}{(2\pi i)^n} \sum_{|\beta| \leq r} \langle D^\beta f_\beta, \prod_{j=1}^{n} \frac{1}{(t_j - z_j)} \rangle \\ &= \frac{1}{(2\pi i)^n} \sum_{|\beta| \leq r} (-1)^{|\beta|} \beta! \langle f_\beta, \prod_{j=1}^{n} \frac{1}{(t_j - z_j)^{\beta_j}} \rangle, \end{aligned} \tag{15}$$

where,

$$|\beta| = \beta_1 + \beta_2 + \cdots + \beta_n \text{ and } \beta! = \beta_1! \beta_2! \cdots \beta_n!.$$

Since, for all $z = (z_1, z_2, \ldots, z_n)$ with $\operatorname{Im} z_i \neq 0$, the function $\prod_{j=1}^{n} (t_j - z_j)^{-\beta_j} \in D_{L_p}(R^n)$, $(1 < p < \infty)$ it follows that the integral

$$\langle f_\beta, \prod_{j=1}^{n} (t_j - z_j)^{-\beta_j} \rangle = \int_{R^n} \frac{f_\beta(t) dt}{\prod_{j=1}^{n} (t_j - z_j)^{\beta_j}} \tag{16}$$

is well defined and convergent. Using the Lemma (10.3) proved by Pandey and Singh [8] we get that

$$\langle f_\beta, \prod_{j=1}^{n} (t_j - z_j)^{-\beta_j} \rangle \to \frac{1}{\beta!} \sum (i\pi)^{n-\ell} H_{j1}.\, H_{j2} \cdots H_{j\ell}[D^\beta f_\beta](x) \quad \text{as } \|y\| = \max_{1 \leq j \leq n} (y_j) \to 0. \tag{17}$$

Therefore, letting $\|y\| = \max_{1 \leq j \leq n} (y_j) \to 0$ in (15) and using (16) and (17) we get

$$\lim_{y \to 0} C(T; x + iy) = \frac{1}{2^n} \sum_{|\beta| \leq r} (-1)^{|\beta|} \sum_{\ell} \frac{1}{(i\pi)^\ell} H_{j1} H_{j2} \cdots H_{j\ell}[D^\beta f_\beta](x)$$

However, the operators $H_{j1}, H_{j2}, \ldots, H_{j\ell}$ and D^β are commutative [8], it follows that

$$\lim_{y \to 0} C(T; x + iy) = \frac{1}{2^n} \sum_{|\beta| \leq r} (-1)^{|\beta|} D^\beta (\sum_{\ell} (\frac{1}{\pi i})^\ell H_{j1} H_{j2} \ldots H_{j\ell}[f_\beta](x)).$$

Since $H_{ji} : L_q(R^n) \to L_q(R^n)$, therefore

$$F_\beta = \sum (\frac{1}{\pi i})^\ell H_{j1} H_{j2} \cdots H_{j\ell}[f_\beta] \in L_q(R^n). \tag{18}$$

Consequently,

$$\lim_{y \to 0} C(T; x + iy) = \frac{1}{2^n} \sum_{|\beta| \leq r} (-1)^{|\beta|} D^\beta F_\beta \in (D_{L_p}(R^n))'.$$

Similarly it can be seen from (17) and (18) that

$$\lim_{y \to 0} C(T; x - iy) = \frac{-1}{2^n} \sum_{|\beta| \le r} (-1)^{|\beta|} D^\beta F_\beta.$$

APPLICATIONS TO PROBABILIY

Every probability density function $f(t), t \in R^n$ can be expressed as the distributional derivative of a probability distribution $\Phi(t)$ i.e.,

$$f(t) = D_t^n \Phi(t), \tag{19}$$

where $D_t^n = \frac{\partial^n}{\partial t_1 \partial t_2 \cdots \partial t_n}$ and $\Phi(t)$ is bounded. If $\Phi(t)$ is a discontinuous function with discontinuities of the first kind, then the probability density $f(t)$ does not exist in the ordinary sense. For example, if we take

$$\begin{aligned} \Phi(t) &= H(t_1, t_2, \ldots, t_n) \\ &= H(t_1)H(t_2) \cdots H(t_n), \end{aligned} \tag{20}$$

where, H is the Heaviside step function, then the probability density does not exist in the ordinary sense. If, however, we admit generalized functions, then

$$f(t) = \delta(t_1)\delta(t_2) \cdots \delta(t_n) \tag{21}$$

is the product of Dirac delta functions.

As a matter of fact, every probability density $f(t)$ defines a continuous linear functional on the space $D_{L_p}(R^n))$, $(1 < p < \infty)$. We prove this fact in the following theorem.

Theorem 3 *Every probability density $f(t), t \in R^n$ defines a generalized function on the space $D_{L_p}(R)$ $(1 < p < \infty)$ of test functions.*

Proof. Let $f(t) = D_t^n \Phi(t)$, where $\Phi(t)$ is a probability distribution. Then, for any $\psi \in (D_{L_p}(R^n))$, we have

$$\begin{aligned} \langle f(t), \psi(t) \rangle &= \langle D_t^n \Phi, \psi(t) \rangle \\ &= (-1)^n \int_{R^n} \Phi(t)[D^n \psi](t) dt \end{aligned} \tag{22}$$

Since the map $\psi \to D^n \psi$ is continuous injection [9] on $D_{L_p}(R^n)$, $(1 < p < \infty)$ and $\Phi(t)$ is bounded, the integral (22) exists, that is the linear map $\psi(t) \to \langle f(t), \psi(t) \rangle$ is well defined. Moreover,

$$\begin{aligned} |\langle f(t), \psi(t) \rangle| &\le \int_{R^n} |D^n \psi(t)| dt \\ &\le C_{p,n} \|\psi\|_{L_p} \qquad [9]. \end{aligned} \tag{23}$$

Therefore, $f(t)$ is bounded. Consequently, $f(t)$ defines a generalized function on $D_{L_p}(R^n)$, $(1 < p < n)$.

Corollary. *For every probability density $f(t)$, $t \in R^n$, the Cauchy representation as defined by (12) exists.*

Example 1. Consider the Poisson probability distribution

$$\Phi(t) = \sum_{j=0}^{\infty} e^{-\lambda} \frac{\lambda^j}{j!} H(t-j).$$

The corresponding probability density is

$$f(t) = \sum_{j=0}^{\infty} e^{-\lambda} \frac{\lambda^j}{j!} \quad \delta(t-j). \tag{24}$$

It should be noted that $\sup f = \{0, 1, 2, 3, \ldots\}$ is not compact and hence $f \notin E'$. However, $f \in (D_{L_p}(R))'$. Therefore, its Cauchy representation exists and is given by

$$C(f; z) = \frac{1}{2\pi i} e^{-\lambda} \sum_{j=1}^{\infty} \frac{\lambda^j}{j!(j-z)}.$$

Example 2: Consider the probability distribution $\Phi(t), \quad t \in R^n$ defined by

$$\Phi(t) = H(t_1)H(t_2)\cdots H(t_n).$$

The corresponding probability density is given by

$$\delta(t) = \delta(t_1)\delta(t_2)\cdots\delta(t_n).$$

Now, $\delta(t) \in (D_{L_p}(R^n))'$ and its Cauchy representation is given by

$$C(\delta; z) = \frac{1}{(-2\pi\, i)^n} \frac{1}{z_1 z_2 \cdots z_n}.$$

ACKNOWLEDGEMENT

The author gratefully acknowledges the support provided for this research by the King Fahd University of Petroleum and Minerals, Dhahran, Saudi Arabia. Private discussions with Professor Munir Ahmad are also appreciated.

REFERENCES

1. M. Aslam Chaudhry, Cauchy representation of distributions and applications to probability, *Bull. Austral. Math. Soc.* 42(1990), 125-131.

2. H. Bremmerman, "Distributions, Complex Variables and Fourier Transform," (Addison-Wesley Publishing Company Inc., 1965).

3. R.D. Carmichael, Representation of distribution in O'_α as boundary values of functions in tube domains, *Glas. Mat. Ser. III*, 7(1972), 213-228.

4. R.D. Carmichael and D. Mitrović, "Distributions and Analytic Functions," (Longman Scientific and Technical, 1980).

5. D. Mitrovic, Plemelj formulas and analytic representation of distributions, *Glas. Mat. Ser. III*, 3(1968), 231-239.

6. ________________, A Hilbert distributional boundary value problem, *Math. Balkanica*, 1(1971), 177-180.

7. J.N. Pandey and M.A. Chaudhry, Hilbert transform of generalized functions and applications, *Canad. J. Math XXXV*, no. 3 (1983), 478-495.

8. O.P. Singh & J.N. Pandey, The n-dimensional Hilbert transform of distributions, its inversion and applications, *Canad. J. Math.*, vol LII, no. 2, 1990,239-258.

9. L. Schwartz, "Theories des Distributions," vol. I, II (Herman, Paris, 1959).

10. A.H. Zemanian, "Generalized Integral Transforms," (Interscience Publishers, New York, 1968).

APPLICATIONS OF GENERALIZED FUNCTIONS TO SHOCKS AND DISCRETE MODELS

Yvonne Choquet-Bruhat
Mecanique Relativiste
Universite Paris 6, France

INTRODUCTION

Generalized functions, also called distributions, have proved useful in many problems related to physics even before their rigorous and general definition. Generalized functions are also a powerful tool to formulate in a unified manner fundamental physical laws. We shall give here two general examples of such an application of generalized functions. The first is to the writing of the equations satisfied by shock waves : we establish in particular the algebraic conditions satisfied across a gravitational shock wave and the propagation of the corresponding discontinuities. In the second example we use a representation of the distribution function of kinetic theory by a sum of discrete measures to find the boltzmann equation for a model with a finite number of velocities, together with propagation equations satisfied by these velocities.

1. GENERALIZED FUNCTION ON A MANIFOLD

To use generalized functions in physics, in particular in the relativistic case it is necessary to have a good definition of the change of variables, and more generally of a generalized function on a differentiable manifold. We recall (cf. for instance Y. Choquet-Bruhat and C.DeWitt - Morette 1982).

Generalized Functions and Their Applications, Edited by
R.S. Pathak, Plenum Press, New York, 1993

Definition Let V be a paracompact C^∞ differentiable manifold. The space $\mathcal{D}'(V)$ of generalized functions on V is the topological dual of $\mathcal{D}(V)$, space of C^∞ functions on V with compact support. We denote, if $f \in \mathcal{D}'(V)$:

$$f : \mathcal{D}(V) \longrightarrow \mathbb{R} \quad \text{by } \varphi \mapsto \langle f, \varphi \rangle$$

The topology of $\mathcal{D}(V)$ is the inductive limit of the Frechet topologies each defined on a compact set K of V by the family of semi norms $p_{k_i, m}(\varphi) \equiv \sup_{x_i \in K_i} |\mathcal{D}^m \varphi_i(x_i)|$, $K_i = u_i(U_i \cap K)$, (U_i, u_i) chart of a locally finite atlas of V with φ_i the representative of φ in this chart and $\mathcal{D}^m$ a partial derivative of order m.

We suppose given on V a volume element η and we define the representative of $f \in \mathcal{D}'(V)$ in the chart U_i of V as the distribution f_i on the open set of $\mathbb{R}^d$ given by

$$\langle f_i, \rho_i \varphi_i \rangle_{\mathbb{R}^d} \equiv \langle f, \varphi \rangle ,$$

where $\langle .,. \rangle_{\mathbb{R}^d}$ denotes the usual duality between $\mathcal{D}'(\mathbb{R}^d)$ and $D(\mathbb{R}^d)$, the representative of η in U_i being

$$\eta_i \equiv \rho_i \, dx^0 dx^1 \dots dx^n , \qquad d = n+1.$$

The representatives in the intersection $U_i \cap U_j$ of two overlapping charts (U_i, u_i) and (U_j, u_j) are linked in $U_i \cap U_j$ by the classical formula, where $\psi \in \mathcal{D}(\Omega)$, $\Omega \equiv u_i \ (U_i \cap U_j)$

$$\langle f_i , \psi \rangle_{\mathbb{R}^d} \equiv \langle f_j , \left| \frac{D(x_i)}{D(x_j)} \right| \psi \circ u_i \circ u_j^{-1} \rangle_{\mathbb{R}^d} .$$

The representatives of partial derivatives obey the usual law by change of variables.

2. GENERALIZED TENSORS

We recall some definitions and properties of generalized, also called distribution-valued, tensors on a C^∞, paracompact, d-dimensional manifold V (cf. Lichnerowicz 1961 or, for instance, Choquet-Bruhat and DeWitt-Morette 1982). The notion of C^∞ tensor fields can be defined without introduction of a particular connection on V. To define topologies on spaces of smooth tensor fields one can proceed through the use of a locally finite atlas along the same lines as for C^∞ functions. The family of semi norms are obtained by the supremum of the absolute value of a partial derivative of arbitrary order of an arbitrary component in a cordinate chart. A linear mapping

$$T : \mathcal{D}_p \mapsto \mathbb{R} \text{ by } \tau \mapsto < T, \tau > ,$$

$\mathcal{D}_p$ space of C^∞ p-contravariant tensor fields on V with compact support, is continous if and only for each τ with compact support in the domain U_i of a chart and each partial derivative operator D^m and component represented by $\tau^{\alpha_1 \ldots \alpha_p}$ there exists a constant C such that

$$| < T,\tau > | \le C \sup_{x_i \in u_i(u_i)} |D^m \tau^{\alpha_1 \ldots \alpha_p}(x_i)| .$$

Different admissible atlases give equivalent definitions of continuity, i.e. equivalent topologies to $\mathcal{D}_p$.

Definition. The space of p-covariant generalized tensors, $\mathcal{D}^{*\prime}_p$, is the topological dual of $\mathcal{D}_p$. We note, if $T \in \mathcal{D}^{*\prime}_p$:

$$T : \mathcal{D}_p \longrightarrow \mathbb{R} \text{ by } \varphi \mapsto < T , \varphi > .$$

Let η be a C^∞ volume element on V. If T is an ordinary locally integrable tensor we identify it with the generalized tensor such that (the integrand is coordinate independent)

$$< T, \tau > \equiv \int_V T.\tau \, \eta \equiv \int_V T_{\alpha_1 \ldots \alpha_p} \tau^{\alpha_1 \ldots \alpha_p} \eta . \qquad (2.1)$$

In a coordinate chart we have

$$\eta \equiv \rho \, dx^0 \, dx^1 \ldots dx^n , \qquad n + 1 = d . \qquad (2.2)$$

The components $T_{\alpha_1 \ldots \alpha'_p}$ locally integrable functions in the image of the chart in $\mathbb{R}^d$, are the representatives in $\mathbb{R}^d$ of generalized functions in the domain of the chart in V, for which we use the same notation, with the property

$$< T_{\alpha_1 \ldots \alpha_p} , \varphi > = < T_{\alpha_1 \ldots \alpha_p} , \rho \, \varphi >_{\mathbb{R}^d} \qquad (2.3)$$

where φ denotes on the left a C^∞ function on V with support in U and on the right its representant in the chart.

Let now T be a generlized p-covariant tensor and let $e_0, \ldots, e_n$ be a basis of vector fields in an open set Ω of V. The components of T in this basis are generalized functions on V defined by ($D(\Omega)$ denotes C^∞ functions on V with compact support in Ω)

$$< T_{\alpha_1 \ldots \alpha_p}, \varphi > \equiv < T, \varphi\, e_{\alpha_1} \otimes \ldots \otimes e_{\alpha_p} >, \; \varphi \in \mathcal{D}(\Omega) \; .$$

It results from the linearity and multiplication law of a generalized function by a C^∞ one that the components of T transform as the components of a p-covariant tensor by change of basis. If Ω is the domain of a chart the representatives of the components of T are generalized functions on an open set of $\mathbb{R}^d$, denoted by the same notation, and satisfying the relation (2.3).

We define the partial derivative $\partial_\beta T_{\alpha_1 \ldots \alpha_p}$ in a coordinate chart as the generalized function such that

$$< \partial_\beta T_{\alpha_1 \ldots \alpha_p}, \varphi > \equiv - < T_{\alpha_1 \ldots \alpha_p}, \rho^{-1} \partial_\beta (\rho\varphi) >. \qquad (2.4)$$

We now suppose that V is endowed with a C^∞ linear connection which leaves invariant the volume element η and has for representative the set of C^∞ functions $\Gamma^\lambda_{\beta\mu}$ in a coordinate chart. We define the covariant derivative of the generalized tensor T by its components in coordinate charts, such that

$$\nabla_\beta T_{\alpha_1 \ldots \alpha_p} \equiv \partial_\beta T_{\alpha_1 \ldots \alpha_p} - \Gamma^\lambda_{\beta\alpha_1} T_{\lambda\alpha_2 \ldots \alpha_p} - \ldots - \Gamma^\lambda_{\beta\alpha_p} T_{\alpha_1 \ldots \alpha_{p-1}\lambda}. \qquad (2.5)$$

It results form the transformation laws of volume elements and linear connections that the quantities (2.5) obey the required transformation law of a (p + 1) - covariant tensor by change of coordinates. The so defined generalized tensor ∇T is such that, if $\tau \in \mathcal{D}_p$

$$< \nabla T, \tau > \equiv - < T, \nabla.\tau >, \text{ with } (\nabla.\tau)^{\alpha_1 \ldots \alpha_p} \equiv \nabla_\beta \tau^{\beta\alpha_1 \ldots \alpha_p}. \qquad (2.6)$$

If the manifold V is endowed with a pseudo riemannian metric with η its volume element and Γ its riemannian connection we find from Lichnerowicz (1961) definition of the covariant derivative of a p generalized tensor.

3. MULTIPLE LAYERS

Particularly useful in applications are some generalized tensors with support in n-dimensional submanifold Σ of V. We suppose now that V is oriented and Σ is defined by an equation $f = 0$, such that $f : V \longrightarrow \mathbb{R}$ is a smooth function with non vanishing gradient in an open neighbourhood Ω of Σ in V. We suppose that Ω is divided by Σ in two disjoint open sets

$$\Omega_+ \equiv \{x\in\Omega,\ f(x) > 0\},\ \Omega_- \equiv \{x\in\Omega,\ f(x) < 0\}.$$

We orient Σ by setting $\Sigma \equiv \partial\Omega_-$. We recall some definitions.

A *Leray form* relative to Σ is an $(n-1)$-form ω such that

$$df \wedge \omega \equiv \eta .$$

It is defined up to the addition of a form vanishing on Σ.

The *Dirac measure* δ_Σ is the generalized function of order zero (measure) with support Σ defined by

$$< \delta_\Sigma ,\ \varphi > \equiv \int_\Sigma \varphi\ \omega$$

The measure δ_Σ depends through ω on the choice of f. Since f has non vanishing gradient we can choose charts on Ω such that, in local coordinates, $f(x) \equiv x^0$ and

$$\omega \equiv \rho\ dx^1 \wedge \ldots \wedge dx^n .$$

Lemma 1. The following formulas hold

$$\nabla Y_+ = -\nabla Y_- = \ell\ \delta_\Sigma , \qquad \ell = \nabla f$$

where Y_+ [resp Y_-] is the locally integrable function equal to 1 in Ω_+ [resp. Ω_-] and to 0 in Ω_- [resp. Ω_+].

Proof. Let $\tau \in \mathcal{D}_1$, then

$$< \nabla Y_+,\tau > = - < Y_+,\nabla.\tau > = - \int_{\Omega_+} \nabla_\alpha \tau^\alpha \eta = - \int_{\partial\Omega_+} \tau^\alpha \ell_\alpha \omega = \int_{\partial\Omega_-} \tau^\alpha \ell_\alpha \omega$$

The result follows from the multiplication law of C^∞ and generalized functions.

Lemma 2. There is a generalized function of order 1 with support Σ, called δ_Σ', such that

$$\nabla \delta_\Sigma = \ell \, \delta_\Sigma' .$$

Proof. Let $\tau \in \mathcal{D}_1$, then

$$< \nabla\delta_\Sigma , \tau > = - < \delta_\Sigma , \nabla . \tau > = - \int_\Sigma \nabla_\alpha \tau^\alpha \omega .$$

Choose local coordinates such that $f(x) \equiv x^0$. Then $x^1, \ldots x^n$ are local coordinates on Σ and, if τ has compact support in the chart

$$\int_\Sigma \nabla_\alpha \tau^\alpha \omega = \int_{x^0=0} \partial_\alpha(\rho\tau^\alpha) dx^1 \ldots dx^n = \int_{x^0=0} \partial_0(\rho\tau^0) dx^1 \ldots dx^n .$$

In the chosen coordinates we have therefore $\nabla_i \delta_\Sigma = 0$ and we set $\delta_\Sigma' = \nabla_0 \delta_\Sigma$. Hence in these coordinates $\nabla\delta_\Sigma = \ell \, \delta_\Sigma'$. This tensorial expression is valid in all coordinates, with δ_Σ' the generalized function of order 1 defined above in particular coordinates.

4. REGURLARLY DISCONTINUOUS TENSORS

Definition. A tensor T is said to be regularly C^k discontinuous across Σ if

1) T is C^k in Ω_+ and Ω_- .

2) $T, \nabla T, \ldots, \nabla^k T$ converge uniformly to tensors denoted $T_+, (\nabla T)_+, \ldots, (\nabla^k T)_+$ when $x \in \Omega_+$ tends to a point of Σ [resp. T_-, $(\nabla T)_-, \ldots, (\nabla^k T)_-$ when $x \in \Omega_-$ tends to Σ].

To $T, \ldots \nabla^k T$, locally integrable tensors are associated generalized tensors. On the other hand the following ordinary tensors are defined and continuous on Σ :

$$[T] \equiv T_+ - T_-, \ldots \; [\nabla^k T] \equiv (\nabla^k T)_+ - (\nabla^k T)_- .$$

and such that $[T]_{\alpha^1 \ldots \alpha^p} = [T_{\alpha^1 \ldots \alpha^p}]$, with an analogous property for the derivatives.

We have therefore

Lemma 1. $[T]\delta_\Sigma, \dots, [\nabla^k T]\delta_\Sigma$ are generalized tensors of order 0 (measures) with support Σ, defined by

$$<[\nabla^k T]\delta_\Sigma \, , \varphi >= \int_\Sigma [\nabla^k T] \cdot \varphi \, \omega, \qquad \varphi \in \mathcal{D}_{p+k} \text{ if T is p-covariant.}$$

Lemma 2. 1) There exists a generalized tensor t of order 1 with support Σ such that

$$\nabla ([T] \delta_\Sigma) - \nabla [T] \delta_\Sigma \equiv \ell \otimes t.$$

2) There exists a generalized tensor v of order 2 such that

$$\nabla(\ell \otimes t) + \ell \otimes \nabla t + \ell \otimes \ell \otimes v \equiv [\nabla\nabla T]\delta_\Sigma$$

Proof. By use of the definitions the components of $\nabla([T]\delta_\Sigma)$ are such that

$$< \partial_\beta ([T]\delta_\Sigma)_{\alpha^1 \dots \alpha^p}, \varphi > = - \int_\Sigma [T_{\alpha^1 \dots \alpha^p}] \, \rho^{-1}\partial_\beta(\varphi\rho) \, \omega,$$

$$\varphi \in \mathcal{D}(\Omega).$$

Hence in adapted coordinates, using integration by parts

$$\partial_\beta ([T]\delta_\Sigma)_{\alpha^1 \dots \alpha^p} = \partial_i ([T])_{\alpha^1 \dots \alpha^p} \delta_\Sigma$$

from which results

$$\nabla_i ([T]\delta_\Sigma)_{\alpha^1 \dots \alpha^p} = [\nabla_i T_{\alpha^1 \dots \alpha^p}] \, \delta_\Sigma$$

and the announced tensorial relation where t is the tensor with components in adapted coordinates such that

$$<t_{\alpha^1 \dots \alpha^p}, \varphi >= \int_{x^0=0} \{[T_{\alpha^1 \dots \alpha^p}]\partial_0(\rho\varphi) - [\nabla_0 T_{\alpha^1 \dots \alpha^p}]\rho\varphi\} dx^1 \dots dx^n$$

The second part is proved analogously.

Corollary. If T is continuous accross Σ and ∇T regularly discontinuous then one has the "Hadamard relation" :

$$[\nabla T]\delta_\Sigma = \ell \otimes t \, ,$$

where t is a tensor valued measure with support Σ and components in adapted coordinates given by

$$t_{\alpha^1 \dots \alpha^p} = - [\nabla_0 T_{\alpha^1 \dots \alpha^p}] \, \delta_\Sigma.$$

We denote by τ the p-tensor defined and continuous on Σ with components $[\nabla_0 T_{\alpha^1 \ldots \alpha^p}]$ in adapted coordinates, then in arbitrary coordinates

$$[\nabla_\beta T_{\alpha^1 \ldots \alpha^p}] = \ell_\beta \tau_{\alpha^1 \ldots \alpha^p}$$

5. SHOCKS IN RELATIVISTIC FLUIDS

The shock equations in mechanics can be obtained by looking for generalized solutions which are regularly discontinuous tensors across a 3 dimensional submanifold Σ of the 4 dimensional space time. We give as an example the case of a perfect relativistic fluid. The dynamical equations are conservation laws

$$\nabla_\alpha T^{\alpha\beta} = 0$$

for the stress energy tensor

$$T_{\alpha\beta} \equiv (r+p)\ u_\alpha u_\beta + pg_{\alpha\beta}$$

r, p specific energy density and pressure, u unit 4-velocity, g space time metric. We say that the fluid undergoes a shock across Σ if T and ∇T are regularly discontinuous accross Σ. We denote by $\nabla_\alpha\{T^{\alpha\beta}\}$ the locally integrable function on Ω equal to $\nabla_\alpha T^{\alpha\beta}$ in $\Omega_+ \cup \Omega_-$.

Lemma 1. If T and ∇T are regularly discontinuous accross Σ the derivative of the generalized tensor T is given by :

$$\nabla_\alpha\ T^{\lambda\mu} \equiv \{T^{\lambda\mu}\} + \ell_\alpha[T^{\lambda\mu}]\delta_\Sigma\ .$$

Proof. Analogous to previous ones.

A consequence of the lemma is the following theorem

Theorem. The stress energy tensor of a fluid which undergoes a shock across Σ satisfies the equations

$$\nabla_\alpha\ \{T^{\alpha\beta}\} = 0$$

and

$$\ell_\alpha[T^{\alpha\beta}] = 0, \text{ the Rankine Hugoniot equations.}$$

6. GRAVITATIONAL SHOCKS

The physical metric g is supposed to be continuous while its partial derivatives are C^2 regularly discontinuous across Σ. In order to have covariant formulas we endow the space time V with a given C^∞ metric e (which can be euclidean metric if permitted by the topology of V) and denote by ∂ the covariant differentiation in the metric e.

The equations satisfied in a generalized sense are Einstein equations in vacuo which read, with H depending smoothly on its arguments

$$R_{\lambda\mu} \equiv -\frac{1}{2} g^{\alpha\beta}\partial^2_{\alpha\beta} g_{\lambda\mu} + \partial_\lambda F_\mu + \partial_\mu F_\lambda + H_{\lambda\mu}(g, \partial g) = 0$$

$$F^\lambda \equiv g^{\alpha\beta}(\Gamma^\lambda_{\alpha\beta} - E^\lambda_{\alpha\beta}), \quad E^\lambda_{\alpha\beta} \text{ Christoffel symbols of e}$$

($F^\lambda = 0$ are the generalized harmonicity conditions).

The results of § 3 show on the one hand that there exists a symmetric space time 2-tensor γ defined and continuous on Σ such that

$$[\partial_\beta g_{\lambda\mu}] = \ell_\beta \gamma_{\lambda\mu} \,.$$

on the other hand they give the formulas

$$\partial^2_{\alpha\beta} g_{\lambda\mu} = \partial^2_{\alpha\beta}\{g_{\lambda\mu}\} + \ell_\alpha \ell_\beta \gamma_{\lambda\mu} \delta_\Sigma \,,$$

$$\partial_\lambda F_\mu = \partial_\lambda\{F_\mu\} + \ell_\lambda [F_\mu] \delta_\Sigma \,,$$

$$R_{\lambda\mu} = \{R_{\lambda\mu}\} + (-\frac{1}{2} g^{\alpha\beta} \ell_\alpha \ell_\beta \gamma_{\lambda\mu} + \ell_\lambda [F_\mu] + \ell_\mu [F_\lambda])\delta_\Sigma$$

The sum of a locally integrable function on V and a measure with support Σ is zero if and only if both are zero. The measure with support Σ in the above equation is zero if and only if the coefficient of δ_Σ is zero. We consider two cases:

1) $g^{\alpha\beta}\ell_\alpha\ell_\beta \neq 0$ on Σ, then $\gamma_{\lambda\mu} = (\ell_\lambda[F_\mu]+\ell_\mu[F_\lambda])(\frac{1}{2}g^{\alpha\beta}\ell_\alpha\ell_\beta)^{-1}$

such a discontinuity in ∂g across Σ is considered as non significant because it can be destroyed by a C^0 regularly C^1 across Σ change of coordinates.

2) $g^{\alpha\beta}\ell_\alpha\ell_\beta = 0$: ℓ is a null vector of the physical metric. Einstein equations reduce then to

$$\{R_{\lambda\mu}\} = 0 \text{ and } [F_\lambda] = 0.$$

The second set of conditions, which are independant of the choice of the backgound metric e, are the "gravitational Rankine Hugoniot equations". They read, using the expressions of F and $[\partial g]$, setting $\ell^{\alpha} = g^{\alpha\beta}\ell_{\beta}$

$$\ell^{\alpha}\theta_{\alpha\beta} = 0 \ , \text{ with } \theta_{\alpha\beta} \equiv \gamma_{\alpha\beta} - \frac{1}{2} g_{\alpha\beta}g^{\lambda\mu}\gamma_{\lambda\mu} \ ,$$

7. PROPAGATION OF GRAVITATIONAL SHOCKS

By the lemma 2 we have, using also the fact that $\ell = \nabla f$ is a smooth covariant vector in a neighbourhood of Σ

$$[\partial^2_{\alpha\beta}g_{\lambda\mu}]\delta_{\Sigma} = (\partial_{\alpha}\ell_{\beta})\gamma_{\lambda\mu}\delta_{\Sigma} + \ell_{\alpha}\partial_{\beta}(\gamma_{\lambda\mu}\delta_{\Sigma}) + \ell_{\beta}\partial_{\alpha}(\gamma_{\lambda\mu}\delta_{\Sigma}) + \ell_{\alpha}\ell_{\beta}V_{\lambda\mu}$$

therefore, since ℓ is a null vector for g in a neibourhood of Σ,

$$g^{\alpha\beta}[\partial^2_{\alpha\beta}g_{\lambda\mu}]\delta_{\Sigma} = 2\ (g^{\alpha\beta}\ \partial_{\alpha}\ \ell_{\beta})\ \gamma_{\lambda\mu}\ \delta_{\Sigma} + \ell^{\alpha}\partial_{\alpha}(\gamma_{\lambda\mu}\delta_{\Sigma}).$$

Use of the fact that $\ell^{\alpha}\partial_{\alpha}$ is a derivation on Σ, with $\ell^{\alpha}\partial_{\alpha}\gamma_{\mu\lambda}$ a continuous function on Σ , and of the formula $\partial_{\alpha}\delta_{\Sigma} = \ell_{\alpha}\delta_{\Sigma}'$, give

$$g^{\alpha\beta}[\partial^2_{\alpha\beta}g_{\lambda\mu}] \equiv \ell^{\alpha}\partial_{\alpha}\ \gamma_{\lambda\mu} + 2\ g^{\alpha\beta}\ \partial_{\alpha}\ell_{\beta}\ \gamma_{\lambda\mu}\ .$$

We have on the other hand, since we have proved that F_{λ} is continuous across Σ ,

$$[\partial_{\lambda}F_{\mu}] \equiv \ell_{\lambda}\ \Phi_{\mu}$$

where Φ is defined and continuous on Σ. We deduce then from $R_{\lambda\mu} = 0$

$$\ell^{\alpha}\partial_{\alpha}\ \gamma_{\lambda\mu} + 2\ g^{\alpha\beta}\partial_{\alpha}\ \ell_{\beta}\ \gamma_{\lambda\mu} - \ell_{\lambda}\ \Phi_{\mu} - \ell_{\mu}\ \Phi_{\lambda} - [H_{\lambda\mu}] = 0.$$

One remarks that, in spite of the discontinuity of ∂g across Σ , the operator $\ell^{\alpha}\nabla_{\alpha}$ is defined over Σ as well as the function $g^{\alpha\beta}\ \nabla_{\alpha}\ \ell_{\beta}$, due to the Rankine Hugoniot equations $[F_{\lambda}] = 0$. It can be proved that the above equation can be written

$$\ell^{\alpha}\nabla_{\alpha}\ \gamma_{\lambda\mu} + 2\ g^{\alpha\beta}\ \nabla_{\alpha}\ \ell_{\beta}\ \ \gamma_{\lambda\mu} - \ell_{\lambda}\ \Phi_{\mu} - \ell_{\mu}\ \Phi_{\lambda} = 0\ .$$

Equations which determine the propagation along the "rays" of the significant part of the discontinuity γ .

8. DISCRETE BOLTZMANN EQUATION

We recall that the usual Boltzmann equation on a pseudo riemannian manifold (V, g) reads

$$\mathcal{L}f \equiv p^{\alpha} \frac{\partial f}{\partial x^{\alpha}} + p^{\alpha} \frac{\partial f}{\partial p^{\alpha}} = \mathcal{J}(f)$$

where x^{α}, p^{α} are local coordinates in the tangent bundle T(V) to V, representing position and 4-momentum of particles; f is the distribution function of particles on T(V); (p,P) is the tangent vector to the trajectories of particles in the phase space T(V) between collisions : the operator $\mathcal{L}$ is the Lie derivative along such a free trajectory, $\mathcal{L}f = 0$ expresses the conservation of f in the absence of collisions. The term $\mathcal{J}(f)$ expresses the loss and gain of particles with momentum p undergoing collisions at the point x.

If no other field is present the particles follow between collisions geodesics of the metric g, in the general case we have :

$$P^{\alpha} \equiv - \Gamma^{\alpha}_{\lambda\mu} \, p_{\lambda} p_{\mu} + Q^{\alpha}$$

where Q is the tangent vector in V to the non gravitational 4-force. If this force originates from an electromagnetic field represented by an exterior 2-form F, then

$$Q^{\alpha} \equiv e \, F^{\alpha}_{\beta} \, p^{\beta} \quad \text{, e the charge of the particles.}$$

The classical discrete Boltzmann equation is obtained by supposing that the velocities of the particles can take only a finite number of fixed given values. Such an hypothesis cannot be made on a curved manifold, neither is it compatible, even in usual euclidean space, with the presence of an external force. Following G. Pichon we shall say that we have a discrete model if the possible velocities, or momenta in a relativistic model, of the particle are finite in number but possibly dependent on the space time point. We shall consider distribution generalized functions of the form

$$f(x,p) \equiv \sum_{I=1}^{N} a_1(x) \, \delta_{B_I(x)} \, (p)$$

where a_1 is a smooth function and B_I a smooth vector field on V while $\delta_{B_I(x)}(p)$ is the measure on T(V) defined by

$$< \delta_{B_I(x)}(p) \ , \ \varphi(x,p) > = \int_V \varphi(x, B_I(x)) \ \eta \ , \quad \varphi \in \mathcal{D}(T(V))$$

with in local coordinates $\eta \equiv \rho dx^0 \ldots dx^n$, with $\rho \equiv |\det g|^{1/2}$ when V is endowed with a pseudo-riemannian metric g. Note that the volume element of T(V) is then $\rho^2 dx^0 \ldots dx^n dp^0 \ldots dp^n$. The partial derivatives of the generalized functions δ_{B_I} on T(V) are such that (we suppress the writing of the dependance in x and p for δ_{B_I} to shorten the writing, but keep it in φ to make the proof more transparent)

$$< \frac{\partial}{\partial p^\alpha} \delta_{B_I} \ , \ \varphi(x,p) > = - < \delta_{B_I} \ , \ \rho^{-2} \frac{\partial(\rho^2 \varphi(x,p))}{\partial p^\alpha} >$$

Hence, since ρ does not depend on p :

$$< \frac{\partial}{\partial p^\alpha} \delta_{B_I} , \ \varphi(x,p) > = - \int_V \left\{\frac{\partial \varphi(x,p)}{\partial p^\alpha}\right\}_{p=B_I(x)} \eta$$

$$< \frac{\partial}{\partial x^\alpha} \delta_{B_I} , \ \varphi(x,p) > \equiv - < \delta_{B_I} , \ \rho^{-2} \frac{\partial}{\partial x^\alpha}(\rho^2 \varphi(x,p)) >$$

We use the property $\rho^{-1} \frac{\partial}{\partial x^\alpha} \rho \equiv \Gamma^\lambda_{\alpha\lambda}$, together with the identity

$$\left\{ \frac{\partial}{\partial x^\alpha} \varphi(x,p) \right\}_{p=B_I} = \frac{\partial}{\partial x^\alpha} \varphi(x,B_I(x)) - \frac{\partial B^\beta_I}{\partial x^\alpha} \left\{\frac{\partial \varphi(x,p)}{\partial p^\beta}\right\}_{p=B_I},$$

the definition of δ_{B_I} and the compactness of the support of φ to obtain

$$\frac{\partial}{\partial x^\alpha} \delta_{B_I} = - \Gamma^\lambda_{\alpha\lambda} \delta_{B_I} - \frac{\partial B^\beta_I}{\partial x^\alpha} \frac{\partial \delta_{B_I}}{\partial p^\beta}$$

Straightforward computation leads then to the following formula

$$p^\alpha \frac{\partial}{\partial x^\alpha} \delta_{B_I} + (Q^\alpha - \Gamma^\alpha_{\lambda\mu} p^\lambda p^\mu) \frac{\partial}{\partial p^\alpha} \delta_{B_I} \equiv (\nabla_\beta B^\beta_I + \frac{\partial Q^\alpha}{\partial p^\alpha}) \delta_{B_I} - \frac{\partial}{\partial p^\alpha}\{ (p^\beta \nabla_\beta B^\alpha_I - Q^\alpha) \delta_{B_I} \}$$

If Q is an electromagnetic force then $\frac{\partial Q^\alpha}{\partial p^\alpha} = 0$. The Boltzmann equation is satisfied in a generalized sense if and only if all independant measures and doublets vanish. This fact leads naturally to the equations, written when $\frac{\partial Q^\alpha}{\partial p^\alpha} = 0$:

$$B_I^\beta \nabla_\beta B_I^\alpha - \{Q^\alpha\}_{p=B_I} = 0 \ , \ I = 1,\ldots,N \qquad (7.1)$$

$$\nabla_\beta (a_I B_I^\beta) = \mathcal{J}_I(a) \ , \qquad I = 1,\ldots,N \qquad (7.2)$$

where $\mathcal{J}_I$ represents the balance of loss and gain of particles with B_I momentum at the point x. For binary collisions $\mathcal{J}_I$ is taken of the form

$$\mathcal{J}_I \equiv \sum_{J,K,L} \{\sigma_{IJ}^{KL} a_K a_L - \sigma_{KL}^{IJ} a_I a_J\}$$

where the σ_{IJ}^{KL} are given positive functions on V.

The system of equations (7.1), (7.2) is, when the metric and the external force Q are given a Leray hyperbolic first order system. When the metric is considered as a field variable satisfying Einstein equations with source the stress energy tensor generated by the distribution generalized function the coupled system is non-strictly in the sense of Leray-Ohya. The same conclusion holds if Q is the Lorentz force corresponding to an electromagnetic field generated by a generalized distribution function of charged particles (cf. Y.C.B. and G.P.)

REFERENCES

1. Y.Choquet(Fourès)-Bruhat, Distributions sur les multiplicites, C.R.A.S. t. 236 (1953).
2. G.De Rham, "Variétés Differentiables" Hermann (1955) Lichnerowicz, A. Propagateurs et Commutateurs en Relativite Generale, publications de l'I.H.E.S. n°10 (1961).
3. G.Pichon, Etude des fluides relarivistes visqueux et charges. These Paris (1964).
4. Y.Choquet-Bruhat,and C. DeWitt-Morette, C. "Analysis, Manifolds and Physics" part I : basics (1982, reprinted 1989).
5. G.Pichon, et M. Huyn-Servet, distribution de Boltzmann discrete sur un fibre tangent, J. *Maths pures et appliquees* (1992).
6. Y.Choquet-Bruhat, and G. Pichon, Plasmas with discrete velocities, in "Discrete Models in Fluid Dynamics" A. Alves ed. World Scientific (1991).

PERTURBATION EXPANSION FOR DISTRIBUTIONS ON SURFACES, LAGRANGE–BÜRMANN THEOREM AND APPLICATIONS TO MECHANICS

F. Constantinescu

Fachbereich Mathematik
Johann Wolfgang Goethe – Universität Frankfurt
Robert–Mayer–Str. 10
6000 Frankfurt am Main/West Germany

Let H_0 be the Hamiltonian of an unperturbed oscillating system and let V be a small perturbation which we take to be analytic in a neighbourhood of the surface $H_0 = E > 0$. If $H = H_0 + \lambda V$ is the total Hamiltonian than there is a convergent perturbation expansion near $\lambda = 0$ of the form

$$\delta(H - E) = \sum_{h=0}^{\infty} \frac{\lambda^n}{n!} V^n \delta^n (H_0 - E) \tag{1}$$

when applied to test function which are analytic near $H_0 = E$. In (1) $\delta(H - E)$ is the Dirac[1] distribution concentrated on the surface $H = E$ and $\delta^{(n)}(H_0 - E)$ are the derivatives of the Dirac distribution on the unperturbed surface $H_0 = E$.

There ist a tight connection of (1) to the following classical Lagrange–Bürmann theorem concerning the expansion of an analytic function in powers of another analytic function[2], the result of the present paper being more general.

Let $\phi(z)$ be a function of z analytic on and inside a contour C surrounding a point a, and let t be such that the inequality

$$|t\phi(z)| > |z - a|$$

is satisfied for all z on C; then the equation

$$\xi = a + t(\zeta)$$

regarded as an equation in ζ, has one root in the interior of C, and further any function $f(\zeta)$ of ζ analytic on and inside C can be expanded as a power series in t by the formula

$$f(\zeta) = f(a) + \sum_{n=1}^{\infty} \frac{t^n}{n!} \frac{d^{n-1}}{da^{n-1}} \left[f'(a)(\phi(a))^n \right]. \tag{2}$$

In particular for $f = 1$

$$\zeta = a + \sum_{n=1}^{\infty} \frac{t^n}{n!} \frac{d^{n-1}}{da^{n-1}} (\phi(a))^n .$$

Generalized Functions and Their Applications, Edited by
R.S. Pathak, Plenum Press, New York, 1993

Several applications to the mechanics of non–linear systems were given by R. Caboz and J.–P. Codaccioni. To give an idea note that the period of a non–linear oscillator defined by the Hamiltonian $H = H_0 + \lambda V$ with H_0 being a quadratic form is given by[3]

$$T(E) = \int \delta(H - E)dpdq \qquad (3)$$

where (p, q) are canonical coordinates. We can apply (1) and get a power series expansion for $T(E)$ with coefficients which in many particular cases can be computed exactely analytically. Other results have been obtained studying nonlinear oscillating systems with several degrees of freedom by applying the generalization of the formula (1) by R. Caboz and J.–P. Codaccioni.

Now let us give precise conditions under which (1) holds[4].

Let H_0 be a positive definite quadratic form in $\mathbb{R}^d, d > 1$, and let V be a C^∞–pertubation of H_0 which can be analytically continued to a given complex neighbourhood of the surface defined by $H_0 = E$, when E varies in an interval I on the positive real axis. Let $H = H_0 + \lambda V$ and suppose that there is a given compact set $K \subset \mathbb{R}^d$ which contains the unperturbed surface $H_0 = E$ as well as a compact connected component of the perturbed surface defined by $H = E$ for all $E \in I$. Let $\nabla H \neq 0$ in K. Then for absolute values of λ small enough the formula (1) holds as convergent Taylor expansion when applied to test functions which are analytic in the given complex neighbourhood of $H_0 = E$.

The idea of the proof is to study the resolvent

$$\frac{1}{H - z} = \sum_{n=0}^{\infty}(-1)^n V^n \frac{1}{(H_0 - z)^{n+1}} \qquad (4)$$

for $z = E \pm i\varepsilon, \varepsilon > 0$ and to represent the δ–distribution concentrated on $H = E$ as difference $\frac{1}{H-E-i\varepsilon} - \frac{1}{H-E+i\varepsilon}$ in the limit $\varepsilon \longrightarrow 0$[5]. The detailed proof, which uses analytic continuation in z requires analyticity of the test functions on which we apply (1). It doesn't work for C^∞–test functions in which case only asymptoticity of the series can be expected.

REFERENCES

1. I. M. GEL'FAND, G. E. SHILOV, *Generalized Functions*, vol. I, Academic Press, New York (1964)

2. E. T. WHITTAKER, G. N. WATSON, *A Course of Modern Analysis*, Cambridge University Press (1952)

3. V. I. ARNOLD, *Mathematical Methods of Classical Mechanics*, Springer, New York – Heidelberg (1978)

4. R. CABOZ, J.–P. CODACCIONI, F. CONSTANTINESCU, *Taylor Series for the Dirac Function on Perturbed Surfaces with Applications to Mechanics*, Math. Meth. in the Appl. Sci. 7 416–425 (1985)

5. F. CONSTANTINESCU, *Distributions and Their Applications in Physics*, Pergamon Press, Oxford (1980)

GENERALIZED D'ALEMBERT FUNCTIONAL EQUATION IN DISTRIBUTIONS

E.Y. Deeba[1] and E.L. Koh[2]

[1]Department of Applied Mathematical Sciences
University of Houston-Downtown
Houston, Texas 77002, U.S.A.
[2]Department of Mathematics and Statistics
University of Regina
Regina, Canada S4S 0A2.

ABSTRACT

In this paper we consider a generalization of D'Alembert's functional equation and study it in the setting of distributions. The method is based on defining certain operators on Schwartz distributions and using these to reformulate the classical equation. It is shown that for regular distributions (i.e. locally integrable functions), the distributional solution reduces to the classical one.

INTRODUCTION

In 1969, Fenyö [1] showed that the most general solution to a functional equation is in a sense a distributional solution. He was answering a problem posed by Hosszu at a conference in Poland in 1967 regarding the functional equation

$$f(x + y - xy) + f(xy) = f(x) + f(y).$$

In recent years there have been several works dealing with functional equations in distributions (see for example [2] to [8]). Recently [9], we studied the distributional analogue of D'Alembert's functional equation

$$f(x + y) + g(x - y) = 2f(x)g(y). \tag{1.1}$$

In this paper, we consider a generalization of (1.1), viz.

$$f(x + y) + g(x - y) = \sum_{k=1}^{n} f_k(x)g_k(y). \tag{1.2}$$

Equation (1.2) has been discussed in [10]. It was shown under appropriate conditions on f, g, f_k and $g_k (k = 1, \cdots, n)$ that f_k and g_k satisfy systems of explicitly second order homogeneous linear equations with constant coefficients and with no first order terms while f and g satisfy the conditions that $f + g$ and $f' - g'$ solve linear homogeneous differential equations of order $2n$ with constant coefficients and all terms of even order.

Our goal is to consider equation (1.2) in distributions, show that for regular distributions (i.e. locally integrable functions), our equation reduces to (1.2) and has indeed the classical solutions. Our approach is functional analytic in the sense that operators are introduced on appropriate function spaces and their duals. These operators generalize the quantities $f(x+g), g(x-y)$ and $f_k(x)g_k(y)$ to distributions. In view of Schwartz's impossibility theorem for multiplication of distributions, we use a tensor product operator.

In solving equation (1.2) (see [10]) it was assumed that the functions f_k and $g_k (k = 1, \cdots, n)$ are linearly independent. Otherwise the number of terms can be reduced. Thus there is no loss of generality in making this assumption. In our context, we shall define linear independence of a set of distributions. This is done in the next section. As well, we shall include some needed background material.

NOTATION AND PRELIMINARY RESULTS

Let $I = (0, \infty) \subset \mathbb{R}$ and $I^2 = I \times I \subset \mathbb{R}^2$. We denote by $\mathcal{D}(I)$ and $\mathcal{D}(I^2)$ the spaces of infinitely differentiable functions with compact support on I and I^2 respectively. Similarly, $\mathcal{E}(I)$ and $\mathcal{E}(I^2)$ are the spaces of infinitely differentiable functions on I and I^2 respectively. The duals of these spaces will be denoted by a prime and we note that $\mathcal{D}(I) \subset \mathcal{E}(I) \subset \mathcal{E}'(I) \subset \mathcal{D}'(I)$ (see [11] or [12]). The second inclusion will be interpreted by identifying the smooth function in $\mathcal{E}(I)$ with the regular distribution it generates in $\mathcal{E}'(I)$. The topologies for these spaces will be the usual convergence concepts for $\mathcal{D}(I)$ and $\mathcal{E}(I)$ as given in [12] and weak topologies for their duals. We denote by $L_{\text{loc}}(I)$ and $L_{\text{loc}}(I^2)$ the spaces of equivalence classes of locally integrable functions on I and I^2 respectively. Two functions are equivalent if they are equal almost everywhere. We shall denote the distribution corresponding to a locally integrable function f by λ_f. If $f \in L_{\text{loc}}(I)$ then

$$< \lambda_f, \varphi >= \int_I f(x)\phi(x)dx$$

for any $\phi \in \mathcal{D}(I)$.

D denotes the differentiation operator on $\mathcal{D}'(I)$ whereas D_1 and D_2 are the partial differentiation operators on $\mathcal{D}'(I^2)$ with respect to the first and second variable from I^2, respectively. These symbols will also denote the differentiation operators on subspaces of $\mathcal{D}'$, e.g. on $\mathcal{D}$. E_1 and E_2 denote integration operators from $\mathcal{D}(I^2)$ into $\mathcal{D}(I)$ given, respectively, for any $\phi \in \mathcal{D}(I^2)$, by

$$E_1[\phi](x) = \int_I \phi(x, y)dy \tag{2.1}$$

and

$$E_2[\phi](y) = \int_I \phi(x, y)dx. \tag{2.2}$$

These are continuous linear operators. Their adjoints $E_i^*(i = 1, 2)$ are continuous linear operators from $\mathcal{D}'(I)$ onto $\mathcal{D}'(I^2)$ and are given by

$$< E_i^*[T], \phi > = < T, E_i(\phi) > \quad i = 1, 2 \tag{2.3}$$

for any $T \in \mathcal{D}'(I)$ and $\phi \in \mathcal{D}(I^2)$.

The next proposition summarizes some properties of the adjoint operators E_i^* (see [5]).

Proposition 2.1.

(a) If $f \in L_{\text{loc}}(I)$, then $g = E_1^*[f] \in L_{\text{loc}}(I^2)$ and $h = E_2^*[f] \in L_{\text{loc}}(I^2)$ where $g(x, y) = f(x)$ and $h(x, y) = f(y)$.
(b) If $\alpha \in \mathcal{E}(I)$, then $E_i^*[\alpha] (i = 1, 2)$ belong to $\mathcal{E}(I^2)$.
(c) If $\alpha \in \mathcal{E}(I)$ and $T \in \mathcal{D}'(I)$, then

$$E_i^*[\alpha T] = E_i^*[\alpha] E_i^*[T], \quad i = 1, 2.$$

(d) If $T \in \mathcal{D}'(I)$, then

$$\begin{aligned} D_1 E_1^*[T] &= E_1^*[DT]; \quad & D_1 E_2^*[T] &= 0 \\ D_2 E_1^*[T] &= 0; & D_2 E_2^*[T] &= E_2^*[DT]. \end{aligned}$$

Let P be the tensor product operator from $\mathcal{D}'(I) \times \mathcal{D}'(I)$ into $\mathcal{D}'(I^2)$ given by

$$\begin{aligned} < P[S; T], \phi > &= < S(x), < T(y), \phi(x, y) >> \\ &= < T(y), < S(x), \phi(x, y) >> \end{aligned}$$

for any $S, T, \in \mathcal{D}'(I)$ and $\phi \in \mathcal{D}(I^2)$. P is a linear operator and that

$$D_1 P[S; T] = P[DS; T]; \quad D_2 P[S; T] = P[S; DT], \tag{2.4}$$

(see [11]). The following results from [5] gives other properties of the operator P.

Proposition 2.2.

(a) If $T \in \mathcal{D}'(I)$, then

$$E_1^*[T] = P[T; 1]; \quad E_2^*[T] = P[1; T].$$

(b) If $\alpha, \beta \in \mathcal{E}(I)$ and $S, T, U \in \mathcal{D}'(I)$, then

$$\begin{aligned} E_1^*[\alpha] P[S; U] + E_1^*[\beta] P[T; U] &= P[\alpha S + \beta T; U] \\ E_2^*[\alpha] P[S; T] + E_2^*[\beta] P[S; U] &= P[S; \alpha T + \beta U]. \end{aligned}$$

(c) If $S, T, U, V \in \mathcal{D}'(I)$ are nonzero distributions and satisfy the equation

$$P[S; T] = P[U; V],$$

then $S = C_1 U$ and $T = C_2 V$ (where C_1 and C_2 are real constants) and conversely.

Let Q_+ and Q_- be linear operators from $\mathcal{D}(I^2)$ into $\mathcal{D}(I)$ given by

$$Q_+[\phi](x) = \int_{\mathbb{R}} \phi(x - y, y) dy = \int_{\mathbb{R}} \phi(y, x - y) dy$$

and

$$Q_-[\phi](x) = \int_{\mathbf{R}} \phi(x+y, y)dy = \int_{\mathbf{R}} \phi(y, y-x)dy.$$

The adjoints of the operators are $Q_\pm^*$ defined from $\mathcal{D}'(I)$ into $\mathcal{D}'(I^2)$ by

$$< Q_\pm^*[T], \phi > = < T, Q_\pm[\phi] >$$

for any $\phi \in \mathcal{D}(I^2)$ and $T \in \mathcal{D}'(I)$. The properties of these operators are given in the following propositions (see [9] for proofs).

Proposition 2.3. Suppose $f \in L_{\text{loc}}(I)$. Then
(a) $Q_+^*[\lambda_f] \in L_{\text{loc}}(I^2)$ and $Q^*[\lambda_f] = f(x+y)$;
(b) $Q_-^*[\lambda_f] \in L_{\text{loc}}(I^2)$ and $Q_-^*[\lambda_f] = f(x-y)$.

Proposition 2.4.

(a) If $\alpha \in \mathcal{E}(I)$, then $Q_\pm^*[\alpha] \in \mathcal{E}(I^2)$.
(b) If $\alpha \in \mathcal{E}(I)$ and $T \in \mathcal{D}'(I)$, then $Q_\pm^*[\alpha T] = Q_\pm^*[\alpha]Q_\pm^*[T]$.
(c) If $T \in \mathcal{D}'(I)$, then

$$\begin{aligned} D_1(Q_+^*[T]) &= D_2(Q_+^*[T]) = Q_+^*[DT]; \\ D_1(Q_-^*[T]) &= Q_-^*[DT]; \quad \text{and} \\ D_2(Q_-^*[T]) &= -Q_-^*[DT]. \end{aligned}$$

Before we proceed to discuss equation (1.2) in distributions, we need the following definition.

Definition 2.1. Let $U_k, k = 1, \cdots, n$ be elements of $\mathcal{D}'(I)$. The set $\{U_k\}$ is said to be linearly independent if there exist test functions $\psi_1, \psi_2, \cdots, \psi_n$ belonging to $\mathcal{D}(I)$ such that the matrix $A = [< u_k, \psi_i >]_{k,i=1}^n$ is invertible.

GENERALIZED D'ALEMBERT EQUATION IN DISTRIBUTIONS

In this section we employ the operators $Q_\pm^*$ and P to write equation (1.2) in distributions and then obtain its solutions.

Let T, U, T_k and $U_k (k = 1, \cdots, n)$ be members in the space $\mathcal{D}'(I)$. We shall refer to the equation

$$Q_+^*[T] + Q_-^*[U] = \sum_{k=1}^{n} P[T_k; U_k] \tag{3.1}$$

as the generalized D'Alembert equation in distributions.

Proposition 3.1. If T, U, T_k and $U_k (k = 1, \cdots, n)$ are regular distributions, that is, there are functions f, g, f_k and g_k in $L_{\text{loc}}(I)$ such that $T = \lambda_f, U = \lambda_g, T_k = \lambda_{f_k}$ and $U_k = \lambda_{g_k}$, then equation (3.1) reduces to equation (1.2).

Proof. In view of Proposition 2.3, we have

$$Q_+^*[\lambda_f] = f(x+y) \quad \text{and} \quad Q_-^*[\lambda_g] = g(x-y).$$

The product operator P as defined implies that for any $\phi \in \mathcal{D}(I^2)$,

$$\sum_{k=1}^{n} < P[\lambda_{f_k};\lambda_{g_k}],\phi > = \sum_{k=1}^{n} < \lambda_{f_k}, < \lambda_{g_k},\phi >>$$
$$= \sum_{k=1}^{n} \int_{I^2} f_k(x)g_k(y)\phi(x,y)dydx$$
$$= \sum_{k=1}^{n} < f_k(x)g_k(y),\phi(x,y) > .$$

If equation (3.1) holds for $\lambda_f, \lambda_g, \lambda_{f_k}$ and λ_{g_k}, then the above calculation implies that

$$\int_{I^2} \left\{ f(x+y) + g(x-y) - \sum_{k=1}^{n} f_k(x)g_k(y) \right\} \phi(x,y)dydx = 0$$

for all $\phi \in \mathcal{D}(I^2)$. Therefore, $f(x+y) + g(x-y) = \sum_{k=1}^{n} f_k(x)g_k(y)$ for almost every $(x,y) \in I^2$. □

In solving equation (3.1) we shall assume without loss of generality that the distributions T_k and $U_k (k = 1, \cdots, n)$ are linearly independent as in definition 2.1.

To solve equation (3.1) we apply the differential operators D_1^2 and D_2^2 to equation (3.1) and use the results of Proposition 2.4(c) and equation (2.4). We obtain the system

$$Q_+^*[D^2T] + Q_-^*[D^2U] = \sum_{k=1}^{n} P[D^2T_k; U_k]$$

and

$$Q_+^*[D^2T] + Q_-^*[D^2U] = \sum_{k=1}^{n} P[T_k; D^2U_k],$$

which in turn yields

$$\sum_{k=1}^{n} P[D^2T_k; U_k] = \sum_{k=1}^{n} P[T_k; D^2U_k]. \tag{3.2}$$

We will first solve for T_k. Since $U_k, k = 1, \cdots, n$, are linearly independent there exist test functions $\psi_1, \psi_2, \cdots, \psi_n$ in $\mathcal{D}(I)$ such that the $n \times n$ matrix $A = [< U_k, \psi_i >]_{k,i=1}^{n}$ is invertible. Let B denote the $n \times n$ matrix $[< D^2U_k, \psi_i >]_{k,i=1}^{n}$.

Now for any ϕ_1, ϕ_2 in the space $\mathcal{D}(I)$, define $\phi_1 \oplus \phi_2$ by

$$(\phi_1 \oplus \phi_2)(x,y) = \phi_1(x)\phi_2(y) \quad \text{for all } (x,y) \in I^2.$$

Clearly $\phi_1 \oplus \phi_2$ is in $\mathcal{D}(I^2)$ and $< P[S;T], \phi_1 \oplus \phi_2 >=< S, \phi_1 >< T, \phi_2 >$ whenever $S, T \in \mathcal{D}'(I)$.

For any $\phi \in \mathcal{D}(I)$, we apply the operator (3.2) on $\phi \oplus \psi_i$ to obtain

$$\sum_{k=1}^{n} < D^2T_k, \phi >< U_k, \psi_i >= \sum_{k=1}^{n} < T_k, \phi >< D^2U_k, \psi_i > .$$

In matrix form, this becomes

$$A \begin{bmatrix} < D^2T_1, \phi > \\ \vdots \\ < D^2T_n, \phi > \end{bmatrix} = B \begin{bmatrix} < T_1, \phi > \\ \vdots \\ < T_n, \phi > \end{bmatrix} \tag{3.3}$$

with A and B being the $n \times n$ matrices $[< U_k, \psi_i >]$ and $[< D^2 U_k, \psi_i >]$ respectively. Equation (3.3) reduces to a second order system of homogeneous linear differential equations with constant coefficients and with no first order terms, namely,

$$\begin{bmatrix} < D^2 T_1, \phi > \\ \vdots \\ < D^2 T_n, \phi > \end{bmatrix} = C \begin{bmatrix} < T_1, \phi > \\ \vdots \\ < T_n, \phi > \end{bmatrix} \tag{3.4}$$

where $C = A^{-1}B$. As a homogeneous system, equation (3.4) may be solved by distributional Laplace transform as in p. 268 of [12] to yield continuous solutions $T_1, \cdots, T_n$. Thus the solutions are precisely the classical ones discussed in [10]. By the same token, one obtains a similar system for the distributions $U_k, k = 1, 2, \cdots, n$. Thus, there exist constants $\alpha_{k,m} \in \mathbb{R}$ such that $T_k = \lambda_{f_k}$ where f_k satisfies the equation

$$f_k'' = \sum_{m=1}^{n} \alpha_{k,m} f_m(x). \tag{3.5}$$

As well, there exist constants $\beta_{k,m} \in \mathbb{R}$ such that $U_k = \lambda_{g_k}$ where g_k satisfies the equation

$$g_k''(x) = \sum_{m=1}^{n} \beta_{k,m} g_m(x). \tag{3.6}$$

We finally need to solve for T and U. Apply the differential operators D_1 and D_2 to equation (3.1) to obtain the system

$$Q_+^*[DT] + Q_-^*[DU] = \sum_{k=1}^{n} P[DT_k; U_k]$$

and

$$Q_+^*[DT] - Q_-^*[DT] = \sum_{k=1}^{n} P[T_k; DU_k].$$

From these equations, we obtain

$$Q_+^*[DT] = \frac{1}{2} \sum_{k=1}^{n} \{P[DT_k; U_k] + P[T_k; DU_k]\}$$

and

$$Q_-^*[DU] = \frac{1}{2} \sum_{k=1}^{N} \{P[DT_k; U_k] - P[T_k; DU_k]\}.$$

For any $\phi \in \mathcal{D}(I^2)$ and using properties of distributions regarding integration with respect to a parameter and translation (see pp. 26, 75 of [12]), we have

$$\begin{aligned} < Q_+^*[DT], \phi > &= < DT(x), \int_{\mathbb{R}} \phi(x - y), y)dy > \\ &= \int_{\mathbb{R}} < DT(x), \phi(x - y, y) > dy \\ &= < DT(x + y), \phi(x, y) > . \end{aligned}$$

On the other hand, by virtue of the fact that T_k and U_k turn out to be regular distributions, we have

$$\begin{aligned} < P[DT_k; U_k], \phi > &= < DT_k(x), < U_k, (y), \phi(x,y) >> \\ &= \int_{I^2} DT_k(x) U_k(y) \phi(x,y) dy dx \\ &= < DT_k(x) U_k(y), \phi(x,y) > . \end{aligned}$$

Similarly,

$$< P[T_k; DU_k], \phi >=< T_k(x) DU_k(y), \phi(x,y) > .$$

Thus,

$$< Q_+^*[DT], \phi >= \frac{1}{2} \sum_{k=1}^{n} \{< P[DT_k; U_k], \phi > + < P[T_k; DU_k], \phi >\}$$

yields

$$< DT(x+y), \phi >= \frac{1}{2} \sum_{k=1}^{n} \{< DT_k(x) U_k(y) + T_k(x) DU_k(x), \phi >\}$$

or

$$DT(x+y) = \frac{1}{2} \sum_{k=1}^{n} \{DT_k(x) U_k(y) + T_k(x) DU_k(y)\}. \tag{3.7}$$

Equation (3.7) can be solved for T since T_k and $U_k, k = 1, \cdots, n$ would have been known. In the same manner, we obtain the following equation for U,

$$DU(x-y) = \frac{1}{2} \sum_{k=1}^{n} \{DT_k(x) U_k(y) - T_k(x) DU_k(y)\} \tag{3.8}$$

which is also solvable.

Alternatively, equations (3.7) and (3.8) give the distributions T and U as C^∞ functions. Thus, one may set $y = 0$ and $U_k(0)$ and $DU_k(0)$ will be well-defined constants. Hence we may add or subtract (3.7) and (3.8) to yield

$$D^2(T+U)(x) = \sum_{k=1}^{n} U_k(0) D^2 T_k(x) \tag{3.9}$$

and

$$D(T-U)(x) = \sum_{k=1}^{n} DU_k(0) T_k(x). \tag{3.10}$$

Equations (3.9) and (3.10) together with (3.5) and (3.6) may now be used to show that $T+U$ and $D(T-U)$ both satisfy linear homogeneous differential equations of order $2n$ with constant coefficients and with all terms of even order. This is done as in pages 231-232 of [10], *mutatis mutandis.* We now have the following theorem.

Theorem 3.1. If T, U, T_k and $U_k (k = 1, \cdots, n)$ belong to $\mathcal{D}'(I)$ and satisfy equation (3.1), and if $\{T_k\}_{k=1,\cdots,n}$ and $\{U_k\}_{k=1,\cdots,n}$ are linearly independent, then T_k and $U_k (k = 1, \cdots, n)$ satisfy systems of second order homogeneous linear differential equations with constant coefficients and without first order terms, while $T+U$ and $D(T-U)$ satisfy linear homogeneous differential equations of order $2n$ with constant coefficients and with all terms of even order.

Remarks

1. We note that T and U can be easily obtained from $T + U$ and $D(T - U)$.

2. Theorems 4.1 and 4.2 of [9] are special cases of Theorem 2.1 above.

3. By means of certain distributional operators we are able to write a classical functional equation in distributions. This new analogue is now solved by properties of distributions.

REFERENCES

1. I. Fenyö, On the general solution of a functional equation in the domain of distributions, *Aequationes Math.* **3**(1969), 236-246.

2. J. Aczél, H. Haruki, M.A. McKiernan, and G.N. Sakovič, General and regular solutions of functional equations characterizing harmonic polynomials, *Aequationes Math.* **1**(1968), 37-53.

3. H. Swiatak, On the regularity of the distributional and continuous solutions of the functional equation $\sum_{i=1}^{k} a_i(x,t)f(x + \phi_i(t)) = b(x,t)$, *Aequationes Math.* **1**(1968), 6-19.

4. M. Neagu, About the Pompeiu equation in distributions, *Inst. Politehn. "Traian Vuia" Timisoara Lucrar. Sem. Mat. Fiz.* (1984), 62-66.

5. E.L. Koh, The Cauchy functional equations in distributions, *Proc. Amer. Math. Soc.* **106**(1989), 641-647.

6. John A. Baker, Functional equations, tempered distributions and Fourier transform, *Trans. Amer. Math. Soc.* **315**(1989), 57-68.

7. John A. Baker, Functional equations, distributions and approximate identities, *Can. J. Math.* **42**(1990), 696-708.

8. E.Y. Deeba and E.L. Koh, The Pexider functional equations in distributions, *Can. J. Math.* **42**(1990), 304-314.

9. E.Y. Deeba and E.L. Koh, D'Alembert functional equations in distributions, *Proc. Amer. Math. Soc.*, to appear.

10. J. Aczél and J. Dhombres, Functional equations in several variables, Cambridge Univ. Press, Cambridge, U.K. 1989.

11. L. Schwartz, Theorie des distributions, Hermann, Paris, 1966.

12. A.H. Zemanian, Distribution theory and transform analysis, McGraw-Hill, Inc., New York, 1965.

RIESZ BASES OF SPECIAL POLYNOMIALS
IN WEIGHTED SOBOLEV SPACES
OF ANALYTIC FUNCTIONS

J. de Graaf

Eindhoven University of Technology
Department of Mathematics and Computing Science
P.O. Box 513
5600 MB Eindhoven, The Netherlands

ABSTRACT

The main subject of this paper is the construction of Riesz bases in Weighted Sobolev Spaces of Analytic Functions on open sets in $\mathbb{C}$. For this two main tools are introduced.
First, starting from weighted Sobolev spaces on a disc or an annulus and supplied with the Taylor basis, we move to an 'arbitrary' open set by means of a conformal mapping. In many cases our weighted Sobolev spaces behave naturally under an analytic pull-back. The analytic functions in our spaces are characterized by the asymptotic behaviour of expansion coefficients and by boundary conditions in ordinary Sobolev spaces.
The second tool is of a completely different nature: By means of upper triangular transition matrices S we construct new Riesz bases out of a given one. Our transition matrices are diagonalizers of a class of given upper triangular matrices. Thus we are able to make estimations in both S and S^{-1} at once. In one of the applications we describe the domains of exponentiated square roots of Jacobi operators in ordinary Sobolev spaces on $[-1, 1]$. This case was left in [GE]. We also relate this to a refinement of Szegö's result on series of Jacobi polynomials on an ellips. (Thm 4.1).

CONTENTS

Generalized Functions and Their Applications, Edited by
R.S. Pathak, Plenum Press, New York, 1993

1. WEIGHTED SOBOLEV SPACES OF ANALYTIC FUNCTION ON THE ANNULUS

For an arbitrary subset Ω of the complex plane $\mathbb{C}$ we denote by $\mathcal{A}(\Omega)$ the set of complex valued functions on Ω which are analytic at each point $z = x + iy \in \Omega$.
Now let Ω be open. For a fixed weight function $\mu \in L_1(\Omega)$, $\mu \geq 0$ and $m \in \mathbb{N}_0 = \mathbb{N} \cup \{0\}$ we define

$$\|\cdot\|_{\Omega,\mu,m} \quad : \quad \mathcal{A}(\Omega) \to [0,\infty] \ ,$$

by

$$f \mapsto \Big(\int_{\Omega} \{|f(z)|^2 + |f^{(m)}(z)|^2\} \ \mu(z) \ dxdy \Big)^{\frac{1}{2}} \ .$$

We introduce the complex vectorspace

$$\mathcal{A}_2(\Omega;\mu,m) = \{f \mid \|f\|_{\Omega,\mu,m} < \infty\}$$

provided with the inner product

$$(f,g)_{\Omega,\mu,m} = \int_{\Omega} \{f(z)\ \overline{g(z)} + f^{(m)}(z)\ \overline{g^{(m)}(z)}\} \ \mu(z) \ dxdy \ .$$

Instead of the triple $(\Omega,\mu,0)$ we denote (Ω,μ). The first special open set Ω that we meet is the *annulus* $A = \{z \mid 0 < a < |z| < b < \infty\}$ for fixed a and b. The *special annulus* A_T is defined by $A_T = \{z \mid e^{-T} < |z| < e^{T}\}$, for some fixed $T > 0$.
The subset $A^\varepsilon \subset A$, $0 < \varepsilon < \frac{1}{2}(b-a)$, is defined by $A^\varepsilon = \{z \mid 0 < a < |z| < a+\varepsilon \ \text{ or } \ b-\varepsilon < |z| < b\}$.
For any given weight function μ on A the new weight function μ_ε on A is defined to be zero on the annulus $A\backslash A^\varepsilon$ and equal to μ on A^ε. So, μ_ε equals μ near the boundary of A.

THEOREM 1.1

Let the weight function μ on the annulus A be such that

- $\forall z \in A \quad \mu(z) = \mu(|z|)$
- $\exists \varepsilon > 0 \ \ \forall z \in A^\varepsilon \quad \mu(z) > 0$ and μ is continuous at z.

Let $m \in \mathbb{N}_0$ be fixed.

(A) $\forall \varepsilon > 0 \ \ \forall j \in \mathbb{N}_0 \ \ \exists M_{\varepsilon,j} > 0 \ \ \forall f \in \mathcal{A}(A)$

$$\sup\{|f^{(j)}(z)| \ |z \in A\backslash A^\varepsilon\} \leq M_{\varepsilon,j}\Big(\int_{A^\varepsilon} |f(z)|^2 \ \mu(z) \ dxdy \Big)^{\frac{1}{2}} \ .$$

(B) The space $\mathcal{A}_2(A;\mu,m)$ is a Hilbert space.
The functions $\{z \mapsto z^n \mid n \in \mathbb{Z}\}$ establish an *orthogonal* basis in $\mathcal{A}_2(A;\mu,m)$.

(C) For $f \in \mathcal{A}(A)$ expressed as a Laurent series $f(z) = \sum_{n=-\infty}^{\infty} a_n \ z^n$ we have

$$f \in \mathcal{A}_2(A;\mu,m) \Longleftrightarrow \sum_{n=-\infty}^{\infty} |a_n|^2 \, \|z^n\|^2_{A,\mu,m} = \sum_{n=-\infty}^{\infty} 2\pi \, \alpha_n^{-2} \, |a_n|^2 < \infty \,.$$

Here $(\alpha_n)^{-2} = \int_a^b r^{2n+1} \mu(r)dr + |(n-m+1)_m|^2 \int_a^b r^{2(n-m)+1} \mu(r)dr$, with Pochhammer's symbol

$$(n-m+1)_m = n \cdot (n-1) \cdot \ldots \cdot (n-m+1) \,.$$

So $(\alpha_n z^n)_{n=-\infty}^{\infty}$ is a Riesz basis in $\mathcal{A}_2(A;\mu,m)$. (That means: each $g \in \mathcal{A}_2$ has a unique expansion $g = \sum_{n=-\infty}^{\infty} a_n(\alpha_n z^n)$, with $(a_n)_{n=-\infty}^{\infty} \in l_2$, which converges in $\mathcal{A}_2$.)

(D) If μ is replaced by ν (not necessarily $\nu(z) = \nu(|z|)$ such that

$$\exists \varepsilon > 0 \; \exists \delta > 0 \; \exists M > 0 \; \forall z \in A^\varepsilon \quad 0 < \delta < \frac{\nu(z)}{\mu(z)} < M < \infty \,,$$

then the spaces $\mathcal{A}_2(A;\mu,m)$ and $\mathcal{A}_2(A;\nu,m)$ have equivalent norms, i.e. they are the same as topological vector spaces. Note that in both these Hilbert spaces the functions $\{z \mapsto \alpha_n \, z^n \mid n \in \mathbb{Z}\}$ establish a Riesz basis.

(E) Let $j \in \mathbb{N}$, $0 < j < m$. Then

$$\forall \varepsilon > 0 \; \exists M^\varepsilon > 0 \; \forall f \in \mathcal{A}_2(A;\mu,m)$$

$$\|f^{(j)}\|^2_{A,\mu} \le M_\varepsilon \|f\|^2_{A,\mu} + \varepsilon \|f^{(m)}\|^2_{A,\mu} \,.$$

(F) Let m distinct point $\underline{z} = (z_0, \ldots, z_{m-1}) \subset A$ be given. The norm $\|\cdot\|_{A,\underline{z},\mu,m}$, defined by

$$\|f\|^2_{A,\underline{z},\mu,m} = \sum_{k=0}^{m-1} |f(z_k)|^2 + \|f^{(m)}\|^2_{A,\mu} \,,$$

is equivalent to the norm $\|\cdot\|_{A,\mu,m}$.

Proof

(A) We use the "mean value property" for analytic functions

$$f^{(j)}(w) = c_j \, \rho^{-(2j+2)} \int_{|w-z|<\rho} f(z) \, (\bar{z} - \bar{w})^j \, dxdy \,, \quad j \in \mathbb{N}_0 \,,$$

with $c_j = (2\pi)^{-1} \, (2j+2)j!$. Take w on one of the circles $|z| = b - \frac{1}{2}\varepsilon$, $|z| = a + \frac{1}{2}\varepsilon$. Take $\rho = \frac{1}{4}\varepsilon$.
Let $\theta > 0$ be such that on $A^{\frac{3}{4}\varepsilon} \backslash A^{\frac{1}{4}\varepsilon}$ one has $\mu(z) > \theta$. Now application of the Cauchy Schwarz inequality to

$$f^{(j)}(w) = c_j \left(\frac{\varepsilon}{4}\right)^{-(2j+2)} \int_{|w-z|<\frac{3}{4}} f(z) \, \frac{(\bar{z}-\bar{w})^j}{\mu} \, \mu \, dxdy$$

leads to

$$|f^{(j)}(w)| \le M_{\varepsilon,j} \|f\|_{A^\varepsilon,\mu}$$

with $M_{\varepsilon,j} = (\pi\theta)^{-\frac{1}{2}} (j+1)^{\frac{1}{2}} \left(\frac{4}{3}\right)^{j+1} j!$.
Because of the maximum principle this result extends to the annular domain between the two circles.

(B) Because of part (A) a Cauchy sequence in $\mathcal{A}_2(A;\mu,m)$ converges uniformly on compact sets in A. So, the limit exists as an analytic function. Further, since $\mu(z) = \mu(|z|)$ it is clear that the functions $\{z^n \mid n \in \mathbb{Z}\}$ are orthogonal. We show that the span of these functions is dense: Let $g \in \mathcal{A}_2(A;\mu,m)$ and suppose $g \perp z^n$ for all $n \in \mathbb{Z}$. Write $g = \sum_{n=-\infty}^{\infty} b_n z^n$, evaluate $(g,z^n)_{A\backslash A^\varepsilon,\mu,m}$ and take the limit $\varepsilon \downarrow 0$. This leads to

$$b_n\{\|z^n\|^2_{A,\mu,m} + |n\cdot(n-1)\cdot\ldots\cdot(n-m+1)|^2\ \|z^{n-m}\|^2_{A,\mu,m}\} = 0\ .$$

Hence the result that all $b_n = 0$.

(C) Let $f = \sum_{n=-\infty}^{\infty} a_n z^n \in \mathcal{A}(A)$. Since the functions $\{z^n \mid n \in \mathbb{Z}\}$ are mutually orthogonal in all spaces $\mathcal{A}_2(A\backslash A^\varepsilon;\mu,m)$ we have

$$\|f\|^2_{A\backslash A^\varepsilon,\mu,m} = \sum_{n=-\infty}^{\infty} |a_n|^2\ \|z^n\|^2_{A\backslash A^\varepsilon,\mu,m}\ .$$

Now let $\varepsilon \downarrow 0$ and observe that the sum is finite iff the integral is finite.

(D) First we show that we arrive at equivalent norms if μ is replaced by μ_ε. On one hand there is the trivial inequality

$$\|f\|^2_{A,\mu_\varepsilon,m} \le \|f\|^2_{A,\mu,m}\ .$$

On the other hand

$$\|f\|^2_{A,\mu,m} = \|f\|^2_{A^\varepsilon,\mu,m} + \|f\|^2_{A\backslash A^\varepsilon,\mu,m}\ .$$

The first term is equal to $\|f\|^2_{A,\mu_\varepsilon,m}$. The second term can be estimated by

$$(M^2_{\varepsilon,0} + M^2_{\varepsilon,m})\ (\int_A \mu(z)\ dxy)\ \|f\|^2_{A,\mu_\varepsilon}\ ,$$

because of (A). Next we compare ν with μ_ε. On one hand we have

$$\|f\|^2_{A,\mu^\varepsilon,m} = \|f\|^2_{A^\varepsilon,\mu,m} \le \frac{1}{\delta}\ \|f\|^2_{A^\varepsilon,\nu,m} \le \frac{1}{\delta}\ \|f\|^2_{A,\nu,m}\ .$$

On the other hand

$$\|f\|^2_{A,\nu,m} = \|f\|^2_{A^\varepsilon,\nu,m} + \|f\|^2_{A\backslash A^\varepsilon,\nu,m} \le$$

$$\le M\ \|f\|^2_{A^\varepsilon,\mu,m} + (M^2_{\varepsilon,0} + M^2_{\varepsilon,m})\ (\int_A \nu(z)dxdy)\ \|f\|^2_{A,\mu_\varepsilon}$$

$$\le \{M + (M^2_{\varepsilon,0} + M^2_{\varepsilon,m}) \int_A \nu(z)dxdy\}\ \|f\|^2_{A^\varepsilon,\mu,m}\ .$$

(E) Write

$$\|f^{(j)}\|^2_{A,\mu} = \sum_{k=-\infty}^{\infty} |a_k|^2\ |k\cdot(k-1)\cdot\ldots\cdot(k-j+1)|^2\ \|z^{k-j}\|^2_{A,\mu}\ .$$

Because of $||z^{k+l}||^2_{A,\mu} \le b^{2l}\ ||z^k||^2_{A,\mu}$, if $l \in \mathbb{N}_0$, we estimate, with $N \in \mathbb{N}$

$$||f^{(j)}||^2_{A,\mu} \le \sum_{|k|\le N} |a_k|^2\ |k\cdot(k-1)\cdot\ldots\cdot(k-j+1)|^2\ ||z^{k-j}||^2_{A,\mu} +$$

$$+b^{2(m-j)} \sum_{|k|>N} \frac{|k\cdot(k-1)\cdot\ldots\cdot(k-m+1)|^2}{|(k-j)\cdot\ldots\cdot(k-m+1)|^2}\ |a_k|^2\ ||z^{k-m}||^2_{A,\mu}\ .$$

Let $\varepsilon > 0$ be given. Take N so large that for all $k \in \mathbb{Z}$, $|k| > N$, $|(k-j)\cdot\ldots\cdot(k-m+1)|^{-2}\ b^{2(m-j)} < \varepsilon$.
Then the second sum can be estimated by $\varepsilon||f^{(m)}||^2_{A,\mu}$.
The first sum can be estimated by $M_\varepsilon||f||^2_{A,\mu}$ if we take

$$M_\varepsilon = \max_{-N\le k\le N} \{|k\cdot\ldots\cdot(k-j+1)|^2\ ||z^{k-j}||^2_{A,\mu}\ ||z^k||^{-2}_{A,\mu}\}\ .$$

(F) Because of part (A) of the theorem there is a constant c_1, which depends only the choice of $z_0, \ldots, z_{m-1}$, such that

$$\sum_{k=0}^{m-1} |f(z_k)|^2 \le c_1||f||^2_{A,\mu}\ .$$

Therefore

$$||f||^2_{A,\underline{z},\mu,m} \le (c_1+1)\ ||f||^2_{A,\mu,m}\ .$$

For the converse inequality, split

$$f = p + f_1 \quad \text{with} \quad p(z) = \sum_{k=0}^{m-1} a_k\ z^k\ .$$

Since p is a polynomial of, at most, degree $m-1$, there is a constant $c_2 > 0$ such that

$$||p||^2_{A,\mu} \le c_2 \sum_{k=0}^{m-1} |p(z_k)|^2 \le$$

$$\le 2\ c_2 \sum_{k=0}^{m-1} |f(z_k)|^2 + 2\ c_2 \sum_{k=0}^{m-1} |f_1(z_k)|^2\ .$$

Again, part (A) of the theorem, there is $c_3 > 0$,

$$||f||^2_{A,\mu} \le 2\ c_2 \sum_{k=0}^{m-1} |f(z_k)|^2 + (2c_2c_3+1)\ ||f||^2_{A,\mu}\ .$$

Finally,

$$||f_1||^2_{A,\mu} = \sum_{k\ne 0,\ldots,m} |a_k|^2\ ||z^k||^2_{A,\mu} \le$$

$$\le b^{2m} \sum_{k=-\infty}^{\infty} |a_k|^2\ |k\cdot(k-1)\cdot\ldots\cdot(k-m+1)|^2\ ||z^{k-m}||^2_{A,\mu} =$$

$$= b^{2m}\ ||f^{(m)}||^2_{A,\mu}\ .$$

(Remind that b is the outer radius of A.) □

Note on part (D) of the theorem. In the estimate $\|\cdot\|^2_{A,\mu,m} \leq C\|\cdot\|^2_{A,\mu_\epsilon,m}$ the best possible constant C is given by

$$C = 1 + \max\{(\int_a^b \mu(r)dr)\,(\int_{b-\epsilon}^b \mu(r)dr)^{-1},\ (\int_a^b \mu(r)dr)\,(\int_a^{a+\epsilon} \mu(r)dr)^{-1}\}$$

as a simple monotonicity argument shows.

In the next theorem we consider important closed linear subspaces of $\mathcal{A}_2(A;\mu,m)$ and Riesz bases for them. We define four linear subspaces $\mathcal{A}(A;+)$, $\mathcal{A}(A;-)$, $\mathcal{A}(A;e)$, $\mathcal{A}(A;o)$ of $\mathcal{A}(A)$ in the following way: $f(z) = \sum_{n=-\infty}^{\infty} a_n\, z^n$ belongs to those spaces if, respectively, $\forall n \in I\!N\ :\ a_{-n} = 0$, $\forall n \in I\!N\ :\ a_n = 0$, $\forall n \in \mathbb{Z}\ :\ a_n = a_{-n}$, $\forall n \in \mathbb{Z}\ :\ a_n = -a_{-n}$.

THEOREM 1.2

Suppose μ satisfies the conditions of Theorem 1.1.

(A) The sets

$$W_+ = \mathcal{A}_2(A;+;\mu,m) = \mathcal{A}_2(A;\mu,m) \cap \mathcal{A}(A;+)$$

$$W_- = \mathcal{A}_2(A;-;\mu,m) = \mathcal{A}_2(A;\mu,m) \cap \mathcal{A}(A;-)$$

are closed linear subspaces in $\mathcal{A}_2(A;\mu,m)$.
The functions in those subspaces can be continued analytically into, respectively, the whole disc $\{z \mid |z| < b\}$, and the set $\{z \mid |z| > a\} \cup \{\infty\}$.

(B) Consider the case $A = A_T$. Define the mapping $Z\ :\ \mathcal{A}(A_T) \to \mathcal{A}(A_T)$ by $(Zf)\,(z) = f(\frac{1}{z})$.
Suppose $\exists C > 0\ \forall z \in A_T\ \mu(\frac{1}{z}) \leq C|z|^4\ \mu(z)$.
Then the operator Z is a continuous bijection on $\mathcal{A}_2(A^T;\mu,m)$. Further, the sets

$$W_e = \mathcal{A}_2(A_T;e;\mu,m) = \mathcal{A}_2(A_T;\mu,m) \cap \mathcal{A}(A;e)$$

$$W_o = \mathcal{A}_2(A_T;o;\mu,m) = \mathcal{A}_2(A_T;\mu,m) \cap \mathcal{A}(A;o)$$

are closed linear subspaces of $\mathcal{A}_2(A_T;\mu,m)$.

(C) With the same additional conditions on μ, the functions

$$\{\beta_0, \beta_n(z^n + z^{-n}),\ \ \beta_n(z^n - z^{-n}) \mid n \in I\!N\}$$

with

$$\beta_n = \alpha_n\ \alpha_{-n}(\alpha_n^2 + \alpha_{-n}^2)^{-\frac{1}{2}}$$

establish a Riesz basis in $\mathcal{A}_2(A_T;\mu,m)$.

Proof

(A) Elementary.

(B) First, we show that Z is well defined.
Suppose $f \in \mathcal{A}_2(A_T;\mu,m)$. If both $\|Z\,f\|^2_{A_T,\mu} < \infty$ and $\|(Z\,f)^{(m)}\|^2_{A_T,\mu} < \infty$ then $Z\,f \in \mathcal{A}_2(A_T;\mu,m)$. We only prove the second inequality. By induction one shows

$$\frac{d^m}{dz^m}\, f\left(\frac{1}{z}\right) = \sum_{j=1}^{m} f^{(j)}\left(\frac{1}{z}\right)\, p_{jm}\left(\frac{1}{z}\right) \, ,$$

where the p_{jm} are polynomials of degree at most $2m$. These p_{jm} do not depend on f. (Cf. Faà di Bruno's formula [AS].)
Put $M = \sup\{|p_{jm}(\frac{1}{z})| \; |z \in A_T,\; 1 \le j \le m\}$. Now

$$\int_{A_T} |(Z\,f)^{(m)}(z)|^2\, \mu(z) dx dy \le m\, M^2 \sum_{j=1}^{m} \int_{A_T} |f^{(j)}\left(\frac{1}{z}\right)|^2\, \mu(z) dx dy \, .$$

Transform the latter integral $z \mapsto \frac{1}{z}$ and use the condition on μ. Then

$$\|(Z\,f)^{(m)}\|^2_{A,\mu} \le m\, M^2 \sum_{j=1}^{m} \int_{A_T} |f^{(j)}(z)|^2\, |z|^{-4}\, \mu\left(\frac{1}{z}\right) dx dy$$

$$\le m\, M^2 C \sum_{j=1}^{m} \int_{A_T} |f^{(j)}(z)|^2\, \mu(z) dx dy \, .$$

After removing the intermediate derivatives with Theorem 1.1 (E), we arrive at

$$\|Z\,f\|^2_{A_T,\mu,m} \le C_1\, \|f\|^2_{A_T,\mu,m}$$

for some $C_1 > 0$ not dependend on f.
Hence Z is a continuous bijection with $Z^2 = I$. The mentioned subspaces are merely the ranges of the respective projection operators $\frac{1}{2}(I + Z)$ and $\frac{1}{2}(I - Z)$.

(C) First we show that from our assumptions follows the existence of constants A_1 and A_2 such that for all $n \in I\!N_0$

$$(*) \qquad 0 < A_1 < \frac{\alpha_{-n}}{\alpha_n} < B_1 \, .$$

Note that, with $b = e^T$,

$$(\alpha_{-n})^{-2} = \int_{b^{-1}}^{b} r^{-2n+1}\, \mu(r) dr +$$

$$+|(-n) \cdot \ldots \cdot (-n - m + 1)|^2 \int_{b^{-1}}^{b} r^{-2(n+m)+1}\, \mu(r) dr \, .$$

Change the variable $r \mapsto \frac{1}{r}$. Take $n > 2(m-1)$

$$(\alpha_{-n})^{-2} = \int_{b^{-1}}^{b} r^{2n-3}\,\mu\left(\frac{1}{r}\right)\,dr +$$

$$+[n(n+1)\cdot\ldots\cdot(n+m-1)]^2 \int_{b^{-1}}^{b} r^{2(n+m)-3}\,\mu\left(\frac{1}{r}\right)\,dr \leq$$

$$\leq C \int_{b^{-1}}^{b} r^{2n+1}\,\mu(r)\,dr + 2\pi\,C\,b^{4m}[n\cdot\ldots\cdot(n+m-1)]^2 \int_{b^{-1}}^{b} r^{2(n-m)+1}\,\mu(r)\,dr$$

$$\leq C(1+b^{4m}\,2^{4m})\,(\alpha_n)^{-2}\,.$$

An inequality of type $(\alpha_n)^2 \leq C_m(\alpha_{-n})^2$ is derived in a similar way.

Now consider $f \in \mathcal{A}_2(A_T; m, \mu)$

$$f(z) = \sum_{n=-\infty}^{\infty} a_n\,\alpha_n\,z^n = a_0\,\alpha_0 + \sum_{n=1}^{\infty} \frac{1}{2}(a_n\,\alpha_n + a_{-n}\,\alpha_{-n})\,(z^n + z^{-n}) +$$

$$+\sum_{n=1}^{\infty} \frac{1}{2}(a_n\,\alpha_n - a_n\,\alpha_{-n})\,(z^n - z^{-n})\,.$$

We write

$$f(z) = a_0\,\alpha_0 + \sum_{n=1}^{\infty} b_n\,\beta_n(z^n + z^{-n}) + \sum_{n=1}^{\infty} c_n\,\beta_n(z^n - z^{-n})$$

with

$$\begin{matrix} b_n \\ c_n \end{matrix} = \frac{1}{2}\,a_n\left(1+\left(\frac{\alpha_n}{\alpha_{-n}}\right)^2\right)^{\frac{1}{2}} \begin{matrix} + \\ - \end{matrix} \frac{1}{2}\,a_{-n}\left(1+\left(\frac{\alpha_{-n}}{\alpha_n}\right)^2\right)^{\frac{1}{2}}\,.$$

Because of (*) the sequences (b_n) and (c_n) are l_2 iff (a_n) is l_2. □

Note that in case μ is of the form

$$\mu(z) = \mu(|z|) = |z|^{-2}\,\chi(|z|)\,\chi(|z|^{-1})\,,$$

with χ arbitrary, and $m = 0$, the basis in Theorem 1.2 (C) is orthonormal up to a factor $\sqrt{2\pi}$.

DEFINITION 1.3

Let ∂A denote the boundary of A. We have $\partial A = \partial_a A \cup \partial_b A$ with $\partial_a A$ and $\partial_b A$ the circles $|z| = a$ and $|z| = b$ respectively.
For $f \in \mathcal{A}(A)$ and $a < r < b$ consider the function

$$f_r\,:\,[-\pi,\pi] \to C\,,\quad t \mapsto f(r\,e^{it})\,.$$

We say that f satisfies a boundary condition in the Sobolev space $H^{\nu}(\partial_b A)$, $\nu \in \mathbb{R}$, if $f_b = \lim_{r\uparrow b} f_r$ exists in $H^{\nu}_{\text{per}}([-\pi,\pi])$, i.e. 2π-periodic.
Similarly we say that $f_a \in H^{\theta}(\partial_a A)$, $\theta \in \mathbb{R}$, if $f_a = \lim_{r\downarrow a} f_r \in H^{\theta}_{\text{per}}([-\pi,\pi])$, i.e. 2π-periodic.

THEOREM 1.4

Consider the Hilbert space $\mathcal{A}_2(A;\mu_{\theta\nu},m)$ with $\mu_{\theta\nu}(|z|) = (b-|z|)^{2\nu-1}\,(|z|-a)^{2\theta-1}$, $\nu,\theta > 0$.
Let $f \in \mathcal{A}(A)$, $f(z) = \sum_{n=-\infty}^{\infty} a_n\, z^n$.
The following three conditions are equivalent

(i) $f \in \mathcal{A}_2(A;\mu_{\theta\nu},m)$

(ii) $(b^n\, n^{m-\nu}\, a_n)_{n=0}^{\infty} \in l_2$ and $(a^n\, |n|^{m-\theta}\, a_n)_{n=-\infty}^{0} \in l_2$

(iii) f satisfies the boundary conditions

$$f_a \in H^{m-\theta}(\partial_a A) \quad \text{and} \quad f_b \in H^{m-\nu}(\partial_b A)\ .$$

So the functions $\{1,\ b^{-n}\, n^{-(m-\nu)}\, z^n,\ a^n\, n^{-(m-\theta)}\, z^{-n} \mid n \in \mathbb{N}\}$ establish a Riesz basis in $\mathcal{A}_2(A;\mu_{\theta\nu},m)$.

Proof

We apply Theorem 1.1 (C), so we have to calculate the asymptotic behaviour for $k \to \pm\infty$ of the integral

$$\gamma_{2k+1} = \int_a^b r^{2k+1}(b-r)^{2\nu-1}\,(r-a)^{2\theta-1}\,dr =$$

$$b^{2\nu-1}\,a^{2\theta-1}\int_{1/b}^{1/a} r^{-(2k+1+2\nu+2\theta)}\left(\frac{1}{a}-r\right)^{2\theta-1}\left(r-\frac{1}{b}\right)^{2\nu-1} dr =$$

$$= b^{2k+1}(b-a)^{2\nu+2\theta-1}\int_0^1 (1-s)^{2\theta-1}\,s^{2\nu-1}\left(1-s\left(1-\frac{a}{b}\right)\right)^{2k+1} ds =$$

$$= \left(\frac{1}{a}\right)^{-(2k+1)}\left(\frac{a}{b}\right)^{2\theta}(b-a)^{2\nu+2\theta-1}\ .$$

$$\cdot\int_0^1 (1-s)^{2\nu-1}\,s^{2\theta-1}\left(1-s\left(1-\frac{a}{b}\right)\right)^{-(2k+1+2\nu+2\theta)} ds\ .$$

Note that, for $0 < B < 1$, $l \in \mathbb{N}$

$$(l\,B)^{2\nu}\int_0^1 (1-s)^{2\theta-1}\,s^{2\nu-1}(1-s\,B)^l\,ds =$$

$$\int_0^1 (1-s)^{2\theta-1}\,(l\,B\,s)^{2\nu-1}\left(1-\frac{1}{l}\,l\,B\,s\right)^l l\,B\,ds =$$

$$= \int_0^{lB} \left(1-\frac{\sigma}{l\,B}\right)^{2\theta-1} \sigma^{2\nu-1}\left(1-\frac{\sigma}{l}\right)^l d\sigma \to$$

$$\to \int_0^{\infty} \sigma^{2\nu-1}\,e^{-\sigma}\,d\sigma = \Gamma(2\nu)\,, \quad \text{as } l\to\infty\,.$$

Since

$$(\alpha_n)^{-2} = \gamma_{2n+1} + |(n-m+1)_m|^2\,\gamma_{2(n-m)+1}$$

we find that the following two limits exist

$$\lim_{n\to\infty} \alpha_n[b^n\,n^{m-\nu}] = L_1 > 0\,,$$

$$\lim_{n\to-\infty} \alpha_n\left[\left(\frac{1}{a}\right)^{-n} |n|^{m-\theta}\right] = L_2 > 0\,.$$

(i) $\Longleftrightarrow$ (ii) We apply Theorem 1.1 (C). Since $b > a$ we have $(\alpha_n^{-1}\,a_n)_{n=-\infty}^{\infty} \in l_2$ iff

$$(b^n\,|n|^{m-\nu}\,a_n)_{n=-\infty}^{\infty} \in l_2 \quad \underline{\text{and}} \quad (a^n\,|n|^{m-\theta}\,a_n)_{n=-\infty}^{\infty} \in l_2$$

(ii) $\Longleftrightarrow$ **(iii)** Note that

$$f_b = \sum_{n=-\infty}^{\infty} a_n\,b^n\,e^{in\phi} \in H^{m-\nu}(\partial_b A)$$

iff $(a_n\,b^n\,n^{m-\nu})_{n=-\infty}^{\infty} \in l_2$.

For f_a the argument is similar. □

As an application of Theorem 1.2 (B) we find

COROLLARY 1.5

In the Hilbert space $\mathcal{A}_2(A_T;\mu_\nu,m)$, with $\mu_\nu(|z|) = (e^T-|z|)^{2\nu-1}\,(|z|-e^{-T})^{2\nu-1}$, $T>0\,,\ \nu>0$, the functions

$$\{1,\ e^{-nT}\,n^{-(m-\nu)}(z^n+z^{-n}),\ \ e^{-nT}\,n^{-(m-\nu)}(z^n-z^{-n}) \mid n\in I\!N\}$$

establish a Riesz basis.

2. WEIGHTED SOBOLEV SPACES OF ANALYTIC FUNCTION ON $\Omega \subset \mathbb{C}$

Consider an open set $\Omega \subset \mathbb{C}$ and a function $\psi \in \mathcal{A}(\Omega)$ which maps Ω bijectively onto an open subset $\psi(\Omega) \subset A$, the annulus of section 1. The linear subspace $\mathcal{A}(\Omega;\psi) \subset \mathcal{A}(\Omega)$ is defined by a pull-back

$$\mathcal{A}(\Omega;\psi) = \{F \mid F(\zeta) = f(\psi(\zeta)),\ \ \zeta = \xi + i\eta \in \Omega,\ \ f \in \mathcal{A}(A)\} .$$

There is a bijective linear correspondence

$$J_\psi \ :\ \mathcal{A}(A) \to \mathcal{A}(\Omega;\psi)\ ,\quad (J_\psi f)\ (\zeta) = f(\psi(\zeta))\ ,$$

so $J_\psi(\mathcal{A}(A)) = \mathcal{A}(\Omega;\psi)$.
Note that $\psi(\Omega) = A$ implies $\mathcal{A}(\Omega;\psi) = \mathcal{A}(\Omega)$. The Hilbert subspace $J_\psi(\mathcal{A}_2(A;\mu,m))$ carries the inner product

$$(F,G)_\psi = (J_\psi^{-1}\, F,\ J_\psi^{-1}\, G)_{A,\mu,m}\ .$$

In this section we investigate the relation between the mentioned "pull back inner product" and an ordinary Sobolev inner product in $\mathcal{A}(\Omega)$. In the next Lemma an auxilliary result concerning the chain rule is presented. The proof is omitted. An explicit expression for the R_{mj} is given by Faà di Bruno's formula [AS].

LEMMA 2.1

Let f and ψ be analytic functions. We have

(i) $(f \circ \psi)^{(m)}(\zeta) = \sum_{j=1}^{m} (f^{(j)} \circ \psi)\ (\zeta) \cdot R_{mj}(\zeta).$

- The R_{mj}, $m \in I\!N$, $1 \le j \le m$, are homogeneous polynomials of degree j in ψ', $\psi^{(2)}, \dots, \psi^{(m-j+1)}$. Put $R_{m,j} = 0$ if $j < 1$ or $j > m$.
- $R_{m+1,j} = R'_{m,j} + \psi' \cdot R_{m,j-1}$, $R_{11} = \psi'$.
- $R_{mm} = (\psi')^m\ ,\quad R_{m1} = \psi^{(m)}$.

(ii) $f^{(m)}(z) = \left(\dfrac{d\psi^{\leftarrow}}{dz}\right)^m \left\{(f \circ \psi)^{(m)}(\psi^{\leftarrow}(z)) - \sum_{j=1}^{m-1} f^{(j)}(z)\ R_{mj}(\psi^{\leftarrow}(z))\right\}$

THEOREM 2.2

Consider an open set $\Omega \subset C\!\!\!\!I$. Let $\psi \in \mathcal{A}(\Omega)$, $\psi\ :\ \Omega \to A$ be injective. Denote $\Omega^\varepsilon = \psi^{\leftarrow}(A^\varepsilon)$. Let $\omega \in L_1(\Omega)$ be a weight function and let $m \in I\!N_0$ be fixed. Finally, let $\mu \in L_1(A)$ be a weight function on A which satisfies the conditions of Theorem 1.1.
Suppose

(a) $\sup_{0 \le q \le m,\ \zeta \in \mathrm{supp}\,\omega} |\psi^{(q)}(\zeta)| = M_1 < \infty$

(b) $\exists \varepsilon > 0\ \exists M_2 > 0\ \exists M_3 > 0\ \forall \zeta \in \Omega^\varepsilon$
$M_2\ \mu(\psi(\zeta))\ |\psi'(\zeta)|^2 \le \omega(\zeta) \le M_3\ \mu(\psi(\zeta)\ |\ |\psi'(\zeta)|^2$

(c) If $m > 0$ then
$\exists \varepsilon > 0\ \exists M_4 > 0\ \forall \zeta \in \Omega^\varepsilon\ |\psi'(\zeta)|^{-1} \le M_4.$

Then

(A) If $\psi(\Omega)$ is dense in A, then on $\mathcal{A}(\Omega;\psi)$ the norms $F \mapsto \|J_\psi^{-1} F\|_{A,\mu,m}$ and $F \mapsto \|F\|_{\Omega,\omega,m}$ are equivalent. So the functions $(\alpha_n(\psi(\zeta))^n)_{n=-\infty}^{\infty}$ establish a Riesz basis in $\mathcal{A}_2(\Omega;\psi;\omega,m)$.

(B) Let W be a closed subspace of $\mathcal{A}_2(A;\mu,m)$ and suppose that the norms $\|\cdot\|_{A,\mu,m}$ and $\|\cdot\|_{\psi(\Omega),\mu,m}$, when restricted to W, are equivalent. Then on the Hilbert space $J_\psi(W) \subset \mathcal{A}(\Omega;\psi;W)$ the norms $F \mapsto \|J_\psi^{-1} F\|_{\psi(\Omega);\mu,m}$ and $F \mapsto \|F\|_{\Omega,\omega,m}$ are equivalent. So $(\sigma_n(z))_{n\in I}$ is a Riesz basis in W iff $(\sigma_n(\psi(\zeta))_{n\in I}$ is a Riesz basis in $\mathcal{A}_2(\Omega;\psi;W;\omega,n)$.

Proof

(A1) Let $F \in \mathcal{A}(\Omega;\psi)$. There exists a unique $f \in \mathcal{A}(A)$ such that $F = f\circ\psi$. For $\varepsilon > 0$ sufficiently small we estimate the two terms of

$$\|F^{(m)}\|^2_{\Omega,\mu,m} = \|F^{(m)}\|^2_{\Omega^\varepsilon,\mu,m} + \|F^{(m)}\|^2_{\Omega\backslash\Omega^\varepsilon,\mu,m}$$

separately. With Lemma 2.1.(i) and the condition on ω we estimate

$$\int_{\Omega^\varepsilon} |(f\circ\psi)^{(m)}(\zeta)|^2\, \omega(\zeta)\, d\xi d\eta \le$$

$$\le c_1 \sum_{j=1}^{m} \int_{\Omega^\varepsilon} |(f^{(j)}\circ\psi)\,(\zeta)|^2\, \mu(\psi(\zeta))\, |\psi'(\zeta)|^2\, d\xi d\eta\ .$$

The constant c_1 depends on m, M_1, M_3 and the shape of the R_{mj}. The latter expression is equal to

$$c_1 \sum_{j=1}^{m} \int_{A^\varepsilon} |f^{(j)}(z)|^2\, \mu(z)\, dxdy \le c_2\, \|f\|^2_{A,\mu,m}\ .$$

The constant c_2 depends on c_1 and on the constants mentioned in Theorem 1.1, part E.
Next we show that

$$\exists c_3 > 0\ \forall\zeta \in (\Omega\backslash\Omega^\varepsilon)\cap \operatorname{supp}\omega\ |F^{(m)}(\zeta)| \le c_3\, \|f\|_{A,\mu}\ .$$

Indeed, for $\zeta \in \operatorname{supp}\,\omega \cap (\Omega\backslash\Omega^\varepsilon)$ write

$$F^{(m)}(\zeta) = (f\circ\psi)^{(m)}(\zeta) =$$

$$= \frac{1}{2\pi i} \sum_{j=1}^{m} (-1)^j\, j!\, R_{mj}(\zeta) \int_K f(z)\, (\psi(\zeta)-z)^{-1-j}\, dz\ .$$

Here K consists of two suitably oriented circles in the interior of $A^{\frac{2}{3}\varepsilon}\backslash A^{\frac{1}{3}\varepsilon}$. On K we have, with the constants of Theorem 1.1.(A), $|f(z)| \le M_{\frac{1}{3}\varepsilon,0}\, \|f\|_{A^{\frac{1}{3}},\mu}$. From this we infer the existence of c_3. We gather that

$$\|F^{(m)}\|^2_{\Omega,\omega} \le (c_2 + c_3\|\omega\|_1)\, \|f\|^2_{A,\mu,m}\ .$$

Since such an estimate also holds for $m = 0$ we arrive at the existence of $c_4 > 0$ such that

$$\|F\|^2_{\Omega,\omega,m} \leq c_4 \, \|f\|^2_{A,\mu,m} \, .$$

(A2) Now let $f \in \mathcal{A}(A)$. With Lemma 2.1.(ii) and conditions (b), (c)

$$\|f^{(m)}\|^2_{A^\epsilon,\mu} \leq c_4 \int\limits_{A^\epsilon} |(f \circ \psi)^{(m)} \, (\psi^{\leftarrow}(z))|^2 \, \mu(z) \, dxdy +$$

$$+ c_5 \sum_{j=1}^{m-1} \int\limits_{A^\epsilon} |f^{(j)}(z) \, R_{mj}(\psi^{\leftarrow}(z))|^2 \, \mu(z) \, dxdy \, .$$

The constant c_5 depends only on M_4 and m.
From the boundedness of the R_{mj} and application of Theorem 1.1.(E) it follows that

$$\|f^{(m)}\|^2_{A^\epsilon,\mu} \leq c_4 \int\limits_{A^\epsilon} |(f \circ \psi)^{(m)} \, (\psi^{\leftarrow}(z)|^2 \, \mu(z) \, dxdy + c_6 \, \|f\|^2_{A^\epsilon,\mu} \, .$$

The constant c_6 depends only on c_5, M_1 and the shape of the R_{mj}. Transform the integral over A^ϵ to an integral over Ω^ϵ and estimate again

$$\|f^{(m)}\|^2_{A^\epsilon,\mu} \leq c_4 \, M^{-1} \, \|F^{(m)}\|^2_{\Omega^\epsilon,\omega} + c_6 \, \|f\|^2_{A^\epsilon,\mu} \, .$$

Since

$$\|f\|^2_{A^\epsilon,\mu} \leq M^{-1} \, \|F\|^2_{\Omega^\epsilon,\omega}$$

we have

$$\|f\|^2_{A,\mu,m} \leq \text{const.} \, \|F\|^2_{\Omega,\omega,m} \, .$$

(B) The proof of part (B) is merely a simple adaptation of the proof of (A). □

REMARK 2.3

The conditions (a) and (c) of Theorem 2.2 are automatically satisfied if $\bar{\Omega}$ is compact and $\psi \in \mathcal{A}(\bar{\Omega})$.

A simple modification of Theorem 2.2 is the following

THEOREM 2.4

Consider an open set $\Omega \subset C$. Let $\psi \in \mathcal{A}(\Omega)$, $\psi : \Omega \to A$ be injective. Let $\omega \in L_1(\Omega)$ be a weight function such that for some $M_1 > 0$, $M_2 > 0$

$$M_1 \, \mu(\psi(\zeta)) \leq \omega(\zeta) \leq M_2 \, \mu(\psi(\zeta)) \, .$$

Let $W \subset \mathcal{A}_2(A; \mu, 1)$ be a closed subspace as in Theorem 2.2.(B). Fix $p \in \Omega$. Then on the Hilbert space $J_\psi(W)$ the norms

$$F \mapsto \|J_\psi^{-1} F\|_{\psi(\Omega);\mu,1} \quad \text{and} \quad F \mapsto (|F(p)|^2 + \|F'\|^2_{\Omega,\omega})^{\frac{1}{2}}$$

are equivalent.
The mapping J_ψ becomes unitary if $\omega(\zeta) = \mu(\psi(\zeta))$ and $\mathcal{A}_2(A;\mu,1)$ carries the norm

$$f \mapsto (|f(\psi(p))|^2 + \|f'\|^2_{A,\mu})^{\frac{1}{2}} .$$

□

In the next theorem we study a class of weight functions on A and Ω which can easily be compared. We need

CONDITION 2.5

Suppose that $\psi : \Omega \to A$ extends continuously to $\psi_e : \bar{\Omega} \to \bar{A}$. Fix a subset $\Delta \subset \partial\bar{\Omega}$. Consider the distance functions $\text{dis}(\zeta,\Delta)$ on Ω and $\text{dis}(z,\psi(\Delta))$ on $\psi(\Omega) \subset A$. Let $\varepsilon > 0$ be fixed. Assume

(i) Each $\zeta \in \Omega^\varepsilon$ can be connected with Δ by means of a differentiable curve within Ω^ε and which has length $\text{dis}(\zeta,\Delta)$.

(ii) Each $z \in \psi(\Omega^\varepsilon)$ can be connected with $\psi_e(\Delta)$ by means of a differentiable curve within $\psi(\Omega^\varepsilon)$ and which has length $\text{dis}(z,\psi(\Delta))$.

THEOREM 2.6

Assume Conditions 2.5. Suppose in addition

$$\exists M > 0\ \forall \zeta \in \Omega^\varepsilon\ |\psi'(\zeta)| \le M \quad \text{and} \quad |\psi'(\zeta)|^{-1} \le M .$$

Let $\theta > -1$. On Ω^ε define $\omega(\zeta) = (\text{dis}(\zeta,\Delta))^\theta$.
On $A^\varepsilon \cap \psi(\Omega)$ define $\mu(z) = (\text{dis}(z,\psi(\Delta)))^\theta$.
Then

$$\forall \zeta \in \Omega^\varepsilon \quad M^{-|\theta|}\ \mu(\psi(\zeta)) \le \omega(\zeta) \le M^{|\theta|}\ \mu(\psi(\zeta))$$

and condition (b) of Theorem 2.2 is satisfied.

Proof

Let $\zeta \in \Omega^\varepsilon$. Take a curve $s \mapsto \chi(s)$ within $\psi(\Omega^\varepsilon)$ which connects $\psi(\zeta)$ and $\psi(\Delta)$ and which has length $L = \text{dis}(\psi(\zeta),\ \psi(\Delta))$. Estimate

$$\text{dis}(\zeta,\Delta) \le \int_0^L \left| \frac{d\psi^{\leftarrow}(\chi(s))}{ds} \right| ds \le ML = M\ \text{dis}(\psi(\zeta),\ \psi(\Delta)) .$$

Similarly we estimate for $z \in \psi(\Omega^\varepsilon)$

$$\text{dis}(z,\psi(\Delta)) \le M\ \text{dis}(\psi^{\leftarrow}(z),\Delta) .$$

Taking the θ-th power leads to the desired inequalities. □

APPLICATION I

Consider the rectangle

$$\Omega = S_T = \{\zeta = \xi + i\eta \mid \xi \in (-\pi,\pi)\,,\quad \eta \in (-T,T)\}\,.$$

Take $\psi(\zeta) = e^{i\zeta}$, $\psi(S_T) = A_T \setminus (-\infty, 0)$. Note that $\psi(S_T)$ is an open dense set in A_T. We find that $\mathcal{A}(S_T;\psi) = \mathcal{A}(S_T; 2\pi\text{-per})$, i.e. the space of functions on S_T which extend to 2π-periodic analytic functions.
For the weight functions on S_T and A_T take

$$\omega_\theta(\zeta) = \omega_\theta(\xi + i\eta) = (T-\eta)^{2\theta-1}(\eta+T)^{2\theta-1}\,,$$

$$\mu_\theta(|z|) = \mu_\theta(|z|) = (e^T - |z|)^{2\theta-1}(|z| - e^{-T})^{2\theta-1}\,.$$

Note that the conditions (a), (b), (c) of Theorem 2.2 are satisfied. Combining Theorem 1.4 and Theorem 2.2.A we find that for all $\theta > 0$ and all $m \in \mathbb{N}_0$

$$J_\psi \;:\; \mathcal{A}_2(A_T;\mu_\theta, m) \to \mathcal{A}_2(S_T; 2\pi\text{–per}; \omega_\theta, m)$$

is a continuous bijection between both Hilbert spaces. As a corollary thereof, the functions

$$(\zeta \mapsto |n|^{-(m-\theta)}\, e^{-|n|T}\, e^{in\zeta})_{n\in\mathbb{Z}}$$

establish a Riesz basis in $\mathcal{A}_2(S_T; 2\pi\text{-per};\omega_\theta, m)$. Note also that $f \in \mathcal{A}(S_T; 2\pi\text{-per})$ belongs to $\mathcal{A}_2(S_T; 2\pi\text{-per};\omega_\theta, m)$ iff the limits of $\xi \mapsto f(\xi + i\eta)$ as $\eta \uparrow T$ or $\eta \downarrow -T$ exist in the Sobolev space $H^{m-\theta}_{\text{per}}(-\pi,\pi)$.
A combination of Theorem 1.2.B and Theorem 2.2.B leads to the Riesz bases

$$(\zeta \mapsto n^{-(m-\theta)}\, e^{-nT}\, \cos n\,\zeta)_{n\in\mathbb{N}_0}$$

$$(\zeta \mapsto n^{-(m-\theta)}\, e^{-nT}\, \sin n\,\zeta)_{n\in\mathbb{N}}$$

for the even and odd subspaces of $\mathcal{A}_2(S_T; 2\pi\text{-per};\omega_\theta, m)$ respectively.

APPLICATION II

Consider the ellips $E_T = \{\zeta = \xi + i\eta \mid 1 - \left(\frac{\xi^2}{\cosh^2 T} + \frac{\eta^2}{\sinh^2 T}\right) > 0\}$.
Take $\Omega = E'_T = E_T \setminus (-\infty, -1] \cup [1, \infty)$.
Take $\psi(\zeta) = e^{i \arccos \zeta}$. Then ψ maps E'_T bijectively onto A_T^+, i.e. the part of A_T which lies in the open upper half plane.
Note that $J_\psi(\mathcal{A}(A_T; e)) = \mathcal{A}(E_T)$ because of its symmetry $f(e^{i \arccos \zeta})$ does not jump at the real axis and extends to an analytic function on the whole of E_T.
Further, since on $\mathcal{A}(A_T; e)$ the norms $\|\cdot\|_{A_T,\mu,m}$ and $\|\cdot\|_{A_T^+,\mu,m}$ are equivalent for any choice of μ which satisfies the conditions of Theorem 1.1, we can apply Theorem 2.2.B.
In particular we apply Theorem 2.6. Take $\Delta = \partial E_T$. On E_T introduce the weight function $\omega_\theta(\zeta) = (\text{dis}(\zeta, \partial E_T))^{2\theta-1}$, $\theta > 0$. On A_T^+ introduce μ_θ as in Application I. We find that for all $\theta > 0$ and all $m \in \mathbb{N}_0$

$$J_\psi \;:\; \mathcal{A}_2(A_T; e; \mu_\theta, m) \to \mathcal{A}_2(E_T; \omega_\theta, m)$$

is a continuous bijection between both Hilbert spaces. As a consequence the Chebyshev polynomials

$$(n^{-(m-\theta)}\, e^{-nT}\, T_n(\zeta))_{n\in \mathbb{N}_0}$$

with $T_n(\zeta) = \cos(n \arccos \zeta)$, $n \in \mathbb{N}$ and $T_0(\zeta) = \pi^{-\frac{1}{2}}$ establish a Riesz basis in $\mathcal{A}_2(E_T;\omega_\theta,m)$.

As a corollary of the general Theorems in sections 1 and 2 we mention the following

Characterization Result

Let $f \in \mathcal{A}([-1,1])$, $f(x) = \sum_{n=0}^{\infty} a_n T_n(x)$, let $\theta > 0$, $m \in \mathbb{N}_0$, then the following three conditions are equivalent:

- $f \in \mathcal{A}_2(E_T;\omega_\theta,m)$
- $(n^{(m-\theta)}\, e^{nT}\, a_n)_{n=0}^{\infty} \in l_2$
- $f \in \mathcal{A}(E_T)$ and $f|_{\partial E_T} \in H^{m-\theta}(\partial E_T)$

In the next section we extend this expansion characterization to a much wider class of suitably normalized polynomials which encompasses all (normalized) Jacobi polynomials. In this way Szegö's result [Sz] is refined in several directions at once.

APPLICATION III

Consider the set $\Omega = B_T = \mathbb{R}\times i\,(-\frac{\pi}{2},\ \frac{\pi}{2}) \setminus \bar{\mathcal{U}}_T$ with $\mathcal{U}_T = \{\xi+i\eta \mid |\sin\xi| > \tanh T\ \cosh\eta\}$. Note that $\bar{\mathcal{U}}_T$ is an infinite set of "ovals" centered at the points $\xi + i\eta = i(k+\frac{1}{2})\pi$, $k \in \mathbb{Z}$. Take $\psi(\zeta) = e^{2i\,\mathrm{arctan}\,\mathrm{sinh}\,\zeta} = \exp 2i \int\limits_0^{\zeta} \frac{1}{\cosh w}\, dw$, $\psi(B_T) = A_T \setminus (-\infty,0)$, which is open and dense in A_T. We find that $\mathcal{A}(B_T;\psi) = \mathcal{A}(B_T; 2\pi i\text{-per},\partial e\infty)$. A function g belongs to this class iff it extends to $\tilde{g} \in \mathcal{A}(\mathbb{C}\setminus\bar{\mathcal{U}}_T)$ such that

(i) $\tilde{g}$ is $2\pi i$-periodic

(ii) $\forall\zeta \in \mathbb{C}\quad \tilde{g}(i\frac{\pi}{2}+\zeta) = \tilde{g}(i\frac{\pi}{2}-\zeta)$

(iii) $\lim\limits_{|\Re\zeta|\to\infty} \tilde{g}(\zeta)$ exists.

There are natural subspaces of $\mathcal{A}(B_T;\psi)$ which exhibit even more symmetry:

- $J_\psi(\mathcal{A}(A_T;e)) = \mathcal{A}(B_T;\pi i\text{-per, even, }\partial e\infty)$ which is the subspace of even πi-periodic functions.
- $J_\psi(\mathcal{A}(A_T;o)) = \mathcal{A}(B_T;2\pi i\text{-per, odd, }\partial e\infty)$ which is the subspace of odd $2\pi i$-periodic functions.

Now, let $\Delta = \partial\mathcal{U}_T \cap \bar{B}_T$. Denote the function

$$\delta^\theta(\zeta) = (\min\{\mathrm{dis}(\zeta,\Delta),1\})^{2\theta-1},\quad \theta > 0\ .$$

For the weight function on B_T we take

$$\omega_\theta(\zeta) = \delta^\theta(\zeta)\ |\cosh\zeta|^{-2}\ .$$

On A_T introduce the weight function μ_θ, as in Application I. For $\varepsilon < (1-e^{-T})$ the conditions of Theorem 2.2 are satisfied (use Theorem 2.6). So we find that for all $\theta > 0$ and all $m \in \mathbb{N}_0$

$$J_\psi \; : \; \mathcal{A}_2(A_T; \mu_\theta, m) \to \mathcal{A}_2(B_T; 2\pi i\text{–per}, \partial e\infty; \omega_\theta, m)$$

is a continuous bijection between both Hilbert spaces. As a consequence we find for the subspace $\mathcal{A}_2(B_T; \pi i\text{-per, even, } \partial e\infty; \omega_\theta, m)$ a Riesz basis

$$\left(n^{-(m-\theta)}\, e^{-nT}\, T_{2n}\left(\frac{1}{\cosh\zeta}\right)\right)_{n=0}^{\infty}$$

and for the subspace $\mathcal{A}_2(B_T; 2\pi i\text{-per, odd, } \partial e\infty; \omega_\theta, m)$ we have the Riesz basis

$$\left(n^{-(m-\theta)}\, e^{-nT}\, (\tanh\zeta)\, U_{2n-1}\left(\frac{1}{\cosh\zeta}\right)\right)_{n=1}^{\infty} .$$

Here the polynomials U en T are defined by $T_n(\cos\theta) = \cos n\,\theta$ and $U_n(\cos\theta) = \frac{\sin(n+1)\theta}{\sin\theta}$. As a corollary of the general theorems in sections 1 and 2 we mention the following

Characterization Result

Let $f \in \mathcal{A}(\mathbb{R})$, $f(x) = \sum_{n=0}^{\infty} a_n T_{2n}\left(\frac{1}{\cosh x}\right) + \tanh x \sum_{n=1}^{\infty} b_n U_{2n-1}\left(\frac{1}{\cosh x}\right)$. Let $\theta > 0$, $m \in \mathbb{N}_0$, then the following three conditions are equivalent:

- $f \in \mathcal{A}_2(B_T; 2\pi i\text{-per}, \partial e\infty; \omega_\theta, m)$
- $(n^{(m-\theta)}\, e^{nT}\, a_n)_{n=0}^{\infty} \in l_2$ and $(n^{(m-\theta)}\, e^{nT}\, b_n)_{n=1}^{\infty} \in l_2$
- $f \in \mathcal{A}(\mathbb{C} \setminus \bar{\mathcal{U}}_T)$, f is $2\pi i$-periodic,

  $$\lim_{|\Re\zeta|\to\infty} f(\zeta) \text{ exists} \quad \text{and} \quad f|_{\partial\mathcal{U}_T} \in H_{loc}^{m-\theta}(\partial\mathcal{U}_T) .$$

Finally it follows that

$$\left(\pi^{-1}(\cosh x)^{-\frac{1}{2}}\, T_{2n-2}\left(\frac{1}{\cosh x}\right),\; \pi^{-1}(\cosh x)^{-\frac{1}{2}}\, \tanh x\, U_{2n-1}\left(\frac{1}{\cosh x}\right)\right)_{n=1}^{\infty}$$

is a complete orthonormal system in $L_2(\mathbb{R})$.
In [BG1] and [BG2] we have characterized the Fourier and Fourier-Jacobi images of some of the spaces mentioned here.

APPLICATION IV

Consider the set $\Omega = R_{ab} = \mathbb{C} \setminus (D_a \cup D_b)$ with $0 < a < 1 < b$. Here D_c, $c > 0$, denotes the closed disc $\{\zeta \mid |\zeta + i\,\frac{c^2+1}{c^2-1}| \le \frac{2c}{|c^2-1|}\}$.
Take $\psi(\zeta) = \frac{i-\zeta}{i+\zeta}$, then $\psi(R_{ab}) = A \setminus \{-1\}$.
We find that $\mathcal{A}(\Omega; \psi) = \mathcal{A}(R_{ab}; \infty)$, i.e. the space of functions which are analytic outside the discs D_a and D_b and at infinity.

Note that if we take $W_+ = \mathcal{A}(A; +)$ then

$$\mathcal{A}(\Omega; \psi; W_+) = \mathcal{A}(\mathbb{C} \setminus D_b; \infty)$$

and, similarly

$$\mathcal{A}(\Omega; \psi; W_-) = \mathcal{A}(\mathbb{C} \setminus D_a; \infty) .$$

Let $\Delta = \partial D_a \cup \partial D_b$. Denote the function

$$\delta^\theta(\zeta) = (\min\{\text{dis}(\zeta, \Delta), 1\})^{2\theta-1}, \quad \theta > 0 .$$

For the weight functions on R_{ab} and A we take

$$\omega_\theta(z) = \delta^\theta(\zeta)\ |i + \zeta|^{-4}$$

$$\mu_\theta(\zeta) = (b - |z|)^{2\theta-1}\ (|z| - a)^{2\theta-1} .$$

For $\varepsilon < \frac{1}{2}(b - a)$ the conditions of Theorem 2.2 are satisfied (via Theorem 2.6). So we find that for all $\theta > 0$ and all $m \in \mathbb{N}_0$

$$J_\psi\ :\ \mathcal{A}_2(A; \mu_\theta, m) \to \mathcal{A}_2(R_{ab}; \infty; \omega_\theta, m)$$

is a continuous bijection between both Hilbert spaces.
As a consequence, for $\mathcal{A}_2(R_{ab}; \infty; \omega_\theta, m)$ we have the Riesz basis

$$\left\{1,\ b^{-n}\, n^{-(m-\theta)} \left(\frac{i-\zeta}{i+\zeta}\right)^n,\ \ a^n\, n^{-(m-\theta)} \left(\frac{i-\zeta}{i+\zeta}\right)^{-n}\right\}_{n\in\mathbb{N}} .$$

Finally, it is easily verified that

$$\left((\pi(1+x^2))^{-\frac{1}{2}}\ \frac{i-x}{i+x}\right)_{n\in\mathbb{Z}}$$

is an orthonormal basis in $L_2(\mathbb{R})$.

Spaces of this example play a key role in [D].

3. THE GENERAL FUNCTIONAL ANALYTIC CLASSIFICATION PROBLEM

Let $K = (K_{mn})_{m,n=0}^{\infty}$ be an infinite matrix with $K_{mn} = 0$ if $m > n$.
We denote $K \in UTM$, i.e. the class of upper triangular matrices. If we suppose that $K_{ii} \neq K_{jj}$ if $i \neq j$ there exists a unique $S \in UTM$ with all $S_{ii} = 1$ such that $S^{-1}\,K\,S = \Delta_k = \text{diag}(K_{00}, K_{11}, \ldots)$. If K has suitable growth properties we construct classes of diagonal matrices $M = \text{diag}(\mu_0, \mu_1, \ldots)$ such that MSM^{-1} and $MS^{-1}M^{-1}$ are l_2-bounded matrices. In general the growth properties of M are related to those of K.

This general result will enable us to construct a great variety of (polynomial) Riesz bases out of a given one, in the weighted Sobolev spaces of the preceding sections. One of the consequences will be a refinement and an extension of Szegö's classical theorem of Jacobi expansions of analytic functions on an ellips.

We start with a separable Hilbert space X and a selected orthonormal basis $(T_n)_{n=0}^{\infty} \subset X$. Associated with a sequence $\mu = (\mu_n)_{n=0}^{\infty}$, $\mu_n > 0$, $\mu_n \to \infty$ as $n \to \infty$, we introduce a Hilbert space X_μ which is dense in X,

$$X_\mu = \{f \mid f \in X, \sum_{n=0}^{\infty} \mu_n^2 \, |(f, T_n)|^2 < \infty\} .$$

Note that $f = \sum_{n=0}^{\infty} a_n T_n \in X_\mu$ iff $\sum_{n=0}^{\infty} \mu_n^2 \, |a_n|^2 < \infty$. Note also that $(\mu_n^{-1} T_n)_{n=0}^{\infty}$ is an orthonormal basis in X_μ. By means of a transition matrix $S \in UTM$ we define a sequence of vectors $(R_n)_{n=0}^{\infty} \subset \text{span}(T_n)$ by $R_n = \sum_{j=0}^{n} S_{jn} T_j$. We take all $S_{nn} = 1$ so that S has an inverse $S^{-1} \in UTM$ of the same type and $T_n = \sum_{j=0}^{n} S_{jn}^{-1} R_j$. Our first question is: When is the sequence $(\mu_n^{-1} R_n)_{n=0}^{\infty}$ a Riesz basis in X_μ?

THEOREM 3.1

Let $M \in UTM$ be defined by $M = \text{diag}(\mu_0, \mu_1, \ldots)$. The sequence $(\mu_n^{-1} R_n)_{n=0}^{\infty}$ is Riesz basis in X_μ iff both matrices MSM^{-1} and $MS^{-1}M^{-1}$ are l_2-bounded. This means that they can be regarded as bounded operators in l_2.

Proof

On $span(T_n)$ define the operators $\mathbf{M}$ and $\mathbf{S}$ by $\mathbf{M} T_n = \mu_n T_n$ and $\mathbf{S} T_n = R_n$, followed by linear extension. On $span(T_n)$ the inverse $\mathbf{S}^{-1}$ exists and $\mathbf{S}^{-1} R_n = T_n$. Observe that $(\mu_n^{-1} R_n)_{n=0}^{\infty}$ is a Riesz basis in X_μ iff $\mathbf{S}$ extends to a continuous bijection on X_μ iff both $\mathbf{S}$ and $\mathbf{S}^{-1}$ extend to continuous mappings on X_μ iff both matrices MSM^{-1} and $MS^{-1}M^{-1}$ are l_2-bounded. The latter equivalence follows from $\|\mathbf{S} f\|_\mu = \|\mathbf{MS} f\| = \|(\mathbf{MSM}^{-1}) \mathbf{M} f\|$.
□

Let $\mathbf{K}$ be a densely defined operator in X which acts invariantly on $span(T_n)$. Suppose that, with respect to the basis $(T_n)_{n=0}^{\infty}$, the operator $\mathbf{K}$ is represented by a matrix $K \in UTM$ with mutually distinct entries on its diagonal. Then there exists a unique $S \in UTM$, with all $S_{ii} = 1$, such that $S^{-1} K S = \Delta_K = \text{diag}(K_{00}, K_{11}, \ldots)$. Consequently, there exists a unique sequence $(T_n^K)_{n=0}^{\infty} \subset span(T_n)$ of eigenvectors of $\mathbf{K}$.
Note that $T_n^K = \sum_{j=0}^{n} S_{jn} T_j$ and $\mathbf{K} T_n^K = K_{nn} T_n^K$. Now we turn $span(T_n)$ into a pre-Hilbert space by declaring the sequence $(\mu_n^{-1} T_n^K)_{n=0}^{\infty}$ to be an *orthonormal* sequence: The Hilbert space completion is then denoted by $X_\mu^{\mathbf{K}}$. Note in particular that $(\mu_n^{-1} T_n^K)_{n=0}^{\infty}$ is a Riesz basis in X_μ iff $X_\mu = X_\mu^{\mathbf{K}}$ as topological vector spaces. Note that $X_\mu = X_\mu^{\mathbf{M}}$. Two problems can now be posed:

- **The classification problem**
  For which pairs $(\mathbf{K}, \mu)$, $(\mathbf{K}', \mu')$ do we have $X_\mu^{\mathbf{K}} = X_\mu^{\mathbf{K}'}$ as topological vector spaces. Or, more generally, when a second Hilbert space Y with a similar construction is involved. When do we have $X_\mu^{\mathbf{K}} = Y_\nu^{\mathbf{L}}$?

- **The characterization problem**
  If X is a function space, describe the elements of $X_\mu^{\mathbf{K}}$ in classical analytic terms.

The characterization problem has been, in fact, the subject of the preceding sections. When dealing with the classification problem we want to apply Theorem 3.1. We solve a problem of the following type: Let there be given an infinite matrix $K \in UTM$ with distinct diagonal elements. Let $S \in UTM$ with all $S_{ii} = 1$ be the diagonalizer of K, so $S^{-1} K S = \Delta_K = \mathrm{diag}(K_{00}, K_{11}, \ldots)$. Find diagonal matrices $M = \mathrm{diag}(\mu_0, \mu_1, \ldots)$ with the property MSM^{-1} and $MS^{-1}M^{-1}$ are l_2-bounded. We consider matrices M of the form $M = e^{t\Lambda}$ with $\mathrm{Re}\, t > 0$ and $\Lambda = \mathrm{diag}(\lambda_0, \lambda_1, \ldots)$. The conditions we impose on K and the conditions on Λ that we look for are growth conditions.

THEOREM 3.2

Let $K \in UTM$. Put $C_n = \max_{1 \le i < j \le n} |K_{ij}|$. Suppose

(i) $\qquad \exists D > 0 \; \forall n > m \quad |K_{mm} - K_{nn}|^{-1} \le \dfrac{D}{n(n-m)}$

(ii) $\qquad \sup_{n \ge 1} \left(\dfrac{1}{n} C_n \right) = \delta < \infty$.

Let $\mu \ge 1$ be fixed. Let $\Lambda = \mathrm{diag}(\lambda_0, \lambda_1, \ldots)$ with

(iii) $\qquad \mathrm{Re}\, \lambda_n = n^\mu (1 + \varepsilon_1(n)), \; \varepsilon_1(n) \to 0 \;$ as $\; n \to \infty$

$$\mathrm{Im}\, \lambda_n = \varepsilon_2(n)\, \mathrm{Re}\, \lambda_n, \; \varepsilon_2(n) \to 0 \;\text{ as }\; n \to \infty$$

$$\mathrm{Re}(\lambda_n - \lambda_m) \ge n^{\mu-1}(n-m)\,(1 + \varepsilon_3(n,m)), \; n > m$$

$$\text{with } \varepsilon_3(n,m) \to 0 \;\text{ as }\; \min(n,m) \to \infty$$

$$\mathrm{Im}(\lambda_n - \lambda_m) = \varepsilon_4(n,m)\, \mathrm{Re}(\lambda_n - \lambda_m)$$

$$\text{with } \varepsilon_4(n,m) \to 0 \;\text{ as }\; \min(n,m) \to \infty .$$

Let $S \in UTM$ be the diagonalizer of K with $S_{ii} = 1, \; 0 \le i < \infty$. The for all $t \in \mathbb{C}, \; \mathrm{Re}\, t > 0$, the matrices

$$e^{t\Lambda} S e^{-t\Lambda} \;\text{ and }\; e^{t\Lambda} S^{-1} e^{-t\Lambda} \;\text{ are } l_2\text{-bounded.}$$

Proof

The proof consists of parts (a)-(g).

(a) Since $K S = S \Delta_K$ and $S^{-1} K = \Delta_K S^{-1}$ we have the following recurrence relations for the entries of S and S^{-1}:

- $$S_{n-q,n} = (K_{n-q,n-q} - K_{nn})^{-1} \sum_{k=0}^{q-1} K_{n-q,n-k}\, S_{n-k,n} ,$$

  $$S_{nn} = 1, \quad 1 \le q \le n$$

- $$S^{-1}_{m,m+q} = (K_{mm} - K_{m+q,m+q})^{-1} \sum_{k=0}^{q-1} K_{m+k,m+q}\, S^{-1}_{m,m+k} ,$$

  $$S^{-1}_{mm} = 1, \quad q \ge 1 .$$

(b) We show that

$$\left.\begin{array}{l} |S_{m,n}| \\ |S^{-1}_{m,n}| \end{array}\right\} \le \exp 2\,\{D\,\delta(n-m)\}^{\frac{1}{2}} .$$

From the recurrence relations for the entries of S we estimate

$$|S_{n-q,n}| \le \frac{D\,C_n}{n\,q} \sum_{k=0}^{q-1} |S_{n-k,n}| , \quad 1 \le q \le n,\ S_{nn} = 1 .$$

By induction, for $1 \le q \le n$

$$|S_{n-q,n}| \le \frac{D\,C_n}{n\,q} \prod_{l=1}^{q-1} \left(1 + \frac{1}{l}\,\frac{D\,C_n}{n}\right) \le \prod_{l=1}^{q} \left(1 + \frac{1}{l}\,D\,\delta\right) \le e^{2\sqrt{qD\delta}} .$$

For the second one

$$|S^{-1}_{m,m+q}| \le \frac{D\,C_{m+q}}{q(m+q)} \sum_{k=0}^{q-1} |S^{-1}_{m,m+k}| , \quad q = 1,2,...,\ S^{-1}_{mm} = 1 .$$

By induction, $q = 1, 2, ...,$

$$|S^{-1}_{m,m+q}| \le \frac{D}{q}\,\frac{C_{m+q}}{m+q} \prod_{l=1}^{q-1} \left(1 + \frac{D}{l}\,\frac{C_{m+l}}{m+l}\right) \le \prod_{l=1}^{q} \left(1 + \frac{D}{l}\,\frac{C_{m+l}}{m+l}\right) \le e^{2\sqrt{qD\delta}} .$$

(c) Put $t = t_1 + i\,t_2$ and $\sigma^t_{mn} = (e^{t\Lambda}\, S\, e^{-t\Lambda})_{mn} = S_{mn}\, e^{-t(\lambda_n - \lambda_m)}$. So with (b)

$$|\sigma^t_{mn}| \le \exp\{2[D\,\delta(n-m)]^{\frac{1}{2}} - t_1 \operatorname{Re}(\lambda_n - \lambda_m) + t_2 \operatorname{Im}(\lambda_n - \lambda_m)\} .$$

(d) We now show that each row in σ^t in an l_2-sequence. So let m be fixed

$$|\sigma^t_{mn}| \le \exp\{t_1 \operatorname{Re}\lambda_m - t_2 \operatorname{Im}\lambda_m\} \cdot \exp\{2[D\,\delta\, n]^{\frac{1}{2}} - t_1\, n^{\mu}(1 + \varepsilon_1(n)) \left(1 - \frac{t_2}{t_1}\, \varepsilon_2(n)\right)\} .$$

Take N so large that for $n > N$ we have

$$2\,n^{-\frac{1}{2}}(D\,\delta)^{\frac{1}{2}} \le \frac{1}{4}\,t_1 \quad \text{and} \quad (1 + \varepsilon_1(n)) \left(1 - \frac{t_2}{t_1}\, \varepsilon_2(n)\right) \ge \frac{1}{2} .$$

For $n > N$ the second factor is smaller than $e^{-\frac{1}{4}t_1 n}$. Hence the result.

(e) Starting from the inequality in (c) we find with conditions (iii)

$$|\sigma^t_{mn}| \le \exp\{2(D\,\delta(n-m))^{\frac{1}{2}} - t_1\, n^{\mu-1}(n-m)\,(1 + \varepsilon_3) \left(1 - \frac{t_2}{t_1}\, \varepsilon_4\right)\} .$$

Below the Mth row, M sufficiently large, we have

$$|\sigma^t_{mn}| \leq \exp\{2(D\,\delta(n-m))^{\frac{1}{2}} - \frac{1}{2}t_1(n-m)\}\ .$$

Therefore each diagonal is a bounded sequence. Finally for $m \geq M$ and $(n-m) \geq \frac{1}{t_1}\,8(D\,\delta)^{\frac{1}{2}}$

$$|\sigma^t_{mn}| \leq e^{-\frac{1}{4}t_1(n-m)}\ .$$

(f) We now split the matrix σ^t in 2 parts:

- The first M rows. They represent a bounded l_2-operator which is Hilbert Schmidt, cf. (d).
- The part below the Mth row also represents a bounded l_2-operator because of the "codiagonal estimate", use

$$\sum_{k=0}^{\infty} \sup_{j-i=k} |\sigma^t_{i,j}| < \infty\ .$$

(g) For $e^{t\Lambda}\,S^{-1}\,e^{-t\Lambda}$ the parts (c)-(g) of the proof apply in exactly the same way. □

REMARK

If $\mu > 1$ the conditions (i) and (ii) can be relaxed considerably, see [GE] and [G]. However in this paper we need the case $\mu = 1$.

THEOREM 3.3

(a) The conditions on Λ in Theorem 3.2 are satisfied if $\lambda_n = n^{\mu}(1+\rho(n))$, $\mu \geq 1$, and

$$\lim_{n\to\infty} \rho(n) = 0 \quad \lim_{m\to\infty} \sup_{n\geq m+1} \left|\frac{\rho(n)\,n - \rho(m)\,m}{n-m}\right| = 0\ .$$

(b) The conditions of (a) are satisfied if ρ is a (complex valued) differentiable function on $(0,\infty)$ with $\lim_{\xi\to\infty} \rho(\xi) = 0$ and $\lim_{\xi\to\infty} \xi\,\rho'(\xi) = 0$.

(c) If $\Lambda = N^{\mu}_a = \operatorname{diag}(\ldots,(n(n+a))^{\frac{1}{2}\mu},\ldots)$, $a \in C$, then condition (b) is satisfied.

Proof

(a) We compute

$$\lambda_n - \lambda_m = (n-m)\,n^{\mu-1}\left\{1 + \left(\frac{1-(\frac{m}{n})^{\mu}}{1-\frac{m}{n}} - 1\right) + \left(\frac{n\,\rho(n) - m\,\rho(m)}{n-m}\right) + \left(\frac{m}{n}\,\frac{1-(\frac{m}{n})^{\mu-1}}{1-\frac{m}{n}}\,\rho(m)\right.\right.$$

On the interval $[0,1)$ one has

$$\mu \geq \frac{1-x^{\mu}}{1-x} \geq 1 \quad \text{and} \quad \frac{1-x^{\mu-1}}{1-x} \leq \max(1,\mu-1) \leq \mu\ .$$

For $\varepsilon_3(n,m)$ take the real parts of the 3rd and 4th terms between { }. We omit the simple verification of the other conditions.

(b) Follows by application of the mean value estimation on $(n-m)^{-1}(\rho(n)\,n-\rho(m)\,m)$.

(c) We have $\rho(x)=\left(1+\frac{a}{x}\right)^{\frac{\mu}{2}}-1$ which satisfies (b). $\square$

Note that sequences like $((\log n)^A\, n^B \exp(t\,n^\mu + Cn + D\sqrt{n}))_{n=1}^\infty$ with $A, B, C, D \in \mathbb{R}$, $t>0$, $\mu \geq 1$ are of type $(\exp(t\,n^\mu(1+\rho(n))))$.
Now we formulate the main result of this section which is a consequence of Theorems 3.1 and 3.2.

THEOREM 3.3. (Classification)

- Consider a Hilbert space X and fix an orthonormal basis $(T_n)_{n=0}^\infty \subset X$.
- Let $\mathcal{K}$ denote the class of all upper triangular matrices $K \in UTM$ which satisfy the conditions (i) and (ii) of Theorem 3.2.
- Let $\mathcal{Y}$ denote the class of sequences $(\lambda_n)_{n=0}^\infty$ which satisfy condition (iii) of Theorem 3.2.
- Let the sequence $\mu = (\mu_n)_{n=0}^\infty$, $\mu_n = |e^{t\lambda_n}|$ with $(\lambda_n)_{n=0}^\infty \in \mathcal{Y}$, be fixed.

Then

(A) For all $K \in \mathcal{K}$ the sequence $(e^{-t\lambda_n}\, T_n^K)_{n=0}^\infty$ is a Riesz basis in X_μ. In other words, the Hilbert spaces X_μ and X_μ^K are the same as topological vector spaces.

(B) If for some positive sequence $\theta = (\theta_n)_{n=0}^\infty$ and some $K \in \mathcal{K}$ the sequence $(\theta_n\, T_n^K)_{n=0}^\infty$ happens to be a Riesz basis in a Hilbert space Y, then for all $K \in \mathcal{K}$ and all $L \in \mathcal{K}$ we have the equality

$$Y_\alpha^K = X_\mu^L\ , \quad \text{with } \alpha = \mu\,\theta^{-1}\ ,$$

as Hilbertizable topological vector spaces.

4. APPLICATION TO THE JACOBI OPERATORS

The results of the preceding section can be applied in the following way. If the, suitably normalized, eigenvectors of some operator establish a Riesz basis in a "scale" of smooth Hilbert spaces then certain perturbed operators will do the same. To put it differently: All perturbed operators when "substituted" into a suitable fixed unbounded function lead to closed operators which all have the same domain.
Such a procedure leads in a straight forward way to long lists of Riesz bases for e.g. the spaces of analytic functions that we discussed in the Applications I-IV in Section 2.
In this final section we carry out such a program for the Jacobi operators which can be considered as perturbations of the Chebyshev operator.

Consider the general Jacobi differentiation operator

$$\mathcal{A}_{\alpha\beta} = -(1-x^2)\,\frac{d^2}{dx^2} + (\alpha+\beta+2)\,x\,\frac{d}{dx} - (\beta-\alpha)\,\frac{d}{dx}\ , \qquad \alpha,\beta \in \mathbb{C},\ \text{fixed}\ .$$

In the special case $\alpha, \beta \in I\!R$, $\alpha > -1$, $\beta > -1$, the differentiation operator $\mathcal{A}_{\alpha\beta}$ can be regarded as a self-adjoint operator in $X_{\alpha\beta} = L_2([-1,1],\ (1-x)^\alpha\ (1+x)^\beta\, dx)$. In this special case for the eigenfunctions we take the normalized (in $X_{\alpha\beta}$) Jacobi polynomials

$$R_n^{\alpha\beta} = \kappa_n^{\alpha\beta}\, P_n^{\alpha\beta} .$$

The explicit expression for the normalizing coefficient $\mathcal{K}_n^{\alpha\beta}$ follows from

$$\lim_{n\to\infty} (2^{\alpha+\beta}\, n^{-1})^{\frac{1}{2}}\, \kappa_n^{\alpha\beta} = 1 .$$

Special cases are the normalized Chebyshev polynomials of the first and second kind: In a local convention we have

$$R_0^{-\frac{1}{2},-\frac{1}{2}}(\cos\theta) = T_0(\cos\theta) = \pi^{-\frac{1}{2}}\ ,\quad R_n^{-\frac{1}{2},-\frac{1}{2}}(\cos\theta) = T_n(\cos\theta) = \left(\frac{2}{\pi}\right)^{\frac{1}{2}}\ \cos n\theta\ ,\quad n \in I\!N$$

$$R_n^{\frac{1}{2},\frac{1}{2}}(\cos\theta) = \left(\frac{2}{\pi}\right)^{\frac{1}{2}} \frac{\sin(n+1)\,\theta}{\sin\theta}\ ,\quad n \in I\!N_0 .$$

Next, let $K = A_{\alpha\beta}$ denote the matrix of $\mathcal{A}_{\alpha\beta}$, $\alpha \in C$, $\beta \in C$, with respect to the basis $(T_n)_{n=0}^\infty$ in the space of all polynomials.
In [GE] we showed that $K_{mn} = (\mathcal{A}_{\alpha\beta} T_n, T_m)_{-\frac{1}{2},-\frac{1}{2}}$, $K_{nn} = n(n+\alpha+\beta+1)$ and $|K_{mn}| \leq (2\,|\beta-\alpha|+|\alpha+\beta+1|)\,n$. Note that this matrix satisfies the conditions (i) and (ii) of Theorem 3.2.
The matrix $S_{\alpha\beta} \in UTM$ is defined by

$$S_{\alpha\beta}^{-1}\, A_{\alpha\beta}\, S_{\alpha\beta} = \operatorname{diag}(\ldots, n(n+\alpha+\beta+1), \ldots) .$$

Via the transition matrix $S_{\alpha\beta}$ we construct the polynomial basis $(T_n^{\alpha\beta})_{n=0}^\infty$ from the Chebyshev basis $(T_n)_{n=0}^\infty$. If $a > -1$, $\beta > -1$, we have $T_n^{\alpha\beta} = r_n^{\alpha\beta}\, R_n^{\alpha\beta}$ where the sequences $(r_n^{\alpha\beta})_{n=0}^\infty$ and $((r_n^{\alpha\beta})^{-1})_{n=0}^\infty$ are bounded.
Now we are in a position to extend the results of Application II in Section 2.

APPLICATION V. (sequel to Applications II)

First we fix some notations.
For all $\alpha > -1$, $\beta > -1$, $m \in I\!N_0$, $\theta > 0$, $t > 0$ we consider the domain $D(\mathcal{A}_{\alpha\beta}^{\frac{1}{2}(m-\theta)}\ \exp t\,\mathcal{A}_{\alpha\beta}^{\frac{1}{2}})$ of the selfadjoint operator $\mathcal{A}_{\alpha\beta}^{\frac{1}{2}(m-\theta)}\ \exp t\,\mathcal{A}_{\alpha\beta}^{\frac{1}{2}}$ in $X_{\alpha\beta} = L_2([0,1]\ (1-x)^\alpha\ (1+x)^\beta\, dx)$. Note that the sequence $\mu = (\mu_n)_{n=0}^\infty$ of eigenvalues of this operator is given by

$$([n(n+\alpha+\beta+1)]^{\frac{1}{2}(m-\theta)}\ \exp t\,[n(n+\alpha+\beta+1)]^{\frac{1}{2}})_{n=0}^\infty .$$

Note also that for fixed m, θ, t the quotient of two such sequences (for different pairs (α,β)) is always bounded. As a corollary of the general theory in Section 3 we obtain

THEOREM 4.1 (Cf. [Sz] Thm 9.1.1)

Let $f \in \mathcal{A}([-1,1])$. For $\alpha, \beta \in C$, $K \in \mathcal{K}$, expand

$$f(x) = \sum_{n=0}^\infty a_n^{\alpha\beta}\, T_n^{\alpha\beta}(x) = \sum_{n=0}^\infty a_n^K\, T_n^K(x) .$$

Denote $a_n^{-\frac{1}{2},-\frac{1}{2}} = a_n$. Let $m \in I\!N_0$, $\theta > 0$, $T > 0$. The following nine conditions are equivalent.

- $f \in \mathcal{A}_2(E_T; \omega_\theta, m)$
- $(n^{(m-\theta)}\, e^{nT}\, a_n)_{n=0}^{\infty} \in l_2$
- $f \in \mathcal{A}(E_T)$ and $f|_{\partial E_T} \in H^{m-\theta}(\partial E_T)$.
- $\exists_{\alpha \in \mathbb{C}}\ \exists_{\beta \in \mathbb{C}}\quad (n^{(m-\theta)}\, e^{nT}\, a_n^{\alpha\beta})_{n=0}^{\infty} \in l_2$
- $\forall_{\alpha \in \mathbb{C}}\ \forall_{\beta \in \mathbb{C}}\quad (n^{(m-\theta)}\, e^{nT}\, a_n^{\alpha\beta})_{n=0}^{\infty} \in l_2$
- $\exists_{\alpha > -1}\ \exists_{\beta > -1}\quad f \in D(\mathcal{A}_{\alpha\beta}^{\frac{1}{2}(m-\theta)}\ \exp T \mathcal{A}_{\alpha\beta}^{\frac{1}{2}})$
- $\forall_{\alpha > -1}\ \forall_{\beta > -1}\quad f \in D(\mathcal{A}_{\alpha\beta}^{\frac{1}{2}(m-\theta)}\ \exp T \mathcal{A}_{\alpha\beta}^{\frac{1}{2}})$
- $\exists K \in \mathcal{K}\quad (n^{(m-\theta)}\, e^{nT}\, a_n^{K})_{n=0}^{\infty} \in l_2$
- $\forall K \in \mathcal{K}\quad (n^{(m-\theta)}\, e^{nT}\, a_n^{K})_{n=0}^{\infty} \in l_2$. □

Note that it is an interesting problem to look for other "known" polynomials which correspond to eigenvectors of a matrix $K \in \mathcal{K}$.

References

[AS] Abramowitz, M. and I.A. Stegun. Handbook of mathematical functions. National Bureau of Standards. Washington 1968.

[BG1] Berkel, C.A.M. van, and J. de Graaf. On a property of the Fourier-cosine transform. Appl. Math. Lett. **1** (3), 307-310, 1988.

[BG2] Berkel, C.A.M. van, and J. de Graaf. The Fourier-Jacobi transform of analytic functions which are (almost) periodic in the imaginary direction. Preprint. Submitted to SIAM J. Math. An.

[D] Duran, A.J. The analytic functionals in the lower half plane as a Gel'fand-Shilov space. To appear.

[G] Graaf, J. de. On approximation of (slowly growing) analytic functions by special polynomials. To appear in "Orthogonal polynomials" Conference Erice 1990.

[GE] Graaf, J. de, and S.J.L. van Eijndhoven. Domains of exponentiated fractional Jacobi Operators: Characterizations, Classifications, Expansion Result. Constr. Approx. **5**, 271-296, 1989.

[Sz] Szegö, G. Orthogonal polynomials, AMS Coll. Publ. XXIII, 1959.

TOEPLITZ OPERATORS ACTING ON GENERALIZED FUNCTIONS

Morisuke Hasumi

Department of Mathematics
Ibaraki University
Mito, Ibaraki 310, Japan

1. INTRODUCTION

There have been extensive studies of Toeplitz operators mainly for Hardy classes and also for Bergman classes. In the present paper we will discuss some preliminary results concerning Toeplitz operators for Bergman classes. More explicitly, we look at the space of boundary functions of elements in the standard Bergman class $B^2(\mathbf{D})$, consisting of functions analytic on the open unit disk $\mathbf{D}$ and square-summable with respect to the area measure. As the functions in $B^2(\mathbf{D})$ usually have boundary values only in the sense of Schwartz distributions, so we are led to the notion of Toeplitz operators acting on distributions.

2. PRELIMINARY DISCUSSION

Let $dA = \dfrac{dxdy}{\pi}$ denote the normalized area measure on the disk $\mathbf{D}$. Then, the standard Bergman class $L_a^2 = L_a^2(\mathbf{D})$ on the unit disk $\mathbf{D}$ is defined as the totality of holomorphic functions $f(z)$ on $\mathbf{D} = \{z \in \mathbf{C} : |z| < 1\}$ satisfying the property

$$\|f\|_B^2 = \iint_{\mathbf{D}} |f(z)|^2 \, dA(z) < \infty,$$

with $z = x + iy$. If we expand such an f into a power series as $f(z) = \sum_{n=0}^{\infty} \hat{f}(n) z^n$, then we have

$$\|f\|_B^2 = \sum_{n=0}^{\infty} \frac{|\hat{f}(n)|^2}{n+1}.$$

Generalized Functions and Their Applications, Edited by
R.S. Pathak, Plenum Press, New York, 1993

From this follows that the sequence $\{\hat{f}(n)\}$ is slowly increasing and therefore that the Fourier series

$$\sum_{n=0}^{\infty} \hat{f}(n)e^{in\theta}$$

represents a (Schwartz) distribution, say f^*, on the unit circle $\mathbf{T}$.

We now denote by $\mathcal{D}' = \mathcal{D}'(\mathbf{T})$ the space of distributions on the unit circle $\mathbf{T}$, which is endowed with the normalized Lebesgue measure $d\sigma = d\theta/2\pi$. Each element S of $\mathcal{D}'$ is expressed as a Fourier series like

$$S = \sum_{n=-\infty}^{\infty} \hat{S}(n)e^{in\theta},$$

with $\hat{S}(n) = \langle S, e^{in\theta}\rangle$, where the bracket operation $\langle\ ,\ \rangle$ is assumed to be linear in the first variable and conjugate-linear in the second. As is well known, a bilateral sequence $\{\hat{S}(n)\}$ of complex numbers represents the Fourier coefficients of a distribution on the circle if and only if it is slowly increasing. A distribution S is called *analytic* if $\hat{S}(n) = 0$ for all $n < 0$. The totality of analytic distributions is denoted by $\mathcal{D}'_A$. Then the projection operator $\widetilde{P} : \mathcal{D}' \to \mathcal{D}'_A$ which is defined by

$$\widetilde{P}\Big(\sum_{n=-\infty}^{\infty} \hat{S}(n)e^{in\theta}\Big) = \sum_{n=0}^{\infty} \hat{S}(n)e^{in\theta}$$

is continuous with respect to the distribution topology.

Now let $\mathcal{D}'_B$ be the space of all distributions S on the circle $\mathbf{T}$ such that its Fourier coefficients $\hat{S}(n) = \langle S, e^{in\theta}\rangle$ satisfy the following:

$$\|S\|_B^2 := \sum_{n=-\infty}^{\infty} \frac{|\hat{S}(n)|^2}{|n|+1} < \infty. \tag{1}$$

We see that the set $\mathcal{D}'_B$ becomes a Banach space under this norm. Let $\mathcal{D}'_{BA}$ be the subspace of $\mathcal{D}'_B$ consisting of those S with $\hat{S}(n) = 0$ for all $n < 0$, i.e. $\mathcal{D}'_{BA} = \mathcal{D}'_B \cap \mathcal{D}'_A$. It is then clear that the projection is a bounded operator of norm one from $\mathcal{D}'_B$ onto $\mathcal{D}'_{BA}$. We also see easily that the correspondence $f \mapsto f^*$ gives an isometric isomorphism of B^2 onto $\mathcal{D}'_{BA}$.

Proposition 1. *Let $f \in B^2(\mathbf{D})$ and put, for $0 < r < 1$,*

$$f_r(e^{i\theta}) := \sum_{n=0}^{\infty} \hat{f}(n)\, r^n e^{in\theta}.$$

Then, $f_r \in \mathcal{D}'_{BA}$ and $f_r \to f^*$ in the norm topology of $\mathcal{D}'_{BA}$ as $r \to 1-0$.

We next denote by $L_h^2(\mathbf{D})$ the totality of complex-valued harmonic functions $u(z)$ on $\mathbf{D}$ such that

$$\|u\|_B^2 = \iint_{\mathbf{D}} |u(z)|^2\, dA(z) < \infty.$$

Since any harmonic function u on $\mathbf{D}$ can be written in the form:

$$u(re^{i\theta}) = \sum_{n=-\infty}^{\infty} \hat{u}(n)\, r^{|n|} e^{in\theta}, \tag{2}$$

we have

$$\|u\|_B^2 = \frac{1}{\pi} \iint_{\mathbf{D}} \Big| \sum_{n=-\infty}^{\infty} \hat{u}(n)\, r^{|n|} e^{in\theta} \Big|^2 r\,dr\,d\theta = \sum_{n=-\infty}^{\infty} \frac{|\hat{u}(n)|^2}{|n|+1}.$$

Using the notation given by (2), we have the following:

Proposition 2. *Let $u \in h^2(\mathbf{D})$. Then, the formal expression*

$$u^*(e^{i\theta}) = \sum_{n=-\infty}^{\infty} \hat{u}(n)\, e^{in\theta}$$

defines, in the form of Fourier series, a distribution belonging to $\mathcal{D}'_B$. The correspondence $u \mapsto u^*$ is an isometric isomorphism of $L^2_h(\mathbf{D})$ onto $\mathcal{D}'_B$. Moreover, put $u_r(e^{i\theta}) := u(re^{i\theta})$ for each $0 < r < 1$. Then, $u_r \in \mathcal{D}'_B$ and $u_r \to u^*$ in the norm topology of $\mathcal{D}'_B$ as $r \to 1 - 0$.

3. DEFINITION OF TOEPLITZ OPERATORS ON THE CIRCLE T

We now define Toeplitz operators on the space $\mathcal{D}'_{BA}$. Formally this is defined as follows: take a square-integrable function φ on $\mathbf{T}$ and consider the product φS for each $S \in \mathcal{D}'_{BA}$, which we assume is well-defined as an element of $\mathcal{D}'_B$. Then, applying the projection $\widetilde{P}$, we get an operator $\widetilde{T}_\varphi$ of $\mathcal{D}'_{BA}$ into itself, i.e.

$$\widetilde{T}_\varphi(S) = \widetilde{P}(\varphi S).$$

The operator $\widetilde{T}_\varphi$ is called a *Toeplitz operator* with symbol φ. Generally speaking, however, a distribution can be multiplied only by C^∞ functions, so that the above "definition" should be made more precise. Since the projection $\widetilde{P}$ is bounded on $\mathcal{D}'_B$, a Toeplitz operator $\widetilde{T}_\varphi$ with symbol φ is bounded provided that the multiplication operator $M_\varphi : S \mapsto \varphi S$ of $\mathcal{D}'_{BA}$ into $\mathcal{D}'_B$ is bounded. On the other hand, since analytic polynomials are dense in the Banach space $\mathcal{D}'_{BA}$, we see that the multiplication operator M_φ is well defined on $\mathcal{D}'_{BA}$ with values in $\mathcal{D}'_B$ if there exists a constant C such that

$$\|\varphi f\|_B \le C\|f\|_B$$

for any analytic polynomial f.

4. OPERATORS WITH ANALYTIC SYMBOLS

We begin with the easier part, in which the symbols are analytic. Take any $\varphi \in H^2(d\sigma)$. Then φf is analytic for any analytic polynomial f, so that $\widetilde{T}_\varphi f = M_\varphi f$. Almost trivially we have

Proposition 3. *The operator $\widetilde{T}_\varphi$ is well-defined as a bounded operator if and only if φ is bounded. If this is the case, then $\|\widetilde{T}_\varphi\| \le \|\varphi\|_\infty$.*

Proof. The multiplication operator M_φ is bounded if and only if φ is bounded in view of a general fact concerning Banach function spaces appearing in [2, Lemma 11], which we sate as a lemma below. The rest is simple and we omit the details. □

Lemma 4. *Let E be a functional Banach space in the sense that*

(a) *E is a Banach space whose elements are complex functions on a set X, with usual pointwise addition and scalar multiplication as algebraic operations,*
(b) *for any $x \in X$ there exists a function $f \in E$ such that $f(x) \ne 0$,*
(c) *for any $x \in X$ the evaluation functional $f \mapsto f(x)$ of E is continuous.*

Then, every multiplier of E is a bounded function on X.

5. A DUALITY IN BERGMAN CLASS

Our next objective is to study Toeplitz operators with conjugate analytic symbols. For this purpose we will use the duality in the Bergman class ($\mathcal{D}'_{BA}$ or B^2). But if the space $\mathcal{D}'_{BA}$, as usual, were viewed as a Hilbert space, its conjugate space would be the same and thus no novelty would happen. So we need some other interpretation. For this purpose we consider the Dirichlet space $D^2 = D^2(\mathbf{D})$, which is defined as the set of all holomorphic functions $g(z)$ on $\mathbf{D}$ such that

$$\sum_{n=0}^{\infty}(n+1)|\hat{g}(n)|^2 < +\infty, \tag{3}$$

where $g(z) = \sum_{n=0}^{\infty} \hat{g}(n)z^n$. It is easy to see that the set D^2 forms a Banach space under the usual pointwise addition and scalar multiplication of functions as algebraic operations and the following norm:

$$\|g\|_D^2 = \sum_{n=0}^{\infty}(n+1)|\hat{g}(n)|^2 = \|g\|_{H^2}^2 + \|g'\|_B^2.$$

Proposition 5. *Let g be holomorphic on $\mathbf{D}$. Then the following are equivalent:*

(a) $g \in D^2$.

(b) $g' \in B^2$.

(c) *The total image (counting multiplicity) of the unit disk under g has finite area.*

Proof. This is clear and well-known. □

Lemma 6. *Let $g \in D^2$ be fixed. Then*

$$\|g\|_D^2 = \sup_Q \left| \frac{1}{2\pi} \int_0^{2\pi} \overline{Q(e^{i\theta})} g(e^{i\theta})\, d\theta \right|, \tag{4}$$

where Q ranges over all polynomials with $\|Q\|_B < 1$.

Proof. Let $Q(z) = \sum_{k=0}^{n} b_k z^k$ be any polynomial with $\|Q\|_B < 1$, i.e.

$$\sum_{k=0}^{n} \frac{|b_k|^2}{k+1} < 1.$$

Then

$$\begin{aligned} \left| \frac{1}{2\pi} \int_0^{2\pi} \overline{Q(e^{i\theta})} g(e^{i\theta})\, d\theta \right|^2 &= \left| \sum_{k=0}^{n} \bar{b}_k \hat{g}(k) \right|^2 \\ &\le \sum_{k=0}^{n} \frac{|b_k|^2}{k+1} \sum_{k=0}^{n} (k+1)|\hat{g}(k)|^2 < \|g\|_D^2, \end{aligned}$$

which implies that the right-hand side of (4) does not exceed the left-hand side.

On the other hand, we set, for each $n \ge 1$, $g_n(z) = \sum_{k=0}^{n} \hat{g}(k)z^k$ and $Q_n(z) = \sum_{k=0}^{n} b_k z^k$ with $b_k = (k+1)\hat{g}(k)/\|g_n\|_D$, $0 \le k \le n$. Then $\|Q_n\|_B = 1$ and

$$\left| \sum_{k=0}^{n} \bar{b}_k \hat{g}(k) \right| = \frac{1}{\|g_n\|_D} \sum_{k=0}^{n} (k+1)\overline{\hat{g}(k)}\hat{g}(k) = \|g_n\|_D$$

This means that the right-hand side of (4) is no smaller than $\|g_n\|_D$ for all n and therefore than $\|g\|_D$. Hence we have the desired equality. □

Proposition 7. *The Bergman space B^2 is regarded as the dual of the Dirichlet space D^2.*

Proof. Let Φ be a bounded linear function on the space D^2 and put $a_n = \overline{\Phi(z^n)}$ for $n = 0, 1, \ldots$. Take any $g = \sum_{k=1}^{\infty} \hat{g}(k) z^k \in D^2$. If we set $g_m = \sum_{k=0}^{m} \hat{g}(k) z^k$ for $m = 0, 1, \ldots$, then

$$\|g - g_m\|_{D^2}^2 = \sum_{k=m+1}^{\infty} (k+1)|\hat{g}(k)|^2 \to 0 \quad \text{as } m \to +\infty,$$

so that $\Phi(g_m) \to \Phi(g)$ as $m \to \infty$. We thus have

$$\Phi(g) = \lim_{m\to\infty} \Phi(g_m) = \lim_{m\to\infty} \Phi\left(\sum_{k=0}^{m} \hat{g}(k) z^k\right) = \lim_{m\to\infty} \sum_{k=0}^{m} \hat{g}(k) \Phi(z^k)$$
$$= \lim_{m\to\infty} \sum_{k=0}^{m} \bar{a}_k \hat{g}(k) = \sum_{k=0}^{\infty} \bar{a}_k \hat{g}(k).$$

Since the function g is chosen arbitrarily in D^2, we have seen that

$$\left|\sum_{k=0}^{\infty} \bar{a}_k b_k\right| < \infty$$

for any sequence $\{b_k\}$ satisfying $\sum_{k=0}^{\infty} (k+1)|b_k|^2 < \infty$. This means that

$$\sum_{k=1}^{\infty} \frac{|a_k|^2}{k+1} < \infty,$$

so that the function f defined by $f(z) = \sum_{k=0}^{\infty} a_k z^k$ belongs to B^2. We now compute the bound

$$\|\Phi\| = \sup\{|\Phi(g)| : \; g \in D^2, \; \|g\|_D < 1\}$$

of the functional Φ. By writing $g(z) = \sum_{k=0}^{\infty} \hat{g}(k) z^k$ we have

$$|\Phi(g)| = \left|\sum_{k=0}^{\infty} \hat{g}(k) \bar{a}_k\right| = \left|\sum_{k=0}^{\infty} \sqrt{k+1}\,\hat{g}(k) \cdot \frac{\bar{a}_k}{\sqrt{k+1}}\right|$$
$$\leq \left(\sum_{k=1}^{\infty} \frac{|a_k|^2}{k+1}\right)^{1/2} \left(\sum_{k=1}^{\infty} (k+1)|\hat{g}(k)|^2\right)^{1/2} = \|f\|_B \|g\|_D.$$

This means that $\|\Phi\| \leq \|f\|_B$. On the other hand, if we put

$$g(z) = \sum_{k=0}^{\infty} \frac{a_k}{k+1} z^k,$$

then $g \in D^2$ because

$$\|g\|_D^2 = \sum_{k=0}^{\infty} (k+1) \left|\frac{a_k}{k+1}\right|^2 = \sum_{k=0}^{\infty} \frac{|a_k|^2}{k+1} = \|f\|_B^2 < \infty.$$

And we have

$$|\Phi(g)| = \left|\sum_{n=0}^{\infty} \bar{a}_n \hat{g}(n)\right| = \left|\sum_{n=0}^{\infty} \bar{a}_n \cdot \frac{a_n}{n+1}\right| = \sum_{n=0}^{\infty} \frac{|a_n|^2}{n+1} = \|f\|_B \|g\|_D,$$

so that $\|\Phi\| \geq \|f\|_B$. Hence we have $\|\Phi\| = \|f\|_B$, as desired. □

It is known that holomorphic functions belonging to B^2 do not necessarily have radial boundary values. Nevertheless, we want to look at boundary values of such functions. From the defining inequality (3) for the space D^2 we deduce the following obvious facts. First, every element g in D^2 belongs to the Hardy class $H^2(\mathbf{D})$ and therefore that the function $g(z)$ has radial boundary values almost everywhere on the unit circle $\mathbf{T}$. Since

$$\|g\|_{H^2}^2 = \sum_{n=0}^{\infty} |\hat{g}(n)|^2 \leq \sum_{n=0}^{\infty} (n+1)|\hat{g}(n)|^2 = \|g\|_D^2,$$

we see that the embedding of D^2 into $H^2(\mathbf{D})$ is continuous. We denote the boundary function by the symbol g^*. Let $D^2(\mathbf{T})$ be the totality of g^* with $g \in D^2$. Then, it is a Banach space in an obvious sense with norm $\|g^*\|_{D^2(\mathbf{T})} = \|g\|_D$. So, the correspondence $g \mapsto g^*$ gives the norm-preserving isomorphism of D^2 onto $D^2(\mathbf{T})$ and therefore the dual, B^2, to D^2 is identified with that of $D^2(\mathbf{T})$, which we may easily identify with $\mathcal{D}'_{BA}$.

6. OPERATORS WITH CONJUGATE ANALYTIC SYMBOLS

Consider the case when the symbol φ is conjugate analytic. So, suppose that the multiplication operator M_φ maps $\mathcal{D}'_{BA}$ into $\mathcal{D}'_B$. Thus, for any $S \in \mathcal{D}'_{BA}$ and $f \in D^2(\mathbf{T})$, we have

$$\langle T_\varphi S, f\rangle = \langle PM_\varphi S, f\rangle = \langle M_\varphi S, f\rangle = \langle S, \bar{\varphi} f\rangle = \langle S, M_{\bar{\varphi}} f\rangle.$$

It follows that the operator $M_{\bar{\varphi}}$ maps $D^2(\mathbf{T})$ into itself. Namely, the analytic function $\bar{\varphi}$ is a multiplier of the Dirichlet space $D^2(\mathbf{T})$ (or D^2). So Stegenga's theorem [6] shows the following:

Proposition 8. *Let $\psi \in H^2(\mathbf{T})$. There exists a constant C such that*

$$\|T_{\bar{\psi}} f\|_B \leq C\|f\|_B$$

for any analytic polynomial f if and only if

$$\int_E (\mathcal{D}_{1/2}\psi)^2 \, d\theta = O(\gamma(E))$$

for any compact subset E of $\mathbf{T}$, where $\gamma(E)$ denotes the logarithmic capacity of E and

$$\mathcal{D}_{1/2}\psi(\theta) = \left\{\int_{-\pi}^{\pi} \frac{|\psi(\theta - t) - \psi(\theta)|^2}{|e^{i\theta} - 1|^2}\, dt\right\}^{1/2}.$$

We note that use of duality in the study of Toeplitz operators appears in Stegenga's paper [5].

7. CLASSICAL TOEPLITZ OPERATORS

We now consider usual Toeplitz operators on the Bergman space $L^2_a(\mathbf{D})$. Let $L^2(\mathbf{D})$ be the space of square-integrable functions on the unit disk $\mathbf{D}$ with respect to the area measure dA. Then the spaces $L^2_a(\mathbf{D})$ and $L^2_h(\mathbf{D})$ are closed subspaces of $L^2(\mathbf{D})$. We denote by P_a and P_h the orthogonal projections of $L^2(\mathbf{D})$ onto $L^2_a(\mathbf{D})$ and $L^2_h(\mathbf{D})$, respectively. A Toeplitz operator is then defined as follows: take a bounded continous function ψ on the unit disk $\mathbf{D}$. For each element $f \in L^2_a(\mathbf{D})$ the pointwise product ψf belongs to the space $L^2(\mathbf{D})$, so that its projection $P_a(\psi f)$ is defined and belongs to $L^2_a(\mathbf{D})$, which we denote by T_ψ.

Proposition 9. *Let $\psi(z)$ be a (not necessarily bounded) continuous function on $\mathbf{D}$. Then it multiplies $L^2_a(\mathbf{D})$ into $L^2_h(\mathbf{D})$, i.e. $f \in L^2_a(\mathbf{D})$ implies $\psi f \in L^2_h(\mathbf{D})$, if and only if it is a bounded analytic function on $\mathbf{D}$.*

Proof. Since the constant function 1 and the coordinate function z belong to $L^2_a(\mathbf{D})$, we see that both $\psi(z)$ and $z\psi(z)$ are harmonic on $\mathbf{D}$. In particular, $\psi(z)$ is twice continuously differentiable. By applying the Laplacian to the harmonic function $z\psi$ we have

$$\frac{\partial^2}{\partial z \partial \bar{z}}(z\psi(z)) = 0.$$

On the other hand, we can find two analytic functions $f(z)$ and $g(z)$ on $\mathbf{D}$ such that $\psi = f + \bar{g}$. It follows that

$$0 = \frac{\partial^2}{\partial z \partial \bar{z}}(zf(z) + z\overline{g(z)}) = \frac{\partial}{\partial z}(z\overline{g'(z)}) = \overline{g'(z)}.$$

This means that $g(z)$ is constant and therefore that ψ is analytic on $\mathbf{D}$.

So, the function ψ is a multiplier of the space $L^2_a(\mathbf{D})$. By Lemma 1 we conclude that ψ is bounded.

As the converse is obvious, this establishes the proposition. □

If we want to have a Toeplitz operator T_ψ on $L^2_a(\mathbf{D})$ which is not a multiplier of $L^2_a(\mathbf{D})$, then the above proposition shows that $\psi L^2_a(\mathbf{D})$ cannot be contained in the space $L^2_h(\mathbf{D})$. So, the Toeplitz operators in the classical sense cannot be discussed within the framework of harmonic functions.

8. FURTHER PROBLEMS

In this section we will explain our motivation of considering Toeplitz operators in spaces of generalized functions. Let us consider a function f, holomorphic on the open unit disk and continuous up to the boundary $\mathbf{T}$. If the boundary function vanishes on a set of positive linear measure, then f vanishes identically on $\mathbf{D}$. So, if two such functions f and g coincide on a set of positive linear measure on $\mathbf{T}$, then they coincide everywhere on the disk $\mathbf{D}$. This means that we can recover, theoretically at least, the original function f from its fragmental data given on any set of positive linear measure on $\mathbf{T}$. The problem is to find a practical formula for reconstruction. The first successful attempt was done by Patil [3; 4], first for H^p-functions with $1 < p < \infty$ on the unit disk $\mathbf{D}$ and subsequently for H^2-functions on polydiscs. The case of H^1-functions have been solved by making use of an idea of Stegenga [5]. But, if $p < 1$, H^p-functions no longer have L^1-boundary functions, so that one has to work with a kind of generalized functions. Such attempts have also been seen in the literature , but, as far as we are aware, no definitive solutions have been achieved so far. This is why we are trying to look into the space $B^2(\mathbf{D})$ in order to see possibility of considering Toeplitz operators on Banach (or Hilbert) spaces consisting of generalized functions. Although we are still far from our goal, there may be some interest in studying such operators.

Acknowledgements

This work was supported in part by Grant-in-Aid for Scientific Research from Ministry of Education, Science and Culture, Japan.

REFERENCES

1. P. L. Duren, "Theory of H^p Spaces," Academic Press, New York and London (1970).
2. P. L. Duren, B. W. Romberg and A. L. Shields, Linear functionls on H^p spaces with $0 < p < 1$, *J. Reine Angew. Math.* 238: 32–60 (1969).
3. D. J. Patil, Representation of H^p-functions, *Bull. Amer. Math. Soc.* 4: 617–620 (1972).
4. D. J. Patil, Recapturing H^2-functions on a polydisc, *Trans. Amer. Math. Soc.* 188: 97–103 (1974).
5. D. A. Stegenga, Bounded Toeplitz operators on H^1 and applications of the duality between H^1 and the functions of bounded mean oscillation, *Amer. J. Math.* 98: 573–589 (1976).
6. D. A. Stegenga, Multipliers of the Dirichlet spaces, *Illinois J. Math.* 24: 113–139 (1980).

NONSTANDARD TREATMENTS OF NEW GENERALISED FUNCTIONS

R. F. Hoskins[1], J. Sousa Pinto[2]

[1]Department of Applied Computing & Mathematics
Cranfield Institute of Technology
Cranfield, England

[2]Departamento de Matemática
Universidade de Aveiro
3800 Aveiro, Portugal

1 INTRODUCTION

Among the most significant developments in recent years in the field of generalised functions has been the appearance of the non-linear theories of Colombeau and others. The new generalised functions of Colombeau [1], [2], [3], [4], include distributions as a special sub-class, together with various non-linear functions of them. In particular the multiplication problem is comprehensively solved and expressions such as δ^2, δ^3 and so on are meaningful. There are, however, difficulties in presenting such theories at an elementary level, particularly to a non-specialist audience of, say, physicists and engineers to whom the material is nevertheless important and valuable. Much of the conceptual difficulty arises from the need to introduce a generalised (complex) number system whose structure is not as readily accessible to intuition as could be desired.

Several authorities have commented on the parallel with the extended number systems used in Nonstandard Analysis, and there already exist nonstandard treatments of Colombeau-type generalised functions by, for example, Todorov [5] and Li-Bang He [6]. Unfortunately these presuppose an acquaintance with relatively advanced non-standard concepts and methods which are not likely to be familiar to a general audience.

We present in this paper a development of a broadly accessible nonlinear theory of generalised functions using only a simple ultrapower model of a hyperreal number system which can easily be introduced at an elementary level. The treatment given here is an extension of (and a modification of) a nonstandard theory of distributions described in an earlier paper [7] which is summarised briefly in the following text. For simplicity we deal throughout with the one-dimensional, real-valued case, and work for the most part in terms of finite-order distributions. No loss of generality is involved however. Such limited material from nonstandard analysis as is required in the present text is given in the following brief description of an elementary hyperreal system.

1.1 Hyperreal Number Systems

A simple model of a hyperreal number system, suitable for non-standard analysis, can be defined as a set of equivalence classes of infinite real sequences. Let $\mathbb{R}^{\mathbb{N}}$ denote as usual the set of all infinite real sequences, $x = (x_n)_{n\in\mathbb{N}}$, and let $\mathcal{U}$ be a free ultrafilter on $\mathbb{N}$. Two real sequences $(x_n)_{n\in\mathbb{N}}$ and $(y_n)_{n\in\mathbb{N}}$ are said to be equivalent if the set $\{n \in \mathbb{N} : x_n = y_n\}$ is a member of $\mathcal{U}$. (Loosely we often say that the sequences are equivalent if $x_n = y_n$ for nearly all $n \in \mathbb{N}$; in particular this will be the case if equality holds for all but finitely many values of n.) Then the quotient ${}^*\mathbb{R} := \mathbb{R}^{\mathbb{N}}/\mathcal{U}$ is a nonarchimedean ordered field which contains the real number system $\mathbb{R}$ as a proper sub-ordered field. The equivalence class containing the real sequence $(x_n)_{n\in\mathbb{N}}$ will be written $[(x_n)_{n\in\mathbb{N}}]$ and will be called a hyperreal number. The real numbers, as members of ${}^*\mathbb{R}$, are often referred to as the standard elements of ${}^*\mathbb{R}$. We have the following classification of the general (standard and nonstandard) elements of ${}^*\mathbb{R}$:

Definition 1

$\xi \in {}^*\mathbb{R}$ is finite if $\xi = [(x_n)_{n\in\mathbb{N}}]$ and $|x_n| < r$ for nearly all n, where r is some positive real number.

$\xi \in {}^*\mathbb{R}$ is infinitesimal if $\xi = [(x_n)_{n\in\mathbb{N}}]$ and $|x_n| < r$ for nearly all n, where r is any positive real number.

$\xi \in {}^*\mathbb{R}$ is infinite if $\xi = [(x_n)_{n\in\mathbb{N}}]$ and $|x_n| \geq r$ for nearly all n, where r is any real positive number.

The set of all finite members of ${}^*\mathbb{R}$ is a ring, and will be denoted by ${}^*\mathbb{R}_b$. Two hyperreal numbers x and y are said to be infinitely close if the difference $(x - y)$ is infinitesimal; we then write $x \approx y$. For each finite $x \in {}^*\mathbb{R}$ there exists a unique real number r such that $x \approx r$; we call r the standard part of x and write $r = st(x)$, or sometimes $r = {}^\circ x$. The set of all hyperreal numbers which are infinitely close to a given real number r is called the monad of r:

$$mon(r) \equiv \{x \in {}^*\mathbb{R} : x \approx r\} = \{x \in {}^*\mathbb{R} : st(x) = r\}.$$

A subset $\mathcal{A}$ of ${}^*\mathbb{R}$ is internal if there exists a sequence $(A_n)_{n\in\mathbb{N}}$ of subsets of $\mathbb{R}$ such that $x = [(x_n)_{n\in\mathbb{N}}] \in \mathcal{A}$ if and only if $x_n \in A_n$ for nearly all n. For convenience we shall then often write $\mathcal{A}$ as $[(A_n)_{n\in\mathbb{N}}]$. In particular, if $A_n = A$ for nearly all n then we write $[(A)_{n\in\mathbb{N}}]$ as ${}^*A$, and call ${}^*A$ the nonstandard extension of the standard set A,

$${}^*A \equiv \{[(x_n)_{n\in\mathbb{N}}] \in {}^*\mathbb{R} : \{n \in \mathbb{N} : x_n \in A\} \in \mathcal{U}\}.$$

All other subsets of ${}^*\mathbb{R}$ are said to be external.

If $(f_n)_{n\in\mathbb{N}}$ is a sequence of functions, $f_n : A \to \mathbb{R}$ where $A \subset \mathbb{R}$, then there exists a function $F : {}^*A \to {}^*\mathbb{R}$ defined by the rule,

$$x = [(x_n)_{n\in\mathbb{N}}] \in {}^*A \to y = F(x) \in {}^*\mathbb{R},$$

where $y = [(y_n)_{n\in N}] = [(f_n(x_n))_{n\in N}]$, i.e.

$$F(x) = [(f_n)_{n\in N}]([(x_n)_{n\in N}]) = [(f_n(x_n))_{n\in N}].$$

Any such function F will be called an internal function, and where necessary will be written as $[(f_n)_{n\in N}]$; all other functions defined on $^*\mathbb{R}$ will be called external. Once again, if $f_n = f$ for nearly all n then we write $[(f)_{n\in N}]$ as $^*f$ and call it the nonstandard extension of the (standard) function f.

The standard part of an arbitrary internal function $F = [(f_n)_{n\in N}]$ will be a well defined function whenever F is finite-valued; it is often written as $^\circ F$. If $(f_n)_{n\in N}$ is a sequence which happens to converge pointwise to a limit function $f : \mathbb{R} \to \mathbb{R}$ then $^\circ F$ certainly exists at standard points since it coincides with the pointwise limit $f = \lim_{n\to\infty} f_n$ on $\mathbb{R}$.

Remarks: (i) This very brief description of *R contains virtually all the material on NSA that we actually need for a theory of generalised functions. (A rather more detailed account is given in [7]). Moreover there is no real difficulty in presenting it at an elementary level. Indeed introductory texts on real analysis now exist which use such hyperreal number systems [8], [9]. The only feature which might be described as in the least degree esoteric is the assumption that a free ultrafilter $\mathcal{U}$ on $\mathbb{N}$ does exist. This does involve an explicit appeal to a transfinite principle (slightly weaker than the Axiom of Choice) which may be viewed with alarm in some quarters. One can avoid even this by working instead with the quotient $\mathbb{R}^N/\mathcal{F}$ where $\mathcal{F}$ denotes the Frechet filter of the cofinite subsets of $\mathbb{N}$. The result is a structure which shares many of the desirable properties of $^*\mathbb{R}$ but which is only a partially ordered ring rather than a totally ordered field. It is effectively what is used by Egorov [10] in a recent treatment of generalised functions which is closely related to the account given here.

(ii) In NSA proper the classification of nonstandard objects as internal or external is of crucial importance. It is the internal objects which inherit the properties (in some appropriately reinterpreted sense) of the corresponding standard ones, and which enlarge the mathematical universe in question. The sequential basis of the ad hoc definitions of internal sets and functions given here makes such transfer of properties transparent, at least in respect of the applications we will be considering in the present context.

2 A NONSTANDARD THEORY OF DISTRIBUTIONS

2.1 Internal Functions and Pre-distributions

The development of a theory of distributions as equivalence classes of internal functions was already envisaged by Abraham Robinson in his fundamental text on Nonstandard analysis in 1966 [11]. Since then such an approach has been adopted, and extended, by a number of workers in the field, most notably perhaps by Detlief Laugwitz [12]. In terms of a simple ultrapower model of the hyperreals (such as that given above) the idea is to use the fact that any distribution μ can be expressed as a limit, in $\mathcal{D}'(\mathbb{R})$, of a sequence of regular distributions

$$< \mu, \varphi >= lim_{n\to\infty} < \mu_n, \varphi > \quad , \quad \forall \varphi \in \mathcal{D}(^*\mathbb{R}).$$

In particular, taking $\mu_n \equiv \delta_n * \mu$ where $(\delta_n)_{n\in N}$ is a delta sequence of functions in $\mathcal{D}(\mathbb{R})$, we may represent μ in terms of a sequence of infinitely differentiable functions.

Then the internal function $F \equiv [(\mu_n)_{n\in\mathbb{N}}]$ is a nonstandard representative of μ, the action of μ on an arbitrary test function φ being given by

$$< \mu, \varphi > = st(< F, {}^\star\varphi >) \equiv st\left[\left(\int_{-\infty}^{\infty} \mu_n(x)\varphi(x)dx\right)_{n\in\mathbb{N}}\right].$$

The principal advantages usually claimed for such a treatment are that distributions do appear as functions (mapping ${}^\star\mathbb{R}$ into ${}^\star\mathbb{R}$) in the proper sense of the word, and that nonstandard arguments are generally shorter and more intuitively appealing than standard ones. The objection that a prior knowledge of NSA is required is not a serious one since the ultrapower model for ${}^\star\mathbb{R}$ is easy to describe and there is no real need for an explicit discussion of the Transfer Principle or other nonstandard techniques at this level. A more powerful objection might appear to be that we cannot identify a distribution μ with a single such internal function F; there are infinitely many nonstandard representatives of a particular $\mu \in \mathcal{D}'(\mathbb{R})$, each of which is referred to as a pre-distribution. What is more, there are internal functions $F = [(f_n)_{n\in\mathbb{N}}]$ where $f_n \in \mathcal{C}^\infty(\mathbb{R})$, which patently do not represent distributions at all.

To counter the first objection it is usual to introduce an equivalence relation for internal functions such as, for example, that described by Laugwitz [12]:

> *Two internal functions F, G will be called distributionally equivalent if an integer k and two internal functions Φ, Ψ exist such that $\Phi(x) \approx \Psi(x)$ for all finite x, and $F = {}^\star D^k\Phi, G = {}^\star D^k\Psi$.*[1]

So far as the second point is concerned, the fact that internal functions may represent something other than distributions ought properly to be regarded as an advantage of the nonstandard approach. It is surely desirable that the existence of the multiplication problem for distributions should be made known to the student at an early stage, and nonstandard representation offers a particularly straightforward way of doing this. If $F = [(f_n)_{n\in\mathbb{N}}]$ and $G = [(g_n)_{n\in\mathbb{N}}]$ are pre-distributions then it is salutary to be able to show that the natural pointwise product $FG = [(f_n g_n)_{n\in\mathbb{N}}]$ does not generally represent a distribution. On the other hand FG will certainly be a representative of a generalised function in some sense. We begin by examining a set of internal functions which contains all pre-distributions and all pointwise products of them.

2.2 The Differential Algebra ${}^\star\mathcal{C}(\mathbb{R})$

Recall the standard definition of continuity for a function $f : \mathbb{R} \to \mathbb{R}$. The function f is continuous at a point $a \in \mathbb{R}$ if, given any positive real r, we can always find a corresponding positive real s (depending, generally, on a and r) such that

$$\forall x \in \mathbb{R} : |x - a| < s \Rightarrow |f(x) - f(a)| < r.$$

If we transfer the standard concept of continuity (suitably reinterpreted) from the context of $\mathbb{R}$ to that of ${}^\star\mathbb{R}$ then we get what is called *-continuity. In general an internal function $F : {}^\star\mathbb{R} \to {}^\star\mathbb{R}$ is said to be *-continuous at a point $a \in {}^\star\mathbb{R}$ if and only if given a positive infinitesimal $\varepsilon \in {}^\star\mathbb{R}$ there exists a corresponding $\delta = \delta(\varepsilon) \in {}^\star\mathbb{R}^+$ such that

$$\forall x \in {}^\star\mathbb{R} : |x - a| < \delta \Rightarrow |F(x) - F(a)| < \varepsilon.$$

[1] The derivation operator ${}^\star D$ is here understood to be in the nonstandard sense as will be explained in the sequel.

For each fixed value of $m(0 \leq m \leq +\infty)$, $\mathcal{C}^m(\mathbb{R})$ denotes as usual the linear space of all m-times continuously differentiable (standard) functions on $\mathbb{R}$. We denote by ${}^*\mathcal{C}^m(\mathbb{R})$ the corresponding set of all internal nonstandard functions $F = [(f_n)_{n\in\mathrm{N}}]$ where $(f_n)_{n\in\mathrm{N}}$ is a sequence of $\mathcal{C}^m(\mathbb{R})$-functions. Any internal function F which belongs to ${}^*\mathcal{C}^0(\mathbb{R})$ is *-continuous on ${}^*\mathbb{R}$. On the other hand if $f : \mathbb{R} \to \mathbb{R}$ is any standard function whose non-standard extension ${}^*f$ is *-continuous at any (standard) point $a \in \mathbb{R}$ then f is continuous in the standard sense at a.

Consider the ${}^*\mathbb{R}_b$-module ${}^*\mathcal{C}^\infty(\mathbb{R})$ consisting of all internal functions $F : {}^*\mathbb{R} \to {}^*\mathbb{R}$ which are such that $F = [(f_n)_{n\in\mathrm{N}}]$ where the standard functions f_n belong to $\mathcal{C}^\infty(\mathbb{R})$. The *-differential operator ${}^*D$ is defined on ${}^*\mathcal{C}^\infty(\mathbb{R})$ by

$$ {}^*DF \equiv [(f_n')_{n\in\mathrm{N}}], $$

and we have immediately that

(a) $F \in {}^*\mathcal{C}^\infty(\mathbb{R}) \Rightarrow {}^*DF \in {}^*\mathcal{C}^\infty(\mathbb{R})$,

(b) $F, G \in {}^*\mathcal{C}^\infty(\mathbb{R}) \Rightarrow [FG \in {}^*\mathcal{C}^\infty(\mathbb{R}) \wedge {}^*D(FG) = ({}^*DF)G + F({}^*DG)]$,

so that ${}^*\mathcal{C}^\infty(\mathbb{R})$ is a differential algebra.

The members of ${}^*\mathcal{C}^\infty(\mathbb{R})$ may be described as internal functions which are infinitely *-differentiable on ${}^*\mathbb{R}$. They include, in particular, all internal functions ${}^*f$ which are the nonstandard extensions of standard infinitely differentiable functions f in $\mathcal{C}^\infty(\mathbb{R})$. But they also include many other nonstandard internal functions which are not the extensions of any standard function. For example, following Colombeau, let us write $\mathcal{A}_0(\mathbb{R})$ for the set of all functions $\theta \in \mathcal{D}(\mathbb{R})$ such that

$$ \int_{-\infty}^{\infty} \theta(x)dx = 1. $$

The sequence $(\theta_n)_{n\in\mathrm{N}}$ obtained by setting $\theta_n(x) \equiv n\theta(nx)$ for $n = 1, 2, \ldots$, is a (delta) sequence of $\mathcal{C}^\infty(\mathbb{R})$ functions, and it defines an internal function δ_θ belonging to ${}^*\mathcal{C}^\infty(\mathbb{R})$. Since θ has compact support it follows that δ_θ must vanish outside $mon(0)$ and therefore that $\delta_\theta(x) = 0$ at every non-zero point x of $\mathbb{R}$. If $\theta(0) = 0$ then $\delta_\theta(0) = 0$ also; if $\theta(0) \neq 0$ then $\delta_\theta(0)$ will be an infinite hyperreal. Hence in neither case can δ_θ be the nonstandard extension of some standard function. For obvious reasons we refer to any such function δ_θ as a delta function.

We also define a *-indefinite integral operator ${}^*\Im_a$, where $a = [(a_n)_{n\in\mathrm{N}}]$, as follows

$$ {}^*\Im_a F(x) \equiv \left[\left(\int_{a_n}^{x_n} f_n(t)dt\right)_{n\in\mathrm{N}}\right] $$

where $x = [(x_n)_{n\in\mathrm{N}}] \in {}^*\mathbb{R}$. Then,

(a) $F \in {}^*\mathcal{C}^\infty(\mathbb{R}) \Rightarrow {}^*\Im_a F \in {}^*\mathcal{C}^\infty(\mathbb{R})$,

(b) ${}^*\Im_a\{{}^*DF\} = F - F(a) \wedge {}^*D\{{}^*\Im_a F\} = F$.

2.3 Nonstandard Distributions

2.3.1 - Finite-Order Distributions Those internal functions in ${}^*\mathcal{C}^\infty(\mathbb{R})$ which are pre-distributions can be characterised most simply by examining another form of continuity appropriate to nonstandard internal functions:

Definition 2

An internal function $F : {}^\star\mathbb{R} \to {}^\star\mathbb{R}$ *is said to be S-continuous at a point* $a \in {}^\star\mathbb{R}$ *if and only if for every point* $x \in {}^\star\mathbb{R}$ *which is infinitely close to* a*,* $F(x)$ *is infinitely close to* $F(a)$*, i.e.*

$$\forall x \in {}^\star\mathbb{R} : [x \approx a \Rightarrow F(x) \approx F(a)].$$

It follows readily that a standard function $f : \mathbb{R} \to \mathbb{R}$ is continuous in the standard sense at a point $a \in \mathbb{R}$ if and only if its nonstandard extension ${}^\star f$ is S-continuous at a. Further, for any internal function F which is S-continuous at $a \in \mathbb{R}$ we can say that its standard part, ${}^\circ F(x) \equiv st([F(x)])$, is a function which is continuous in the standard sense at a. However, not every internal function $F \in {}^\star\mathcal{C}^\infty(\mathbb{R})$ will be S-continuous on ${}^\star\mathbb{R}$, even if the standard part of F happens to be continuous in the standard sense at every standard point. We denote by ${}^s\mathcal{C}(\mathbb{R})$ the ${}^\star\mathbb{R}_b$-submodule of ${}^\star\mathcal{C}^\infty(\mathbb{R})$ consisting of all those functions which are S-continuous at every standard point of ${}^\star\mathbb{R}$. Then for any $F \in {}^s\mathcal{C}(\mathbb{R})$ and any $a \in {}^\star\mathbb{R}_b$ we have that ${}^\star\Im_a F \in {}^s\mathcal{C}(\mathbb{R})$. But it is not generally true that ${}^\star DF \in {}^s\mathcal{C}(\mathbb{R})$ for $F \in {}^s\mathcal{C}(\mathbb{R})$.

Remark : A function $F \in {}^s\mathcal{C}(\mathbb{R})$ which is such that ${}^\star DF \in {}^s\mathcal{C}(\mathbb{R})$ may be described as **S-differentiable**. If ${}^\star D^j F \in {}^s\mathcal{C}(\mathbb{R})$ for some $j \in \mathbb{N}_0$ then F is said to be j-times S-differentiable. A function $F \in {}^s\mathcal{C}(\mathbb{R})$ such that F is j-times S-differentiable for every $j \in \mathbb{N}_0$ is said to be infinitely S-differentiable, and is convenient to denote by ${}^s\mathcal{C}^\infty(\mathbb{R})$ the ${}^\star\mathbb{R}_b$-submodule of ${}^s\mathcal{C}(\mathbb{R})$ comprising all infinitely s-differentiable functions. ${}^s\mathcal{C}^\infty(\mathbb{R})$ is an ${}^\star\mathbb{R}_b$-differential sub-algebra of ${}^\star\mathcal{C}^\infty(\mathbb{R})$.

We now denote by ${}^\star D^j\{{}^s\mathcal{C}(\mathbb{R})\}$ the set of all internal functions (in ${}^\star\mathcal{C}^\infty(\mathbb{R})$) which are the j-th *-derivatives of functions in ${}^s\mathcal{C}(\mathbb{R})$. If we define

$$\begin{aligned} {}^\star D^\infty\{{}^s\mathcal{C}(\mathbb{R})\} &\equiv \cup_{j=0}^\infty {}^\star D^j\{{}^s\mathcal{C}(\mathbb{R})\} \\ &= \left\{F \in {}^\star\mathcal{C}^\infty(\mathbb{R}) : [\exists j \in \mathbb{N}_0, \Phi \in {}^s\mathcal{C}(\mathbb{R}) : F = {}^\star D^j\Phi]\right\} \end{aligned}$$

then we have the following (strict) inclusions:

$${}^s\mathcal{C}^\infty(\mathbb{R}) \subset {}^s\mathcal{C}(\mathbb{R}) \subset {}^\star D^\infty\{{}^s\mathcal{C}(\mathbb{R})\} \subset {}^\star\mathcal{C}^\infty(\mathbb{R}).$$

The members of ${}^\star D^\infty\{{}^s\mathcal{C}(\mathbb{R})\}$ are the nonstandard representatives of (finite order) Schwartz distributions on $\mathbb{R}$. If F and G are internal functions in ${}^\star D^\infty\{{}^s\mathcal{C}(\mathbb{R})\}$ then we say that F and G are **distributionally equivalent**, and write $F \Xi G$, if and only if there exists some integer $m \in \mathbb{N}_0$ and a polynomial p_m of degree $< m$, with coefficients in ${}^\star R$, which is such that

$${}^\star\Im_a^m(F - G) \approx p_m$$

for any $a \in {}^\star\mathbb{R}_b$. (A polynomial of degree < 0 is understood here to mean an infinitesimal constant.)

Clearly this will agree with the definition of distributional equivalence used by Laugwitz [12]. Then for any $F \in {}^\star D\{{}^s\mathcal{C}(\mathbb{R})\}$ we denote by $\mu \equiv {}^\Xi[F]$ the equivalence class containing F, and call it a **finite order** ${}^\Xi$**-distribution**. The set of all such equivalence classes determined on ${}^\star D^\infty\{{}^s\mathcal{C}(\mathbb{R})\}$ by the equivalence relation Ξ will be denoted by

$${}^\Xi\mathcal{C}_\infty(\mathbb{R}) \equiv {}^\star D^\infty\{{}^s\mathcal{C}(\mathbb{R})\}/\Xi.$$

${}^\Xi\mathcal{C}_\infty(\mathbb{R})$ is a nonstandard model for the axiomatic definition of finite order distributions on $\mathbb{R}$ proposed by J.Sebastião e Silva [13], in which the concept of standard continuous function on $\mathbb{R}$ is replaced by that of Ξ-equivalence class of nonstandard S-continuous ${}^\star\mathcal{C}^\infty(\mathbb{R})$-internal functions.

2.3.2 - Global-distributions The extension to global-distributions, as discussed in Hoskins and Pinto [7], can be summarised briefly as follows: Let $\{\mathbb{K}_\alpha\}_{\alpha\in A}$ be the family of all compact subintervals of $\mathbb{R}$ with more than one point. Denote by ${}^\pi\mathcal{C}(\mathbb{R})$ the ${}^\star\mathbb{R}_b$ sub-module of all ${}^\star\mathcal{C}^\infty(\mathbb{R})$-functions F such that for each $\alpha \in A$ there exists $r_\alpha \in \mathbb{N}_0$ and $F_\alpha \in {}^s\mathcal{C}(\mathbb{K}_\alpha)$ such that

$$\forall x \in {}^\star\mathbb{K}_\alpha : F(x) = {}^\star D^{r_\alpha} F_\alpha(x).$$

The equivalence relation extends to ${}^\pi\mathcal{C}(\mathbb{R})$ as follows. Two functions $F, G \in {}^\pi\mathcal{C}(\mathbb{R})$ are distributionally equivalent, and we write again $F \Xi G$, if and only if for each $\alpha \in A$ there exists an integer $m_\alpha \in \mathbb{N}_0$ and a polynomial p_{m_α} of degree $< m_\alpha$ such that

$$\forall x \in {}^\star\mathbb{K}_\alpha : {}^\star\Im^{m_\alpha}_{a_\alpha}(F - G)(x) \approx p_{m_\alpha}(x)$$

where, for each $\alpha \in A$, a_α is an arbitrarily fixed point in $\mathbb{K}_\alpha$. We write

$${}^\Xi\mathcal{C}_\pi(\mathbb{R}) \equiv {}^\pi\mathcal{C}_\infty(\mathbb{R})/\Xi$$

and call each equivalence class ${}^\Xi[F]$ a **global ${}^\Xi$-distribution**.

Remarks: This nonstandard theory of distributions is relatively concise and simple to describe. It has the usual conceptual appeal of such theories in that the representation of distributions as functions avoids some of the apparent paradoxes which standard theories involve. For example, a standard real function $f : \mathbb{R} \to \mathbb{R}$ is said to be a null function if it vanishes almost everywhere. The standard part of a delta function such as

$$\frac{1}{\pi}\frac{\omega}{1+\omega^2x^2}$$

is a null function in this sense since it is zero for all real $x \neq 0$. This illustrates the fact that, so far as distributions are concerned, the archimedean property of $\mathbb{R}$ and the resulting absence of infinitely large and infinitely small elements is a disavantage. However, it is also easy to show that a general theory of multiplication is still not possible for Ξ-distributions, although internal functions can be multiplied freely.

3 NONSTANDARD GENERALISED FUNCTIONS

3.1 Growth Characterisation of Pre-distributions

To develop a theory of generalised functions which includes distributions and products of distributions we need to define an appropriate sub-algebra of ${}^\star\mathcal{C}^\infty(\mathbb{R})$. The limitations of the set ${}^\star D^\infty\{{}^s\mathcal{C}(\mathbb{R})\}$ in this respect are easily seen. So far as the Schwartz product is concerned there is of course no problem. If $F \in {}^\star D^\infty\{{}^s\mathcal{C}(\mathbb{R})\}$ and $\theta \in {}^s\mathcal{C}^\infty(\mathbb{R})$ then it is easily shown that $\theta F \in {}^\star D^\infty\{{}^s\mathcal{C}(\mathbb{R})\}$. Further, if F and G belong to ${}^\star D^\infty\{{}^s\mathcal{C}(\mathbb{R})\}$ and $F \Xi G$, then for any $\theta \in {}^s\mathcal{C}^\infty(\mathbb{R})$ we have $\theta F \Xi \theta G$. Hence if $\mu \equiv {}^\Xi[F]$ and if we denote by θ the (regular) ${}^\Xi$-distribution ${}^\Xi[\theta]$ defined by a function $\theta \in {}^s\mathcal{C}^\infty(\mathbb{R})$ then the product $\theta\mu$ is the ${}^\Xi$-distribution defined by $\theta\mu \equiv {}^\Xi[\theta F]$. This still holds good if we take a regular ${}^\Xi$-distribution defined by a function $\theta \in \mathcal{C}^r(\mathbb{R})$, where r is some non-negative integer, provided that μ is a ${}^\Xi$-distribution of finite order (in the sense of J.S.Silva) not greater than r.

On the other hand consider the delta function defined by

$$\delta(x) \equiv \frac{1}{2\pi}\frac{\omega}{1+\omega^2x^2},$$

where ω denotes the infinite positive hyperreal $[(n)_{n\in\mathrm{N}}]$. Clearly δ belongs to ${}^\star D^\infty\{{}^s\mathcal{C}(\mathbb{R})\}$ for we have,

$$\delta(x) = {}^\star D\left({}^\star\Im_0\delta(x)\right) = {}^\star D\left(\pi^{-1}\arctan(\omega x)\right) = {}^\star D^2(\Phi(x)),$$

where $\Phi(x)$ belongs to ${}^s\mathcal{C}(\mathbb{R})$. However the function

$$\delta^2(x) = \frac{1}{\pi^2}\frac{\omega^2}{(1+\omega^2x^2)^2}$$

is not the m-th derivative of a function in ${}^s\mathcal{C}(\mathbb{R})$ for any $m \in \mathbb{N}_0$ (except at the origin). ${}^\star D^\infty\{{}^s\mathcal{C}(\mathbb{R})\}$ is not an algebra.

We return once more to the fact that any finite-order distribution μ is for some $r \in \mathbb{N}_0$ the r-th (distributional) derivative of a continuous function and examine the growth conditions which this entails. For any function f defined and continuous on $\mathbb{R}$, we define

$$f_n(x) \;=\; (f * \theta_n)(x) \;\equiv\; \int_{-\infty}^{\infty} f(t)\theta_n(x-t)dt \;\;,\;\; n = 1, 2, \ldots,$$

where $\theta_n(x) = n\theta(nx)$ and θ is a function in $\mathcal{A}_0(\mathbb{R}) \subset \mathcal{D}(\mathbb{R})$. Then $(f_n)_{n\in\mathrm{N}}$ is a sequence of $\mathcal{C}^\infty(\mathbb{R})$ functions which converges uniformly on compacts in $\mathbb{R}$ to f. Further, if the support of θ is contained in the interval $[-a, a]$ then we have

$$\begin{aligned}|f_n(x)| \;&\le\; \left\{\sup_{x-a/n\le t\le x+a/n}|f(t)|\right\}\cdot\int_{x-a/n}^{x+a/n} n|\theta(n(x-t))|dt\\ &=\; C_\theta\left\{\sup_{x-a/n\le t\le x+a/n}|f(t)|\right\},\end{aligned}$$

where

$$C_\theta \;=\; \int_{-a}^{a}|\theta(\tau)|d\tau.$$

Hence if $\mathbb{K}$ is any compact subset of $\mathbb{R}$ then we have

$$sup_{x\in\mathrm{K}}|f_n(x)| \le C_\theta\left\{sup_{\mathrm{K}+[-a,a]}|f(t)|\right\},$$

where the right-hand side is independent of n. It follows that if F is the internal function $F \equiv [(f_n)_{n\in\mathrm{N}}] \in {}^s\mathcal{C}(\mathbb{R})$ then for each compact $\mathbb{K} \sqsubset \mathbb{R}$ there exists a positive constant $C \in {}^\star\mathbb{R}_b$ such that

$$sup_{x\in{}^\star\mathrm{K}}|F(x)| \le C.$$

Moreover if G is any other internal function in ${}^s\mathcal{C}(\mathbb{R})$ such that ${}^\circ G = f$ then $G \Xi F$ and $G - F \approx 0$. It follows that every internal function in ${}^s\mathcal{C}(\mathbb{R})$ is bounded by a constant in ${}^\star\mathbb{R}_b$ on the nonstandard extension of each compact of $\mathbb{R}$.

More generally if f and θ are as before and if r is any non-negative integer, then for any $x \in \mathbb{R}$ we have

$$\begin{aligned}|f_n^{(r)}(x)| \;&\le\; \left\{\sup_{x-a/n\le t\le x+a/n}|f(t)|\right\}\int_{x-a/n}^{x+a/n} n^{r+1}|\theta^{(r)}(n(x-t))|dt\\ &\le\; C_{\theta,r}\cdot n^r\left\{\sup_{x-a/n\le t\le x+a/n}|f(t)|\right\},\end{aligned}$$

where

$$C_{\theta,r} \;=\; \int_{-a}^{a}|\theta^{(r)}(\tau)|d\tau.$$

Hence, if $\mathbb{K}$ is any compact in $\mathbb{R}$ then we have

$$\sup_{x\in\mathbb{K}} |f_n^{(r)}(x)| \leq C_{\theta,r} \cdot n^r \left\{ \sup_{\mathbb{K}+[-a,a]} |f(t)| \right\} \leq C' \cdot n^r,$$

where C' is independent of n. It follows that the function $F \equiv [(f_n)_{n\in\mathbb{N}}]$ is a member of ${}^s\mathcal{C}(\mathbb{R})$ such that for any $r \in \mathbb{N}_0$ and any compact $\mathbb{K} \sqsubset \mathbb{R}$ there exists an integer $m = m(r) \leq r$ such that

$$sup_{x\in{}^*\mathbb{K}} |{}^*D^r F(x)| \leq C \cdot \omega^m,$$

where $C = C(\mathbb{K})$ is a positive constant in ${}^*\mathbb{R}_b$ and ω is the infinite hyperreal number, $\omega = [(n)_{n\in\mathbb{N}}]$.

Accordingly we adopt the following definition:

Definition 3

A function $F \in {}^*\mathcal{C}(\mathbb{R})$ is said to be of ${}^\omega$-growth on ${}^*\mathbb{R}_b$ if, for every compact $\mathbb{K} \sqsubset \mathbb{R}$, there exists a non-negative integer m such that

$$sup_{t\in{}^*\mathbb{K}} |F(t)| \leq C \cdot \omega^m,$$

for some C in ${}^*\mathbb{R}_b$.[2]

3.2 The Differential Algebra ${}^\omega\mathcal{G}(\mathbb{R})$

We denote by ${}^\omega\mathcal{C}^0(\mathbb{R})$ the ${}^*\mathbb{R}_b$ sub-module of ${}^*\mathcal{C}^\infty(\mathbb{R})$ consisting of all functions of ${}^\omega$-growth on ${}^*\mathbb{R}_b$; then ${}^s\mathcal{C}(\mathbb{R}) \subset {}^\omega\mathcal{C}^0(\mathbb{R})$. If F is any function in ${}^\omega\mathcal{C}^0(\mathbb{R})$ then it is easily shown that, for any $a \in {}^*\mathbb{R}_b$, the function ${}^*\Im_a F$ also belongs to ${}^\omega\mathcal{C}^0(\mathbb{R})$. Therefore, for all $F \in {}^*\mathcal{C}^\infty(\mathbb{R})$, if ${}^*DF \in {}^\omega\mathcal{C}^0(\mathbb{R})$ then F itself must also belong to ${}^\omega\mathcal{C}^0(\mathbb{R})$. On the other hand it is not generally true that $F \in {}^\omega\mathcal{C}^0(\mathbb{R})$ implies ${}^*DF \in {}^\omega\mathcal{C}^0(\mathbb{R})$. We define accordingly a modified concept of differentiability as follows:

Definition 4

An internal function F which belongs to ${}^*\mathcal{C}^\infty(\mathbb{R})$ is said to be ${}^\omega$-differentiable on ${}^*\mathbb{R}_b$ if and only if the function ${}^*DF$ is of ${}^\omega$-growth on ${}^*\mathbb{R}_b$.

We denote by ${}^\omega\mathcal{C}^1(\mathbb{R})$ the set of all ${}^\omega$-differentiable functions in ${}^*\mathcal{C}(\mathbb{R})$; then ${}^\omega\mathcal{C}^1(\mathbb{R}) \subset {}^\omega\mathcal{C}^0(\mathbb{R})$. More generally, for any $k \in \mathbb{N}_0$ we say that $F \in {}^*\mathcal{C}^\infty(\mathbb{R})$ is k-times ${}^\omega$-differentiable if ${}^*D^kF$ belongs to ${}^\omega\mathcal{C}^0(\mathbb{R})$, and write

$${}^\omega\mathcal{C}^k(\mathbb{R}) \equiv \{F \in {}^*\mathcal{C}^\infty(\mathbb{R}) : {}^*D^kF \in {}^\omega\mathcal{C}^0(\mathbb{R})\},$$

for the set of all such functions. If ${}^*D^kF \in {}^\omega\mathcal{C}^0(\mathbb{R})$ then for all $j = 0, 1, 2, \ldots, k$, we have ${}^*D^jF \in {}^\omega\mathcal{C}^0(\mathbb{R})$, and so

$${}^\omega\mathcal{C}^k(\mathbb{R}) \subset {}^\omega\mathcal{C}^{k-1}(\mathbb{R}) \subset \ldots {}^\omega\mathcal{C}^0(\mathbb{R}) \subset {}^*\mathcal{C}^\infty(\mathbb{R}).$$

Definition 5

An internal function F in ${}^*\mathcal{C}^\infty(\mathbb{R})$ is said to be infinitely ${}^\omega$-differentiable or to be an ${}^\omega$-moderate function if, for all $k \in \mathbb{N}_0$ we have that F is in ${}^\omega\mathcal{C}^k(\mathbb{R})$. Then

$${}^\omega\mathcal{C}^\infty(\mathbb{R}) \equiv \cap_{k=0}^\infty {}^\omega\mathcal{C}^k(\mathbb{R})$$

is the ${}^*\mathbb{R}_b$ sub-module of all ${}^\omega$-moderate functions in ${}^*\mathcal{C}^\infty(\mathbb{R})$; it is easy to see that ${}^\omega\mathcal{C}^\infty(\mathbb{R})$ is in fact a differential sub-algebra of ${}^*\mathcal{C}^\infty(\mathbb{R})$.

[2]Note that in general both the numbers m and C depend on the compact $\mathbb{K} \sqsubset \mathbb{R}$.

The restriction to $^{\omega}$-moderate internal functions ensures that we confine attention to functions which do not assume infinite values which are, in some sense, *too large.* In the same way we choose to eliminate functions which are *too small.* It is most convenient to do this in terms of a field structure which is intermediate between $\mathbb{R}$ and $^*\mathbb{R}$. First, define the following two subsets of $^*\mathbb{R}$:

$$^{\omega}\mathbf{fin}(^*\mathbb{R}) \equiv \{x \in {}^*\mathbb{R} : [\exists m \in \mathbb{N} : |x| < \omega^m]\}$$

$$^{\omega}\mathbf{inf}(^*\mathbb{R}) \equiv \{x \in {}^*\mathbb{R} : [\forall m \in \mathbb{N} : |x| < \omega^{-m}]\}.$$

Both $^{\omega}\mathbf{fin}(^*\mathbb{R})$ and $^{\omega}\mathbf{inf}(^*\mathbb{R})$ are rings with respect to the algebraic operations of $^*\mathbb{R}$, and $^{\omega}\mathbf{inf}(^*\mathbb{R})$ is a maximal ideal of $^{\omega}\mathbf{fin}(^*\mathbb{R})$. The relation $\simeq$ defined by

$$\forall x, y \in {}^{\omega}\mathbf{fin}(^*\mathbb{R}) : [x \simeq y \Leftrightarrow x - y \in {}^{\omega}\mathbf{inf}(^*\mathbb{R})]$$

is an equivalence relation on $^{\omega}\mathbf{fin}(^*\mathbb{R})$. Then $^{\omega}\mathbf{fin}(^*\mathbb{R})/^{\omega}\mathbf{inf}(^*\mathbb{R})$ is a field, which we call the Robinson field and denote by $^{\omega}\mathbb{R}$. It was introduced by Abraham Robinson in [14] and is studied intensively in the text by Lightstone and Robinson [15].

As usual we write $^{\omega}[x]$ for the equivalence class in $^{\omega}\mathbb{R}$ which contains the element $x \in {}^*\mathbf{fin}(^*\mathbb{R}) \subset {}^*\mathbb{R}$. An element $^{\omega}[x]$ of $^{\omega}\mathbb{R}$ is said to be positive in case x is a positive hyperreal; this defines an ordering relation on $^{\omega}\mathbb{R}$ which extends it to an ordered field. By construction the field $^{\omega}\mathbb{R}$ lies between the fields $\mathbb{R}$ and $^*\mathbb{R}$. In fact $^{\omega}\mathbb{R}$ is an extension of $\mathbb{R}$ that includes all sufficiently large infinitesimals and all sufficiently small infinite numbers, but which excludes those infinitesimals (other than zero) which are too small and those infinite numbers which are too large. Each member $\imath$ of $^{\omega}\mathbf{inf}(^*\mathbb{R})$ is an infinitesimal which is too small in this sense, and for $\imath \neq 0$ its inverse $\imath^{-1}$ is an infinite number which is too large. We call each member of $^{\omega}\mathbf{inf}(^*\mathbb{R})$ an **iota**, and the inverse of any non-zero iota a **mega**. Thus $\imath \in {}^*\mathbb{R}$ is an iota if and only if $|\imath| < \omega^{-m}$ for all $m \in \mathbb{N}$, and $\nu \in {}^*\mathbb{R}$ is a mega if and only if $|\nu| > \omega^n$ for all $n \in \mathbb{N}$.

Definition 6

A function $F \in {}^{\omega}\mathcal{C}^{\infty}(\mathbb{R})$ is said to be iota-null on $^*\mathbb{R}_b$ if and only if $F(x) \simeq 0$ for all $x \in {}^*\mathbb{R}_b$; that is, if and only if $|F(x)| < \omega^{-m}$ for every $m \in \mathbb{N}_0$.

The set $^{\omega}\mathcal{N}(\mathbb{R})$ of all functions $F \in {}^{\omega}\mathcal{C}^{\infty}(\mathbb{R})$ such that $^*D^kF$ is iota-null on $^*\mathbb{R}_b$ for all $k \in \mathbb{N}_0$ is an ideal of $^{\omega}\mathcal{C}^{\infty}(\mathbb{R})$; further we have $^*D\{^{\omega}\mathcal{N}(\mathbb{R})\} \subset {}^{\omega}\mathcal{N}(\mathbb{R})$. We define

$$^{\omega}\mathcal{G}(\mathbb{R}) \equiv {}^{\omega}\mathcal{C}^{\infty}(\mathbb{R})/^{\omega}\mathcal{N}(\mathbb{R}),$$

and denote the equivalence class containing $F \in {}^{\omega}\mathcal{C}^{\infty}(\mathbb{R})$ by $^{\omega}[F]$.

$^{\omega}\mathcal{G}(\mathbb{R})$ is a differential algebra under the following operations:

$$^{\omega}[F] + {}^{\omega}[G] \equiv {}^{\omega}[F+G]\,,$$

$$^{\omega}[F] \cdot {}^{\omega}[G] \equiv {}^{\omega}[FG]\,,$$

$$a \cdot {}^{\omega}[F] \equiv {}^{\omega}[\alpha F]\,,$$

$$D\{^{\omega}[F]\} \equiv {}^{\omega}[^*DF]\,,$$

where $a = {}^{\omega}[\alpha]$ and α may be any point in $^{\omega}\mathbf{fin}(^*\mathbb{R})$.

3.3 Association

The algebra ${}^{\omega}\mathcal{G}(\mathbb{R})$ contains the subalgebra

$${}^{\omega}\mathcal{E}(\mathbb{R}) \equiv {}^{s}\mathcal{C}^{\infty}(\mathbb{R})/{}^{\omega}\mathcal{N}(\mathbb{R})$$

whose members are in some sense nonstandard representatives of $\mathcal{C}^{\infty}(\mathbb{R})$-functions. It also contains the (non-differential) algebra

$$({}^{s}\mathcal{C}(\mathbb{R}) \cap {}^{\omega}\mathcal{C}^{\infty}(\mathbb{R}))\ /{}^{\omega}\mathcal{N}(\mathbb{R}).$$

However the correspondence

$$\mathcal{C}^{\infty}(\mathbb{R}) \to {}^{\omega}\mathcal{E}(\mathbb{R}) \subset {}^{\omega}\mathcal{G}(\mathbb{R})$$

is not well defined. If $f \in \mathcal{C}^{\infty}(\mathbb{R})$ and $\theta \in \mathcal{A}_0(\mathbb{R})$, and we write $F_\theta \equiv [(f * \theta_k)_{k\in\mathbb{N}}]$ then the equivalence classes ${}^{\omega}[{}^{\star}f]$ and ${}^{\omega}[F_\theta]$ do not generally coincide in ${}^{\omega}\mathcal{G}(\mathbb{R})$. Nevertheless they do satisfy a weaker form of equality which we call $\mathcal{D}$-association.

Let ${}^{\omega}f \equiv {}^{\omega}[F] \in {}^{\omega}\mathcal{G}(\mathbb{R})$, where $F = [(f_n)_{n\in\mathbb{N}}]$. For any $\varphi \in \mathcal{D}(\mathbb{R})$ we have that

$$< F, {}^{\star}\varphi > = \left[\left(\int_{-\infty}^{\infty} f_n(t)\varphi(t)dt\right)_{n\in\mathbb{N}}\right]$$

belongs to ${}^{\omega}\mathbf{fin}({}^{\star}\mathbb{R})$. If F and F_1 are any two representatives of ${}^{\omega}f$ then they differ by at most an iota-null function on ${}^{\star}R_b$, and so

$$< F_1, {}^{\star}\varphi > = < F, {}^{\star}\varphi > + < F_1 - F, {}^{\star}\varphi >$$

where

$$< F_1 - F, {}^{\star}\varphi > = \int_{{}^{\star}\mathbb{R}} [F_1(t) - F(t)]^{\star}\varphi(t)\ dt$$

which is an iota since the integration is only over a finite range. Hence it makes sense to define

$$< {}^{\omega}f, \varphi > \equiv {}^{\omega}[< F, {}^{\star}\varphi >],$$

where F is an arbitrary representative of ${}^{\omega}f$.

Denote by ${}^{\omega}mon(0)$ the set of ${}^{\omega}$-infinitesimals in ${}^{\omega}\mathbb{R}$ defined by

$${}^{\omega}mon(0) = mon(0)/{}^{\omega}\mathbf{inf}({}^{\star}\mathbb{R}).$$

Two elements $\xi, \eta \in {}^{\omega}\mathbb{R}$ are said to be ${}^{\omega}$-infinitesimally close, and we write $\xi \cong \eta$, if and only if $\xi - \eta \in {}^{\omega}mon(0)$.

Definition 7

${}^{\omega}f, {}^{\omega}g \in {}^{\omega}\mathcal{G}(\mathbb{R})$ are said to be $\mathcal{D}$-associated if and only if

$$< {}^{\omega}f, \varphi > \cong < {}^{\omega}g, \varphi >$$

for all $\varphi \in \mathcal{D}(\mathbb{R})$, and then we write ${}^{\omega}f\ \Xi\ {}^{\omega}g$.

Clearly Ξ is an equivalence relation over ${}^{\omega}\mathcal{G}(\mathbb{R})$, and we show below that it is consistent with the relation of distributional equivalence defined earlier (as the notation suggests).

The quotient

$${}^{\Xi}\mathcal{G}(\mathbb{R}) \equiv {}^{\omega}\mathcal{G}(\mathbb{R})/\Xi$$

is a linear space over ${}^\omega\mathbb{R}$ but not an algebra. An element ${}^\Xi[{}^\omega f]$ of ${}^\Xi\mathcal{G}(\mathbb{R})$ is called a ${}^\Xi$-generalised function.

Let ${}^\omega[F] \,\Xi\, {}^\omega[G]$, where $F, G \in {}^\omega\mathcal{C}^\infty(\mathbb{R})$. Then since we have

$$< {}^\star D(F-G), \varphi > = < F-G, -\varphi' >$$

it follows that ${}^\omega[{}^\star DF] \;\Xi\; {}^\omega[{}^\star DG]$. Hence, we may define the ${}^\Xi$-derivation operator ${}^\Xi D : {}^\Xi\mathcal{G}(\mathbb{R}) \to {}^\Xi\mathcal{G}(\mathbb{R})$ by setting for any ${}^\Xi$-generalised function $\mu = {}^\Xi[{}^\omega f]$,

$${}^\Xi D\mu \;\equiv {}^\Xi[D\{{}^\omega f\}] \in {}^\Xi(\mathbb{R}),$$

where $D\{{}^\omega f\} \equiv {}^\omega[{}^\star DF]$ for any $F \in {}^\omega\mathcal{C}^\infty(\mathbb{R})$ such that ${}^\omega[F] = {}^\omega f$.

Theorem 1

$F, G \in {}^\omega\mathcal{C}^\infty(\mathbb{R})$ are distributionally equivalent if and only if the equivalence classes ${}^\omega f \equiv {}^\omega[F]$ and ${}^\omega g \equiv {}^\omega[G]$ are $\mathcal{D}$-associated.

Proof: We have that $F \Xi G$ if and only if for some $m \in \mathbb{N}_0$ there exists a polynomial p_m of degree $< m$ such that

$${}^\star\Im_0^m(F-G) \approx p_m$$

on ${}^\star\mathbb{R}_b$. Hence

$$\begin{aligned} < F-G, {}^\star\varphi > &= < {}^\star\Im_0^m(F-G), (-1)^{m\star}\varphi^{(m)} > \\ &\approx < p_m, (-1)^{m\star}\varphi^{(m)} > \;=\; 0 \end{aligned}$$

and so

$$< {}^\omega f, \varphi > \;\cong\; < {}^\omega g, \varphi >,$$

for all $\varphi \in \mathcal{D}(\mathbb{R})$.

The only if part is immediate. □

Corollary 1

The inclusion ${}^\Xi\mathcal{C}_\infty(\mathbb{R}) \subset {}^\Xi\mathcal{G}(\mathbb{R})$ is valid, that is, every ${}^\Xi$-distribution on $\mathbb{R}$ of finite order is a ${}^\Xi$-generalised function on $\mathbb{R}$.

In the same way $\mu \equiv {}^\Xi[{}^\omega f]$ belongs to ${}^\Xi\mathcal{C}_\pi(\mathbb{R})$ if and only if ${}^\omega f$ may be represented by an internal function $F \in {}^\omega\mathcal{C}^\infty(\mathbb{R})$ such that for every compact $\mathbb{K}$ there exists $r = r(\mathbb{K}) \in \mathbb{N}_0$ and $\Phi \in {}^\pi\mathcal{C}(\mathbb{R})$ such that

$$\forall x \in {}^\star\mathbb{K} : F(x) = {}^\star D^r\Phi(x),$$

and it follows that

$${}^\Xi\mathcal{C}_\pi(\mathbb{R}) \subset {}^\Xi\mathcal{G}(\mathbb{R}).$$

4 COLOMBEAU-BIAGIONI GENERALISED FUNCTIONS

The simplified form of Colombeau's theory of New Generalised Functions, due to Hebe Biagioni [1], can be summarised as follows:

Denote by $\mathcal{E}_s(\mathbb{R})$ the set of all mappings

$$\begin{aligned} T : (0,\infty) \times \mathbb{R} &\to \mathbb{R} \\ (\varepsilon, x) &\to T(\varepsilon, x) \end{aligned}$$

which are infinitely differentiable with respect to $x \in \mathbb{R}$ for each fixed $\varepsilon > 0$. Let $\mathcal{E}_{M,s}(\mathbb{R})$ be the subset of $\mathcal{E}_s(\mathbb{R})$ which consists of all mappings T that are moderate in the following sense: for each compact $\mathbb{K} \subset \mathbb{R}$ and each non-negative integer $r \in \mathbb{N}_0$ there exist numbers $q \in \mathbb{N}_0, c > 0$ and $\eta > 0$ such that

$$|D^r T(\varepsilon, x)| \leq c \cdot \varepsilon^{-q}$$

for all $x \in \mathbb{K}$ and all ε such that $0 < \varepsilon < \eta$. Further, let $\mathcal{N}_s(\mathbb{R})$ be the subspace of $\mathcal{E}_s(\mathbb{R})$ consisting of all T in $\mathcal{E}_s(\mathbb{R})$ such that given any compact $\mathbb{K} \subset \mathbb{R}$ and $r \in \mathbb{N}_0$ there exists $p \in \mathbb{N}$ such that for all $q \geq p$ there are $c > 0$ and $\eta > 0$ with

$$|D^r T(\varepsilon, x)| \leq c \cdot \varepsilon^q$$

for all $x \in \mathbb{K}$ and all ε such that $0 < \varepsilon < \eta$.

$\mathcal{E}_{M,s}(\mathbb{R})$ is a sub-algebra of the algebra $\mathcal{E}_s(\mathbb{R})$ and $\mathcal{N}_s(\mathbb{R})$ is an ideal of $\mathcal{E}_{M,s}(\mathbb{R})$. The Colombeau-Biagioni algebra of New Generalised Functions is defined to be the quotient

$$\mathcal{G}_s(\mathbb{R}) \equiv \mathcal{E}_{M,s}(\mathbb{R})/\mathcal{N}_s(\mathbb{R}).$$

In what follows we understand $\mathcal{G}_s(\mathbb{R})$ to have been defined with the variable ε restricted to the discrete values $\varepsilon = 1/n$, where $n \in \mathbb{N}$. Let T be an element of $\mathcal{E}_{M,s}(\mathbb{R})$ and for each $n \in \mathbb{N}$ define T_n by writing

$$T_n(x) \equiv T(1/n, x).$$

Then $T_n \in \mathcal{C}^\infty(\mathbb{R})$ and for any compact $\mathbb{K} \sqsubset \mathbb{R}$ and $r \in \mathbb{N}_0$ there exist $q \in \mathbb{N}_0$ and $c > 0$ such that the inequality

$$sup_{x \in \mathbb{K}} |T_n^{(r)}(x)| \leq c \cdot n^q$$

holds for all sufficiently large $n \in \mathbb{N}$. Hence the internal function $F_T \equiv [(T_n)_{n \in \mathrm{N}}]$ belongs to ${}^\omega\mathcal{C}^\infty(\mathbb{R})$. We may define the (injective) mapping $\xi : \mathcal{E}_{M,s}(\mathbb{R}) \to {}^\omega\mathcal{C}^\infty(\mathbb{R})$ by setting $\xi(T) = F_T$ for every T in $\mathcal{E}_{M,s}(\mathbb{R})$. Now, if $T \in \mathcal{N}_s(\mathbb{R})$ then given $\mathbb{K} \sqsubset \mathbb{R}$ and r in $\mathbb{N}_0$ there exists $p \in \mathbb{N}$ such that for all $q \in \mathbb{N}_0$ with $q \geq p$ there exists $c > 0$ such that for all sufficiently large n

$$sup_{x \in \mathbb{K}} |T_n^{(r)}(x)| \leq c \cdot n^{-q}.$$

Hence for every $r \in \mathbb{N}$ it follows that

$${}^\star D^r F_T \equiv [(T_n^{(r)})_{n \in \mathrm{N}}]$$

is an iota and so $\xi(\mathcal{N}_s(\mathbb{R})) \subset {}^\omega\mathcal{N}(\mathbb{R})$, giving a homomorphism of algebras.

References

[1] H.A.Biagioni, "A Nonlinear Theory of Generalized Functions," Springer-Verlag, Berlin - New York - London (1990).

[2] J. F. Colombeau, "New Generalized Functions and Multiplication of Distributions," North-Holland, Amsterdam - New York - Oxford (1984).

[3] J. F. Colombeau, "Elementary Introduction to New Generalized Functions," North-Holland, Amsterdam - New York - Oxford (1985).

[4] E. E. Rosinger, "Generalized Solutions of Partial Differential Equations," North-Holland, Amsterdam - New York - Oxford (1987).

[5] T. D. Todorov, Colombeau's generalized functions and nonstandard analysis, in: "Proc. Int. Conf. on Generalized Functions and their applications," Dubrovnik (1987).

[6] Li Bang-He and Li Ya-qing, New generalized functions in nonstandard framework, (to appear).

[7] R. F. Hoskins and J. Sousa Pinto, A nonstandard realization of the J.S.Silva axiomatic theory of distributions, PortugaliæMath., 48 : 2 (1991).

[8] J. M. Henle and E. M. Kleinberg, "Infinitesimal Calculus," MIT Press, Cambridge Mass - London (1979).

[9] R. F. Hoskins, "Standard and Nonstandard Analysis," Ellis Horwood, New York - London -Tokyo (1991).

[10] Yu. V. Egorov, A contribution to the theory of generalized functions, Uspekhi Mat. Nauk 45 : 5 (1990).

[11] A. Robinson, "Nonstandard Analysis," North-Holland, Amsterdam - New York - Oxford (1966).

[12] D. Laugwitz, Infinitesimals in Physics, in: "Papers dedicated to Prof. I.Iliev's 70th birthday," Sofia (1984).

[13] J. Sebastião e Silva, Sur l'axiomatique des distributions et ses possible modeles, in: "Obras de J.S.Silva, vol. III," INIC, Lisboa (1982).

[14] A. Robinson, Function theory on some non-archimedean fields, Amer. Math. Monthly, 80 Supplement 87-109 (1973).

[15] A. H. Lightstone and A. Robinson, "Non-archimedean Fields and Asymptotic Expansions," North-Holland, Amsterdam - New York - Oxford (1975).

INTERPLAY BETWEEN SINGULARITIES ASYMPTOTICS AND GENERALIZED FUNCTIONS

Ram P. Kanwal

Pennsylvania State University
University Park, Pennsylvania, 16802, U.S.A.

Abstract

We present various aspects of generalized functions and their applications to wave propagation, asymptotic expansions and singular integral equations.

We start with the basic concepts of the sifting property of the delta function and the fact that it is the derivative of the Heaviside function. Then we build up the derivatives of the generalized functions across a surface of discontinuity. We demonstrate that this information is sufficient to derive the Leibniz integral formula and the transport theorems of wave fronts and interfaces. The theory built up in the process is found very useful in the study of various other aspects of the wave propagation. As an illustration we present the analysis of the delta function concentrated on the light cone.

Our next study is related to the asymptotic expansions. We demonstrate that the concept of the moment expansion of generalized functions makes the subject of asymptotic series extremely simple. We illustrate our theory by deriving the Watson lemma, the Laplace formula and the related results.

Finally, we discuss the distributional solution of integral equations. We introduce this subject with the help of the Wiener-Hopf integral equation.

1. SINGULAR SURFACES

We begin with the very basic formulas such as the sifting property of delta function $\delta(x-\xi)$:

$$\int_{-\infty}^{\infty} \delta(x-\xi)\phi(x)dx = \phi(\xi) \tag{1}$$

and the generalized derivative of the Heaviside function $H(x-\xi)$:

$$\frac{\overline{d}H(x-\xi)}{dx} = \delta(x-\xi), \tag{2}$$

where the bar stands for the generalized derivative.

Generalized Functions and Their Applications, Edited by
R.S. Pathak, Plenum Press, New York, 1993

The first elegant application of the above formula is the Leibniz formula

$$\frac{d}{dt}\int_{a(t)}^{b(t)} f(x,t)dx = \int_{a(t)}^{b(t)} \frac{\partial f(x,t)}{\partial t}dx$$
$$+f(b(t),t)b'(t) - f(a(t),t)a'(t), \tag{3}$$

where $f(x,t)$ is a continuous function in a closed region $a(t) \le x \le b(t)$. We extend this function to all (x,t) region by setting $f(x,t) = 0$, for the interval $x < a(t)$, and $x > b(t)$. This extended function $F(x,t)$ has jump discontinuities at $x = a(t)$ and $x = b(t)$ so that we can write it as

$$F(x,t) = f(x,t)H(x - a(t))H(b(t) - x). \tag{4}$$

Then our problem reduces to the evaluation of the quantity

$$\frac{d}{dt}\int_{-\infty}^{\infty} F(x,t)dx = \frac{d}{dt}\langle F(x,t), 1\rangle_x, \tag{5}$$

where by $\langle\ \rangle_x$ we imply that integration is with respect to x so that

$$\frac{d}{dt}\int_{-\infty}^{\infty} F(x,t)dx = \left\langle \frac{\overline{\partial}}{\partial t}(f(x,t)H(x - a(t)H(b(t) - x)), 1\right\rangle_x$$
$$= \left\langle H(x - a(t))H(b(t) - x)\frac{\partial f(x,t)}{\partial t} + \right.$$
$$\left. f(a(t),t)a'(t)\delta(x - a(t)) + f(b(t),t)b'(t)\delta(x - b(t)), 1\right\rangle_x,$$

which, with the help of the sifting property, becomes the required formula (1).

Now let $F(x)$ be a function which is smooth everywhere except a $x = \xi$ where it has a jump discontinuity of magnitude $a = [F] = F|_{\xi+} - F|_{\xi-}$. then the generalized derivative of $F(x)$ is

$$\frac{\overline{d}}{dt}F(x) = \frac{dF}{dx} + [F]\delta(x - \xi), \tag{6}$$

which is obtained in precisely the same fashion as we did in the previous example. This result extends to higher dimensions very easily. Let us restrict ourselves to the space $\mathbb{R}^3$ and let this space be separated by a two-dimensional curved surface $\Sigma(\underset{\sim}{x},t)$ in two parts which we call positive and negative. Here $\underset{\sim}{x} \in \mathbb{R}^3$ and t is the time. The basic distribution concentrated on this moving and deforming wave front is the delta function $\delta(\Sigma)$ whose action on a test function $\phi(\underset{\sim}{x},t) \in \mathcal{D}$ (the space of C^∞ functions with compact support) is

$$\langle \delta(\Sigma(\underset{\sim}{x},t)), \phi(\underset{\sim}{x},t)\rangle = \int_{-\infty}^{\infty}\int_{\Sigma(\underset{\sim}{x},t)} \phi(\underset{\sim}{x},t)dS(\underset{\sim}{x})dt, \tag{7}$$

where dS is the surface element on Σ. Thus, $\delta(\Sigma)$ is the single layer distribution. The second surface distribution is $\delta'(\Sigma)$ defined as

$$\delta'(\Sigma) = n_i\frac{\overline{\partial}}{\partial x_i}(\delta(\Sigma), \tag{8}$$

where n_i are the components of the unit normal $\underset{\sim}{n}$ which is oriented from the negative side to the positive side of Σ. More interesting and useful surface distribution is the normal derivative operator $d_n\delta(\Sigma)$ given by

$$\langle d_n\delta(\Sigma), \phi\rangle = -\int_{-\infty}^{\infty}\int_{\Sigma}\frac{d\phi}{dn}dS(\underset{\sim}{x})dt. \tag{9}$$

This is the dipole layer distribution. These three distributions are related as

$$d_n\delta(\Sigma) = \delta'(\Sigma) - 2\Omega\delta(\Sigma), \tag{10}$$

where Ω is the mean curvature of the surface Σ.

The derivative formulas for a vector function $\underset{\sim}{F}(\underset{\sim}{x},t)$ which has a jump discontinuity across $\Sigma(\underset{\sim}{x},t)$ have the forms similar to the one-dimensional formula (6). Indeed,

$$\frac{\overline{\partial}\underset{\sim}{F}}{\partial t} = \frac{\partial \underset{\sim}{F}}{\partial t} - G[\underset{\sim}{F}]\delta(\Sigma), \tag{11}$$

$$\overline{\text{grad}\underset{\sim}{F}} = \text{grad}\underset{\sim}{F} + \underset{\sim}{n}[\underset{\sim}{F}]\delta(\Sigma), \tag{12}$$

$$\overline{\text{div}\underset{\sim}{F}} = \text{div}\underset{\sim}{F} + \underset{\sim}{n}\cdot[\underset{\sim}{F}]\delta(\Sigma), \tag{13}$$

$$\overline{\text{curl}\underset{\sim}{F}} = \text{curl}\underset{\sim}{F} + \underset{\sim}{n}\times[\underset{\sim}{F}]\delta(\Sigma), \tag{14}$$

where $[\underset{\sim}{F}] = F_+ - \underset{\sim}{F}_-$ is the jump across the wave front Σ and G is the normal speed of this front.

Let us use this information to derive the famous transport equation for the wave fronts. Let $\Sigma(\underset{\sim}{x},t)$ be given locally by a smooth function $u(\underset{\sim}{x},t)=0$, then $n_i = U^{-1}(\partial u/\partial x^i)$ and $-G = U^{-1}(\partial u/\partial t)$, where $U = |\nabla u| = [(\partial u/\partial x^i)(\partial u/\partial x^i)]^{1/2}$ and i is summed. Now let $f(\underset{\sim}{x},t)$ be a smooth function defined in a moving solid $V(t)$ and assume that its boundary γ is a smooth moving hypersurface $\Sigma(t)$. We consider the function

$$F(t) = \int_{V(t)} f(\underset{\sim}{x},t)d\underset{\sim}{x} \tag{15}$$

and set $V(t) = V_+$, $\mathbb{R}^n\backslash\overline{V}(t) = V_-$, so that $\underset{\sim}{n}$ is the outward normal. Extend $f(\underset{\sim}{x},t)$ to all $\mathbb{R}^n$ by setting $f(\underset{\sim}{x},t)=0$ in V_-. Just as in the Leibniz formula, we write (15) as

$$F(t) = \langle f(\underset{\sim}{x},t), 1\rangle_{\underset{\sim}{x}}$$

and use formula (11) to get

$$\begin{aligned}\frac{dF}{dt} &= \langle\frac{\overline{\partial}}{\partial t}f(\underset{\sim}{x},t),1\rangle_{\underset{\sim}{x}}\\ &= \langle\frac{\partial f(\underset{\sim}{x},t)}{\partial t} - G[f]\delta(\Sigma),1\rangle_{\underset{\sim}{x}}.\end{aligned} \tag{16}$$

Then (15) and (16) yield the transport equation

$$\frac{d}{dt}\int_{V(t)} f(\underset{\sim}{x},t)dt = \int_{V(t)}\frac{\partial f(\underset{\sim}{x},t)}{\partial t}d\underset{\sim}{x} + \int_{\Sigma} G\,f\,dS. \tag{17}$$

The next concept that we want to discuss is the propagation of singular surfaces which carry the charge and current densities concentrated on them. As a first step we mention the derivatives of the delta function $\delta(\Sigma)$:

$$\frac{\bar{\partial}}{\partial t}(\delta(\Sigma(\underset{\sim}{x},t))) = -G\,\delta'(\Sigma(\underset{\sim}{x},t)), \tag{18}$$

and

$$\frac{\bar{\partial}}{\partial x^i}(\delta(\Sigma(\underset{\sim}{x},t))) = n_i\delta'(\Sigma(\underset{\sim}{x},t)), \tag{19}$$

where $\delta'(\Sigma)$ is defined by (8). Next, suppose that the wave front $\Sigma(\underset{\sim}{x},t)$ carries a single layer of strength $f(\underset{\sim}{x},t)$, i.e. $f\delta(\Sigma)$. Then

$$\frac{\bar{\partial}}{\partial t}(f\delta(\Sigma)) = \frac{\delta f}{\delta t}\delta(\Sigma) - G\,f\,\delta'(\Sigma), \tag{20a}$$

$$\frac{\bar{\partial}}{\partial x_i}(f\delta(\Sigma) = \frac{\delta f}{\delta x_i}\delta(\Sigma) + n_i f\delta'(\Sigma), \tag{20b}$$

where

$$\frac{\delta f}{\delta t} = \frac{\partial f}{\partial t} + G\frac{df}{dn} \tag{21}$$

is the time derivative as apparent to an observer moving with the front Σ. Similarly,

$$\frac{\delta f}{\delta x^i} = \frac{\partial f}{\partial x^i} - n_i\frac{df}{dn}, \tag{22}$$

is the surface derivative with respect to the Cartesian coordinates of the surrounding space. If the wave front has an edge or a curve of discontinuity Δ, then we need the generalized derivative $\bar{\delta}/\delta t$. In that case $\delta(\Delta)\delta(\Sigma) = \delta(\Delta)$.

The use of the above formulas enables us to derive the first order derivatives of the single layer $f\delta(\Sigma)$. Indeed,

$$\begin{aligned}\frac{\bar{\partial}}{\partial t}(f\delta(\Sigma)) &= \frac{\bar{\delta} f}{\delta t}\delta(\Sigma) - G\,f\,\delta'(\Sigma)\\ &(\frac{\bar{\delta} f}{\delta t} - 2\Omega\,f)\delta(\Sigma) - G\,f\,d_n\delta(\Sigma),\end{aligned} \tag{23}$$

$$\frac{\bar{\partial}}{\partial x_i}(f\delta(\Sigma)) = \left(\frac{\bar{\delta} f}{\delta x_i} + 2\,\Omega\,f\right)\delta(\Sigma) + n_i f\,d_n\delta(\Sigma). \tag{24}$$

The formulas for the $\overline{\mathrm{div}}(f\,\delta(\Sigma))$ and $\overline{\mathrm{curl}}(f\,\delta(\Sigma))$ can be easily read off from (24).

This analysis for the single and double layers can be readily extended to the general multilayers $f\,d_n^P\delta(\Sigma)$, where $d_n^P\delta(\Sigma)$ is the P-th order normal derivative operator

$$\langle d_n^P\delta(\Sigma),\ \phi(\underset{\sim}{x},t)\rangle = (-1)^m\int_{-\infty}^{\infty}\int_{\Sigma(t)}\frac{d^P\phi}{dn\,P}dS(\underset{\sim}{x})dt. \tag{25}$$

These results are expressed in terms of the second fundamental form $\underset{\sim}{\mu} = \mu_{ij} = (\delta n_i/\delta x_j)$

and its powers $\mu_{ij}^{(P)} = \mu_{ik}^{(P-1)}\mu_{kj}$. Observe that $-\mu_{ii} = -\frac{\delta n_i}{\delta x_i} = 2\Omega$, where i is summed. Moreover, $\mu_{it} = \left(\frac{\delta n_i}{\delta t}\right) = -\left(\frac{\delta G}{\delta x_i}\right)$, $\mu_{tt} = -(\delta G/\delta t)$ and similar results for higher order time derivatives.

An interesting application of these concepts is found in the study of the delta function concentrated on the light cone. For this purpose we consider the length in the Minkowski space $\mathbb{R}^4(\underset{\sim}{x}, t)$ and set $\underset{\sim}{y} = (\underset{\sim}{x}, t)$. Then

$$\langle \underset{\sim}{y}, \underset{\sim}{y} \rangle = t^2 - |\underset{\sim}{x}|^2 = t - r^2 = (t-r)(t+r). \tag{26}$$

Then $\delta(\underset{\sim}{y}, \underset{\sim}{y})$ is the delta function with support on the light cone and we have

$$\delta(\langle \underset{\sim}{y}, \underset{\sim}{y} \rangle) = \delta((t-r)(t+r)) = \frac{\delta(t-r)}{2r} + \frac{\delta(t+r)}{2r}. \tag{27}$$

The values of the relevant geometric quantities for the light cone $t - r = 0$ are

$$\left.\begin{array}{l} n_i = x_i/t, \quad \mu_{ij}^P = t^{-P}(\delta_{ij} - n_i n_j), \\ \Omega = -1/t, \quad G = 1, \quad \mu_{it} = \mu_{tt} = 0. \end{array}\right\} \tag{28}$$

The generalized derivative for the multilayers $f\, d_n^P \delta(t-r)$ are

$$\frac{\overline{\partial}}{\partial t}\left(f\, d_n^P \delta(t-r)\right) = -f\, d_n^{P+1}\delta(t-r) + \left(\frac{\delta f}{\delta t} + \frac{2f}{t}\right)\delta_n^P \delta(t-r), \tag{29}$$

and

$$\frac{\overline{\partial}}{\partial x_i}\left(f\, d_n^P \delta(t-r)\right) = f\, n_i d_n^{P+1}\delta(t-r)$$
$$+\sum_{M=0}^{P} \frac{P!}{M!}\frac{1}{t^{P-M}}\left(\frac{\delta f}{\delta x_i} - \frac{2n_i f}{t}\right) d_n^M \delta(t-r). \tag{30}$$

Many more results also follow.

2. ASYMPTOTICS

Our second topic is the distributional theory of asymptotic expansions. To appreciate the application of generalized functions we recall the behavior of the Laplace integral

$$I(\lambda) = \int_a^b \phi(x) e^{-\lambda h(x)} dx, \tag{31}$$

as $\lambda \to \infty$. If x_0 is the only global minimum of $h(x)$, then we have the Laplace formula

$$I(\lambda) \sim \left(\frac{2\pi}{\lambda h''(x_0)}\right)^{\frac{1}{2}} \phi(x_0) e^{\lambda h(x_0)}. \tag{32}$$

Now observe that if we can prove that $e^{-\lambda h(x)}$ has an asymptotic series of delta functions so that

$$e^{-\lambda h(x)} \sim \left(\frac{2\pi}{\lambda\, h''(x_0)}\right) e^{\lambda h(x_0)} \delta(x - x_0), \tag{33}$$

then we can substitute (33) in (31). The sifting property of the delta function then yields (32). This concept has helped us to build a distributional theory for the asymptotics.

To present our theory we first define the concept of moments. The moments μ_n of a function $F(x)$ are defined as

$$\mu_n = \langle F(x), x^n \rangle = \int_{-\infty}^{\infty} F(x) x^n \, dx. \tag{34}$$

Now consider a test function $\phi(x)$ whose Taylor series is

$$\phi(x) = \sum_{n=1}^{\infty} \phi^{(n)}(0) \frac{x^n}{n!}. \tag{35}$$

Then it follows from (34) that

$$\begin{aligned} \langle F(x), \phi(x) \rangle &= \langle F(x), \sum_{n=0}^{\infty} \phi^{(n)}(0) \frac{x^n}{n!} \rangle \\ &= \sum_{n=0}^{\infty} \phi^{(n)}(0) \frac{\mu_n}{n!}. \end{aligned} \tag{36}$$

Since $\phi^{(n)}(0) = (-1)^n \langle \delta^{(n)}(x), \phi(x) \rangle$, the above relation can be written as

$$\langle F(x), \phi(x) \rangle = \langle \sum_{n=0}^{\infty} (-1)^n \frac{\mu_n}{n!} \delta^{(n)}(x), \phi(x) \rangle, \tag{37}$$

and we find that

$$F(x) \sim \sum_{n=0}^{\infty} (-1)^n \mu_n \frac{\delta^{(n)}(x)}{n!}. \tag{38}$$

This is precisely what we are looking for in (33). We are now ready to present the basic results of asymptotic expansions:

(i) Watson's Lemma. Let us take $F(x) = H(x)e^{-x}$. Then

$$\mu_n = \int_0^{\infty} e^{-x} x^n \, dx = n!. \tag{39}$$

Substituting these values in (38) we get

$$H(x)e^{-x} \sim \sum_{n=0}^{\infty} (-1)^n n! \frac{\delta^{(n)}(x)}{n!} = \sum_{n=0}^{\infty} (-1)^n \delta^{(n)}(x). \tag{40}$$

Next we use the formula $\delta^{(n)}(\lambda x) = (1/\lambda^{n+1})\delta^{(n)}(x)$ and find from (40) that

$$H(x)e^{-\lambda x} \sim \sum_{n=0}^{\infty} (-1)^n \frac{\delta^{(n)}(x)}{\lambda^{n+1}}, \quad \text{as } \lambda + \infty. \tag{41}$$

For a smooth function $\phi(x)$, the above formula yields

$$\begin{aligned} \int_0^{\infty} e^{-\lambda x} \phi(x) dx &\sim \sum_{n=0}^{\infty} (-1)^n \langle \frac{\delta^{(n)}(x)}{\lambda^{n+1}}, \phi(x) \rangle \\ &= \frac{\phi(0)}{\lambda} + \frac{\phi'(0)}{\lambda^2} + \frac{\phi''(0)}{\lambda^2} + \cdots + \frac{\phi^{(N)}(0)}{\lambda^{N+1}} + \cdots, \end{aligned} \tag{42}$$

as $\lambda \to \infty$. This is precisely Watson's lemma.

(ii) The Laplace Formula. This formula is derived by starting with the function $f(x) = e^{-x^2}$. The moments of this function are

$$\mu_n = \int_{-\infty}^{\infty} e^{-x^2} x^n dx = \begin{cases} \Gamma(\frac{n+1}{2}), & n \text{ even} \\ 0 \quad , & n \text{ odd}. \end{cases} \tag{43}$$

so that

$$e^{-\lambda x^2} \sim \sum_{n=0}^{\infty} \frac{\Gamma\left(\frac{2n+1}{2}\right) \delta^{(2n)}(x)}{(2n)!\, \lambda^{\left(\frac{2n+1}{2}\right)}}, \quad \text{as } \lambda \to \infty. \tag{44}$$

Suppose now that a smooth function $h(x)$ has a minimum at $x = x_0$ so that $h'(x_0) = 0$, $h''(x_0) > 0$ and consider the Laplace integral

$$I(\lambda) = \int_{-\infty}^{\infty} e^{-\lambda h(x)} \phi(x) dx = \langle e^{-\lambda h(x)}, \phi(x) \rangle. \tag{45}$$

where $\phi(x)$ is so chosen that the support of $\phi(x)$ is a small enough neighborhood of x_0 so that it contains no other critical point of $h(x)$. Accordingly, we can find an increasing smooth function $\psi(x)$ with $\psi(x_0) = 0$, $\psi'(x_0) > 0$ for all $x \in \mathbb{R}$ such that

$$h(x) = h(x_0) + (\psi(x))^2,$$

in the support of $h(x)$. Then $h'(x) = 0 + 2\psi(x)\psi'(x)$ and $h'(x_0) = 2\psi(x_0)\psi'(x_0) = 0$. Therefore we set

$$u = \psi(x), \quad du = \psi'(x)dx \text{ or } dx = du/\psi'(x)$$

Then (45) becomes

$$I(\lambda) = e^{-\lambda h(x_0)} \int_{-\infty}^{\infty} e^{-\lambda(\psi(x))^2} \phi(x) \frac{1}{\psi'(x)} \psi'(x) dx$$

or

$$I(\lambda) = e^{-\lambda h(x_0)} \int_{-\infty}^{\infty} e^{-\lambda u^2} \phi_1(u) du, \tag{46}$$

where

$$\phi_1(u) = \frac{\phi(u)}{\psi'(u)}.$$

For $e^{-\lambda u^2}$ in the integral (46) we substitute the moment formula (44) and we have

$$I(\lambda) \sim e^{-\lambda h(x_0)} \int_{-\infty}^{\infty} \sum_{n=0}^{\infty} \frac{\Gamma(\frac{2n+1}{2})}{(2n)!} \frac{\delta^{2n}(u)}{\lambda^{\frac{2n+1}{2}}} \phi_1(u) du \tag{47}$$

as $\lambda \to \infty$.

The first term of the formula (47) is

$$\begin{aligned} I(\lambda) &\sim e^{\lambda h(x_0)} \int_{-\infty}^{\infty} \Gamma(\frac{1}{2}) \frac{\delta(u)}{\sqrt{\lambda}} \phi_1(u) du \\ &= e^{-\lambda h(x_0)} \sqrt{\pi} \frac{\phi_1(0)}{\sqrt{\lambda}} \\ &= e^{-\lambda h(x_0)} \sqrt{\pi} \frac{1}{\sqrt{\lambda}} \frac{\phi(x_0)}{\psi'(x_0)}. \end{aligned} \tag{48}$$

When we substitute these values in (38) we get

$$e^{-x^2} \sim \sum_{n=0}^{\infty} \frac{\Gamma\left(\frac{2n+1}{2}\right)\delta^{(2n)}(x)}{(2n)!},$$

To put this formula in the form (32) of the Laplace formula, we observe that

$$h''(x) = 2(\psi'(x))^2 + 2\psi(x)\psi''(x),$$

so that

$$h''(x_0) = 2(\psi'(x_0)^2.$$

When we substitute this value of $\psi'(x_0)$ in (48) we recover the formula (32), namely

$$I(\lambda)\lambda \sim e^{-\lambda h(x_0)} \left(\frac{2\pi}{h''(x_0)}\right)^{\frac{1}{2}} \frac{\phi(x_0)}{\sqrt{\lambda}}.$$

(iii) Oscillatory Integrals. In this case we have to evaluate the integral

$$I(\lambda) = \int_{-\infty}^{\infty} e^{i\lambda h(x)}\phi(x)dx, \tag{49}$$

as $\lambda \to \infty$. This integral is processed in the same fashion as the Laplace integral. Indeed, the steps leading (49) to the relation

$$I(\lambda) = \int_{-\infty}^{\infty} e^{i\lambda u^2}\phi_1(u)du \tag{50}$$

are the same. The slight difference arises in the values of the moments μ_n which now are

$$\mu_n = \int_{-\infty}^{\infty} e^{i\lambda u^2} u^n du = \begin{cases} \Gamma\left(\frac{n+1}{2}\right) e^{\pi i\left(\frac{2n+1}{4}\right)} & , \quad n \text{ even}, \\ 0 & , \quad n \text{ odd}. \end{cases} \tag{51}$$

This yields the moment expansion of $e^{i\lambda u^2}$ as

$$e^{i\lambda u^2} \sim \sum_{n=0}^{\infty} \frac{\Gamma(\frac{2n+1}{2})e^{\pi i(\frac{2n+1}{4})}}{(2n)!} \frac{\delta^{(2n)}(u)}{\lambda^{\frac{2n+1}{2}}}. \tag{52}$$

When we substitute this value for $e^{i\lambda u^2}$ in the integral (50) we obtain

$$\begin{aligned} I(\lambda) \sim e^{-i\,h(x_0)} \int_{-\infty}^{\infty} \sum_{n=0}^{\infty} & \frac{\Gamma(\frac{2n+1}{2} e^{-\pi i(\frac{2n+1}{4})}}{(2n)!} \\ & \times \frac{\delta^{(2n)}(u)}{\lambda^{\frac{2n+1}{2}}} \phi_1(u)dn. \end{aligned} \tag{53}$$

The first term of this formula yields

$$I(\lambda) \sim e^{i(\lambda h(x_0) + \frac{\pi}{4})} \left(\frac{2\pi}{h''(x_o)}\right)^{\frac{1}{2}} \frac{\phi(x_0)}{\sqrt{\lambda}}. \tag{54}$$

Classically this formula is derived by the method of stationary phase.

The interesting feature of this approach is that we get the entire asymptotic expansions right away. Then we can take as many terms as are needed.

At this stage we mention the kinds of spaces that we have used in our studies.

1. The Schwartz space $S = S(\mathbb{R}^n)$ is the space of smooth function $\phi(\underset{\sim}{x})$ on $\mathbb{R}^n$ such that $x^\alpha D^\beta \phi(x)$ vanishes at ∞ for all α and β. Its dual space S' is the space of tempered distributions.
2. The space of all smooth functions on $\mathbb{R}^n$ is denoted by $\mathcal{E} = \mathcal{E}(\mathbb{R}^n)$ and its dual space $\mathcal{E}'$ of distributions of compact support on $\mathbb{R}^n$.
3. Intermediate spaces $\mathcal{O}_M$ is defined as the space of smooth functions that, together with all their derivatives, are of polynomial growth at infinity. $\mathcal{O}_C$ is the subspace of $\mathcal{O}_M$ for which the bounding polynomials are of the same degree for all derivatives. $\mathcal{O}'_C$ and $\mathcal{O}'_M$ are the corresponding dual spaces.

Then there are some more suitable subspaces which are useful in number theory and other applications.

3. WIENER-HOPF INTEGRAL EQUATION

The integral equation of the type

$$\int_0^\infty k(t-s)g(s)ds = \lambda\, g(t) + f(t), \quad 0 < t < \infty, \tag{55}$$

is called the Wiener-Hopf integral equation. Its distinguishing features are the difference kernel and the semi-infinite interval.

Observe that to evaluate the integral on the left for positive values of s we need the value of the kernel $k(t)$ for $-\infty < t < \infty$. Secondly, to apply Fourier transform, we must extend the integral equation for all values of its argument. But then we must do something compatible about the right side of (55). This is achieved as follows.

We set

$$\int_{-\infty}^\infty k(t-s)g(s)ds = \begin{cases} \lambda g(t) + f(t), & 0 < t < \infty\,, \\ u(t) \quad , & -\infty < t < 0. \end{cases} \tag{56}$$

Now we extend the definition of the function g, f and u so that we have

$$g = 0,\ t < 0; \quad f = 0,\ t < 0; \quad u = 0,\ t > 0.$$

This enables us to write (56) as

$$\int_{-\infty}^\infty k(t-s)g_+(s)ds = \lambda g_+(t) + f_+(t) + u_-(t), \ -\infty < t < \infty, \tag{57}$$

where the subscripts $\pm$ indicate the $\pm$ half-line on which the function is non-vanishing.

Let us see now the way the generalized functions make their way into this analysis. The Wiener-Hopf integral equation

$$\int_0^\infty e^{-|t-s|} g_+(s)ds = 1, \qquad t \geq 0, \tag{58}$$

does not have a classical solution. We have proved that its distributional solution is

$$g_+(s) = \frac{1}{2}H(s) - \frac{1}{2}\delta'(s) + C(\delta(s) + \delta'(s)), \tag{59}$$

where C is an arbitrary constant.

For the general equation

$$\int_0^\infty e^{-|t-s|} g_+(s) = f_+(t) \quad , \quad t \geq 0, \tag{60}$$

we have proved that the solution is

$$g_+(s) = \frac{1}{2}(f_+(s) - f''_+(s)) + C(\delta(s) + \delta'(s)). \tag{61}$$

Motivated by these results we have presented a general distributional theory of the Wiener-Hopf integral and integro-differential equations. Earlier we have presented the distributional solutions of various other kinds of singular single and dual integral equations.

REFERENCES

COPSON, E. T., 1965, Asymptotic Expansions, Cambridge University Press, Cambridge.

ESTRADA, R. AND KANWAL, R. P., 1985, Distributional solutions of singular integral equations, J. Integ. Eqns., 8:41–85.

ESTRADA, R. AND KANWAL, R. P., 1985, Distributional solutions of dual integral equations of Cauchy, Abel and Titchmarsch types, J. Integ. Eqns., 9:272–305.

ESTRADA, R. AND KANWAL, R. P., 1987, Higher order fundamental forms of a surface and their applications to wave propagation and generalized derivatives, Circo Mat. Palermo, Ser. 2, 36:27–62.

ESTRADA, R. AND KANWAL, R. P., 1988, An analysis for the delta function with support on the light cone, J. Phys. A: Math. Gen., 21:2667–2675.

ESTRADA, R. AND KANWAL, R. P., 1990, A distributional theory for asymptotic expansions, Proc. Roy. Soc. Lond. A, 428, 399–430.

ESTRADA, R. AND KANWAL, R. P., 1991, Non-classical derivation of the transport theorems for wave fronts, J. Math. Analy. Applic., 159:290–297.

ESTRADA, R. AND KANWAL, R. P., 1992, The asymptotic expansion of certain multi-dimensional generalized functions, J. Math. Anal. Applics., 163:264–283.

ESTRADA, R. AND KANWAL, R. P., *1991, Distributional solution of the Wiener-Hopf integral and integro-differential equations, J. Integ. Eqns. Applic., 3:489–514.*

KANWAL, R. P., 1983, Generalized Functions, Theory and Technique, Academic Press, New York.

WIENER, N. AND HOPF, E., 1931, Übewr eine klausse singuläres Integral gleiching, 30/32:696–706.

HYPERFUNCTIONS, ULTRADISTRIBUTIONS

AND MICROFUNCTIONS

Hikosaburo Komatsu

Department of Mathematics, Faculty of Science
University of Tokyo
Hongo, Tokyo, 113 Japan

1. DISTRIBUTIONS, ULTRADISTRIBUTIONS AND HYPERFUNCTIONS

As exemplified in the introduction of L. Schwartz' book [24] individual generalized functions had been known before he introduced the distributions $\mathcal{D}'(\Omega)$ in 1948-51. Dirac's δ function and Hadamard's finite part of $x_+^\alpha, \alpha < -1$, are famous. More generally P. Lévy [16] had determined the dual of $C^m([a, b])$. Yet the true history of generalized functions started with Schwartz.

He defined the space $\mathcal{D}'(\Omega)$ of all *distributions* on an open set Ω in $\mathbf{R}^n$ as the dual

$$\mathcal{D}'(\Omega) := (\mathcal{D}(\Omega))', \tag{1}$$

i.e. the space of all continuous linear functionals on the locally convex space

$$\mathcal{D}(\Omega) := \{\varphi(x) \in C^\infty(\Omega) \ ; \ \mathrm{supp}\ \varphi \ \text{compact}\}. \tag{2}$$

By this definition mathematicians had the power of manipulating generalized functions just like functions or more easily. The characteristic feature is that the distributions $\mathcal{D}'(\Omega), \Omega \subset \mathbf{R}^n$, form a sheaf on which differential operators $P(x, \partial) = \sum_{|\alpha| \leqq m} a_\alpha(x)\partial^\alpha$ with C^∞ coefficients act as sheaf homomorphisms.

A *sheaf* $\mathcal{F}$ is by definition a collection of linear spaces $\mathcal{F}(\Omega)$ defined for all open sets Ω together with linear mappings $\mathcal{F}(\Omega) \to \mathcal{F}(\Omega_1)$ defined for all pairs $\Omega \supset \Omega_1$ such that we have for any open covering $\{\Omega_\lambda\}$ of Ω

$$f \in \mathcal{F}(\Omega), \ \ f|_{\Omega_\lambda} = 0 \ \text{ for all } \lambda \Rightarrow f = 0, \tag{S.1}$$

$$f_\lambda \in \mathcal{F}(\Omega_\lambda), f_\lambda|_{\Omega_\lambda \cap \Omega_\mu} = f_\mu|_{\Omega_\lambda \cap \Omega_\mu} \Rightarrow \exists f \in \mathcal{F}(\Omega) \ f_\lambda = f|_{\Omega_\lambda}. \tag{S.2}$$

Generalized Functions and Their Applications, Edited by
R.S. Pathak, Plenum Press, New York, 1993

A *sheaf homomorphism* $h : \mathcal{F} \to \mathcal{G}$ is a collection of linear mappings $h(\Omega) : \mathcal{F}(\Omega) \to \mathcal{G}(\Omega)$ compatible with the restriction mappings.

These properties exhibit the local nature of the distributions and the action on them of linear differential operators.

Ten years later in 1958-60 M. Sato [20], [21] introduced the sheaf $\mathcal{B}$ of hyperfunctions. His definition was completely different. Let Ω be an open set in $\mathbf{R}^n$ and V an open set in $\mathbf{C}^n$ containing Ω as a closed set. Then the space $\mathcal{B}(\Omega)$ of *hyperfunctions* on Ω is defined to be the n-th relative cohomology group with coefficient in the sheaf $\mathcal{O}$ of holomorphic functions over $\mathbf{C}^n$:

$$\mathcal{B}(\Omega) := H^n_\Omega(V, \mathcal{O}) = H^n(V, V \backslash \Omega ; \mathcal{O}). \tag{3}$$

At approximately same time Roumieu [19] and Beurling (c.f. [2]) introduced an infinitely many classes of generalized functions between $\mathcal{D}'$ and $\mathcal{B}$. They are called *ultradistributions*. The most important among them are the Gevrey classes of ultradistributions. We will restrict ourselves to these classes for the sake of simplicity. Let $1 < s < \infty$. Then the space $\mathcal{D}^{(s)\prime}(\Omega)$ (resp. $\mathcal{D}^{\{s\}\prime}(\Omega)$) of ultradistributions of class (s) (resp. class $\{s\}$) on $\Omega \subset \mathbf{R}^n$ is defined to be the dual of the locally convex space

$$\mathcal{D}^{(s)}(\Omega) := \{\varphi(x) \in \mathcal{D}(\Omega);\, \sup \frac{|\partial^\alpha \varphi(x)|}{h^{|\alpha|}|\alpha|!^s} < \infty \quad \text{for all} \quad h > 0\} \tag{4}$$

(resp.

$$\mathcal{D}^{\{s\}}(\Omega) := \{\varphi(x) \in \mathcal{D}(\Omega);\, \sup \frac{|\partial^\alpha \varphi(x)|}{h^{|\alpha|}|\alpha|!^s} < \infty \quad \text{for some} \quad h\}). \tag{5}$$

These classes of generalized functions form sheaves under the natural restriction mappings. We have the following inclusion relations for $1 < s < t < \infty$:

$$\mathcal{D}' \subset \mathcal{D}^{(t)\prime} \subset \mathcal{D}^{\{s\}\prime} \subset \mathcal{D}^{(s)\prime} \subset \mathcal{B}. \tag{6}$$

Let $\mathcal{E}^{(s)}(\Omega)$ (resp. $\mathcal{E}^{\{s\}}(\Omega)$) be the corresponding classes of smooth functions without support condition, and let $\mathcal{E}$ and $\mathcal{A}$ be the sheaves of infinitely differentiable and real analytic functions respectively. Then we have the opposite inclusions:

$$\mathcal{E} \supset \mathcal{E}^{(t)} \supset \mathcal{E}^{\{s\}} \supset \mathcal{E}^{(s)} \supset \mathcal{A}. \tag{7}$$

It is easy to prove the existence of hyperfunction solutions u to linear differential equations

$$P(x, \partial)u(x) = f(x) \tag{8}$$

with real analytic coefficients $a_\alpha(x)$ (e.g. [12], [4]). We are interested in the problem of regularity of a given solution. For example, it is known that every solution u in $\mathcal{D}^{\{2\}\prime}$ of the heat equation $\partial_t u = \Delta_x u$ is in $\mathcal{E}^{\{2\}}$ but there are solutions in $\mathcal{D}^{(2)\prime}$ which are not smooth.

To solve this problem we need a representation of the solution u independent of the class of generalized functions. This was not easy because the definition of hyperfunctions was totally different from the other.

2. HYPERFUNCTIONS AS BOUNDARY VALUES OF HARMONIC FUNCTIONS

The background of Sato's definition is his understanding of hyperfunctions as sums of (not necessarily topological) boundary values of holomorphic functions. This is a very natural approach but we have technical difficulties in the higher dimensional case. Therefore we will instead employ harmonic functions as in Real Analysis [14], [15].

We denote by $\mathcal{P}$ the sheaf of harmonic functions on $\mathbf{R}^{n+1}$ and by $\mathcal{P}_0$ its subsheaf of all harmonic functions vanishing on $\mathbf{R}^n$.

Definition 1. We define the space $\mathcal{B}(\Omega)$ of *hyperfunctions* on an open set Ω in $\mathbf{R}^n$ as the quotient space

$$\mathcal{B}(\Omega) := \mathcal{P}(V)/\mathcal{P}_0(\widetilde{V}), \tag{9}$$

where V is an open set in $\mathbf{R}^{n+1}_+ := \{(x,t); x \in \mathbf{R}^n,\ t > 0\}$ such that $\widetilde{V} = V \cup \Omega \cup \overline{V}$ is open in $\mathbf{R}^{n+1}$ and contains Ω as a closed set. Here $\overline{V} := \{(x,t) \in \mathbf{R}^{n+1};\ (x,-t) \in V\}$ is the mirror image of V.

If $H(x,t) \in \mathcal{P}(V)$, we denote its class in $\mathcal{B}(\Omega)$ by $H(x,+0)$.

In order that this definition makes sense we have to prove its independence of V. This follows from the following Mittag-Leffler theorem for harmonic functions.

Lemma 1. *Let $\{V_\lambda\}$ be a family of open sets in $\mathbf{R}^{n+1}$. If harmonic functions $H_{\lambda\mu} \in \mathcal{P}(V_\lambda \cap V_\mu)$ satisfy*

$$H_{\lambda\mu} + H_{\mu\nu} + H_{\nu\lambda} = 0 \tag{10}$$

for all $V_\lambda \cap V_\mu \cap V_\nu \neq \phi$, then there are $H_\lambda \in \mathcal{P}(V_\lambda)$ such that

$$H_{\lambda\mu} = H_\mu - H_\lambda \quad \text{on} \quad V_\lambda \cap V_\mu. \tag{11}$$

This is proved in the same way as the Mittag-Leffler theorem for holomorphic functions of one variable (see e.g. [9]). The following is also an easy consequence of Lemma 1.

Theorem 1. *$\mathcal{B}(\Omega)$, $\Omega \subset \mathbf{R}^n$, form a flabby sheaf, i.e. the restriction mappings are surjective.*

Given an element $f \in \mathcal{F}(\Omega)$ of a sheaf $\mathcal{F}$, there is the maximal open subset Ω_1 of Ω on which it vanishes by Axiom (S.1). Its complement is called the *support* of f and denoted by supp f. We use the notation

$$\mathcal{F}_S(\Omega) := \{f \in \mathcal{F}(\Omega);\ \operatorname{supp} f \subset S\}. \tag{12}$$

The general hyperfunctions can not be defined by duality but hyperfunctions with compact support can. Let K be a compact set in an open set Ω in $\mathbf{R}^n$. If $f(x) = H(x,+0) \in \mathcal{B}_K(\Omega)$ and $\varphi(x) \in \mathcal{A}(K)$, we define the pairing

$$\langle \varphi, f \rangle = \int_\Omega \varphi(x) f(x) dx$$

as follows. Solve the Cauchy problem

$$\begin{cases} (\Delta_x + \partial_t^2)\Phi(x,t) = 0, \\ \Phi(x,0) = 0, \\ \partial_t\Phi(x,0) = \varphi(x). \end{cases} \tag{13}$$

Then

$$\int_\Omega \varphi(x)f(x)dx := \int_S \left(\frac{\partial\Phi}{\partial n}H - \Phi\frac{\partial H}{\partial n}\right) dS, \tag{14}$$

where S is a surface in the intersection of V and the domain of Φ with the boundary ∂S in $\Omega\backslash K$. It does not depend on H nor on S by Green's formula.

Under this pairing we have the following duality theorem.

Theorem 2.

$$\mathcal{B}_K(\Omega) \cong (\mathcal{A}(K))'. \tag{15}$$

Here the space $\mathcal{A}(K)$ of the germs of real analytic functions on a neighborhood of K is regarded as a locally convex space by its inductive limit representation

$$\mathcal{A}(K) = \varinjlim \mathcal{O}(U) \tag{16}$$

as U ranges over the open neighborhoods of K in $\mathbf{C}^n$.

Theorem 2 is derived from the Grothendieck duality [8]:

Lemma 2. *If K is a compact set in an open set $\widetilde{V}$ in $\mathbf{R}^{n+1}$, then*

$$(\mathcal{P}(K))' \cong \mathcal{P}(\widetilde{V}\backslash K)/\mathcal{P}(\widetilde{V}). \tag{17}$$

For any hyperfunction $f \in \mathcal{B}_K$ with compact support a defining harmonic function $H(x,t)$ is obtained as its *Poisson integral*

$$H(x,t) = \int P_t(x-y)f(y)dy, \tag{18}$$

where

$$P_t(x) = \frac{2}{\omega_{n+1}} \frac{t}{(x^2+t^2)^{\frac{n+1}{2}}} = \frac{1}{(2\pi)^n}\int_{\mathbf{R}^n} e^{-t|\xi|+ix\xi}d\xi. \tag{19}$$

If we topologize $\mathcal{A}(\Omega)$ for open Ω by

$$\mathcal{A}(\Omega) = \varprojlim_{K\Subset\Omega} \mathcal{A}(K), \tag{20}$$

then we have the duality

$$(\mathcal{A}(\Omega))' \cong \mathcal{B}_{\text{cp}}(\Omega), \tag{21}$$

where

$$\mathcal{B}_{\text{cp}}(\Omega) := \{f \in \mathcal{B}(\Omega); \text{supp } f \text{ compact}\}. \tag{22}$$

Theorems 1 and 2 characterize the sheaf $\mathcal{B}$ of hyerfunctions. Hence our hyperfunctions are isomorphic to Sato's.

3. CHARACTERIZATION OF ULTRADISTRIBUTIONS

Since the imbedding $\mathcal{A}(\Omega) \hookrightarrow \mathcal{E}^*(\Omega)$, where $* = \phi$ or (s) or $\{s\}$, is a continuous linear operator with dense image, we have by duality the imbedding $\mathcal{D}^{*\prime}_{\text{cp}}(\Omega) \hookrightarrow \mathcal{B}_{\text{cp}}(\Omega)$. Since this preserves the support, it is uniquely extended to a sheaf isomorphism $\mathcal{D}^{*\prime} \hookrightarrow \mathcal{B}$.

Theorem 3. *Let $\Omega \subset \mathbf{R}^n$ and $V \subset \mathbf{R}^{n+1}_+$ be as in Definition 1. The following conditions are equivalent for $H(x,t) \in \mathcal{P}(V)$:*

(a) $$H(x,+0) \in \mathcal{D}^{*\prime}(\Omega);$$

(b) $H(x,t)$ *converges in* $\mathcal{D}^{*\prime}(\Omega)$ *as* $t \to 0$,

(c) $\{H(x,t);\ 0 < t < \varepsilon\}$ *is bounded in* $\mathcal{D}^{*\prime}(\Omega)$;

(d) *For any compact set K in Ω there are a growth function $G(t)$ of class $*$ and constants C and r such that*

$$\sup_{x \in K} |H(x,t)| \leqq C \exp G(t), \quad 0 < t < r. \tag{23}$$

Then $H(x,t)$ converges to $H(x,+0)$ in $\mathcal{D}^{*\prime}(\Omega)$ as t tends to zero.

Here the growth function $G(t)$ is defined by

$$G(t) := \begin{cases} -h \ \log t \quad \text{for some } h \text{ if } * = \phi, \\ (ht)^{-\frac{1}{s-1}} \quad \text{for some } h \text{ if } * = (s), \\ \sup \log \dfrac{p!}{t^p H_p p!^s} \quad \text{with } H_p = h_1 \cdots h_p \\ \qquad \text{for an } h_p \nearrow \infty \text{ if } * = \{s\}. \end{cases} \tag{24}$$

In the one dimensional case Sato's definition of hyperfunctions reduces to the quotient representation

$$\mathcal{B}(\Omega) = \mathcal{O}(V \backslash \Omega)/\mathcal{O}(V) \tag{25}$$

for an (and any) open set V in $\mathbf{C}$ containing Ω as a closed set. Then we have given in [13] a similar characterization theorem of ultradistributions in terms of defining holomorphic functions. Theorem 3 can be proved by modifying it slightly [14].

Similarly we have

Theorem 4. *Let Ω and V be as above. The following conditions are equivalent for $H(x,t) \in \mathcal{P}(V)$:*

(a) $$H(x,+0) \in \mathcal{E}^*(\Omega);$$

(b) $H(x,t)$ *converges in* $\mathcal{E}^*(\Omega)$ *as* $t \to 0$;

(c) $\{H(x,t); 0 < t < \varepsilon\}$ *is bounded in* $\mathcal{E}^*(\Omega)$;

(d) *For any ultradifferential operator $P(\partial)$ of class $*$ $P(\partial_x)H(x,t)$ is locally bounded near Ω.*

Here

$$\mathcal{P}(\partial) = \sum_{|\alpha|=0}^{\infty} a_\alpha \partial^\alpha, \quad a_\alpha \in \mathbf{C}, \tag{26}$$

is called an *ultradifferential operator of class* $*$ if

$$a_\alpha = 0 \text{ for sufficiently large } |\alpha| \text{ if } * = \phi; \tag{27}$$

$$|a_\alpha| \leqq C\, k^{|\alpha|}/|\alpha|!^s \tag{28}$$

with constants k and C if $* = (s)$, for any $k > 0$ with a constant C if $* = \{s\}$.

4. MICROFUNCTIONS AS SINGULARITIES OF HOLOMORPHIC FUNCTIONS

In 1969 Sato [22] introduced microfunctions as a sheaf $\mathcal{C}$ over the cotangent sphere bundle

$$S^*\mathbf{R}^n := (T^*\mathbf{R}^n \backslash \mathbf{R}^n)/\mathbf{R}_+ \cong \mathbf{R}^n \times S^{n-1} \tag{29}$$

such that

$$0 \longrightarrow \mathcal{A} \longrightarrow \mathcal{B} \overset{\text{sp}}{\longrightarrow} \pi_*\mathcal{C} \longrightarrow 0 \tag{30}$$

is exact, where $\pi : S^*\mathbf{R}^n \longrightarrow \mathbf{R}^n$ is the canonical projection, so that we have the isomorphisms

$$\mathcal{B}(\Omega)/\mathcal{A}(\Omega) \cong \mathcal{C}(\pi^{-1}\Omega), \ \Omega \subset \mathbf{R}^n. \tag{31}$$

Sato-Kawai-Kashiwara [23] defined it as the n-th derived sheaf

$$\mathcal{C} := \mathcal{H}^n_{S^*\mathbf{R}^n}(\pi^{-1}\mathcal{O})^a \tag{32}$$

over the comonoidal transformation $(\mathbf{C}^n \backslash \mathbf{R}^n) \coprod S^*\mathbf{R}^n$, which is endowed with a non-Hausdorff topology. a stands for the antipodal mapping in $S^*\mathbf{R}^n$.

We denote by $SS\,u$ the support $\operatorname{supp} \operatorname{sp} u$ of the image of $u \in \mathcal{B}(\Omega)$ under the spectral mapping sp, and call it the *singularity spectrum.* If u is a distribution, $SS\,u$ coincides essentially with the analytic wave front set $WF_A(u)$ of Hörmander [10]:

$$SS\,u = \operatorname{supp} \operatorname{sp} u = WF_A(u)/\mathbf{R}_+\,. \tag{33}$$

We will give here a much more elementary definition of microfunctions. Let

$$D\mathbf{R}^n := \{x + iy \in \mathbf{C}^n; |y| < 1\} \tag{34}$$

be the tube domain with the unit ball as its base. We identify the cotangent sphere bundle $S^*\mathbf{R}^n$ with its boundary with the antipodal coordinates :

$$S^*\mathbf{R}^n := (\partial D\mathbf{R}^n)^a = \{x - i\omega; x \in \mathbf{R}^n, |\omega| = 1\}. \tag{35}$$

Definition 2. We define the sheaf $\mathcal{C}$ of microfunctions by the exact sequence

$$0 \longrightarrow \mathcal{O}\,|_{[D\mathbf{R}^n]} \longrightarrow \mathcal{O}\,|_{D\mathbf{R}^n} \longrightarrow \mathcal{C}^a \longrightarrow 0 \tag{36}$$

of sheaves, where $[\mathbf{DR}^n]$ denotes the closure $\mathbf{DR}^n \cup (S^*\mathbf{R}^n)^a$ of $\mathbf{DR}^n$, and $\mathcal{O}\mid_A$ the sheaf which assigns to each open set W in $\mathbf{C}^n$ the space $\mathcal{O}(W\cap A) = \varinjlim_{V\supset W\cap A} \mathcal{O}(V)$ for a locally closed set A.

Namely a microfunction near $(x,\omega)\in S^*\mathbf{R}^n$ is the singularity of a holomorphic function defined on the intersection of $\mathbf{DR}^n$ and a neighborhood of $x-i\omega$ modulo the holomorphic functions on a neighborhood of $x-i\omega$.

The following theorem is an easy consequence of the fact that $\mathbf{DR}^n$ is a strictly pseudo-convex domain [1], [14], [15].

Theorem 5. *The sheaf $\mathcal{C}$ of microfunctions is flabby.*

In order to define the spectral mapping $\mathrm{sp}:\mathcal{B}\longrightarrow \pi_*\mathcal{C}$, we make use of the following result due to Siciak [25], [15].

Lemma 3. *Let Ω be an open set in $\mathbf{R}^n$. If we take the open set V in Definition 1 so large that every closed unit ball in $\mathbf{R}^{n+1}_+ \cup \mathbf{R}^n$ osculating at an $x\in\Omega$ is included in V except for x, then every $H(x,t)\in\mathcal{P}(V)$ has an analytic extension $H(x+iy,1)$ to $D\Omega=\{x+iy;\ x\in\Omega, |y|<1\}$. The extension $H(x+iy,1)$ is holomorphic on $[D\Omega]=\{x+iy;\ x\in\Omega,|y|\leqq 1\}$ if and only if the hyperfunction $H(x,+0)$ is real analytic on Ω.*

Definition 3. The class $[H(x+iy,1)]$ in $\mathcal{C}^a(\pi^{-1}\Omega)$ is defined to be the image $\mathrm{sp}\, f$ of $f(x)=H(x,+0)\in\mathcal{B}(\Omega)$:

$$\mathrm{sp}\, H(x,+0)\mid_{(x,\omega)} := [H(x+iy,1)]_{y=-\omega}\,. \tag{37}$$

Theorem 6. *Sato's sequence*

$$0\longrightarrow \mathcal{A}\longrightarrow \mathcal{B}\xrightarrow{\mathrm{sp}} \pi_*\mathcal{C}\longrightarrow 0 \tag{30}$$

is exact.

Lemma 3 proves the theorem except for the surjectivity of the spectral mapping sp. We proved it in [14] by showing that if $K(z)\in\mathcal{O}(\mathbf{DR}^n)$ and if $\Omega_0\supset\Omega$ is a sufficiently large relatively compact domain in $\mathbf{R}^n$, then the Poisson integral

$$H(z,t) := \int_{\Omega_0} P_{t-1}(z-w)K(w)dw$$

has an analytic continuation to a harmonic function in $\mathcal{P}(V)$ for an open set V in $\mathbf{R}^{n+1}_+$ such that $V\cup\Omega\cup\overline{V}$ is an open set containing Ω as a closed set.

Definition 4. We define the sheaf $\mathcal{C}^*$ of *microfunctions of class* $*$ (resp. $\mathcal{C}_*$ of *microfunctions of regular class* $*$) by the exact sequence of sheaves

$$0\longrightarrow \mathcal{O}\mid_{[\mathbf{DR}^n]}\longrightarrow \mathcal{O}^*\longrightarrow (\mathcal{C}^*)^a\longrightarrow 0 \tag{38}$$

$$(\text{resp. } 0\longrightarrow \mathcal{O}\mid_{[\mathbf{DR}^n]}\longrightarrow \mathcal{O}_*\longrightarrow (\mathcal{C}_*)^a\longrightarrow 0), \tag{39}$$

where $\mathcal{O}^*$ (resp. $\mathcal{O}_*$) is the sheaf over $\mathbf{C}^n$ of holomorphic functions on $D\mathbf{R}^n$ with the growth condition (d) of class $*$ in Theorem 3 (resp. with the regularity condition (d) of Theorem 4) as $x + iy$ approaches the boundary $(S^*\mathbf{R}^n)^a$. We write $\mathcal{C}^{(\infty)}$ and $\mathcal{C}_{(\infty)}$ if $* = \phi$ to avoid confusion.

Then we have the following theorem by the proof of the previous theorem.

Theorem 7. *We have the exact sequence of sheaves:*

$$0 \longrightarrow \mathcal{A} \longrightarrow \mathcal{D}^{*\prime} \longrightarrow \pi_*\mathcal{C}^* \longrightarrow 0, \tag{40}$$

$$0 \longrightarrow \mathcal{A} \longrightarrow \mathcal{E}^* \longrightarrow \pi_*\mathcal{C}_* \longrightarrow 0. \tag{41}$$

The sheaves $\mathcal{C}^{(\infty)}$ and $\mathcal{C}_{(\infty)}$ had been introduced by Bony [3] and Bengel-Schapira [1]. De Roever [5] and Eida [6], [7] considered the other cases.

A sheaf $\mathcal{F}$ is said to be *supple* [1] if for any open set Ω and closed sets F_1 and F_2 in Ω we have

$$\mathcal{F}_{F_1 \cup F_2}(\Omega) = \mathcal{F}_{F_1}(\Omega) + \mathcal{F}_{F_2}(\Omega). \tag{42}$$

The following theorem is due to Bengel-Schapira [1] if $* = \phi$ and to Eida [6] in the other cases.

Theorem 8. *The sheaves $\mathcal{C}^*$ and $\mathcal{C}_*$ are supple.*

5. BOUNDARY VALUES OF HOLOMORPHIC FUNCTIONS ON TUBOIDS

Let Ω be an open set, and Γ a proper convex cone in $\mathbf{R}^n$. An open set W in $\mathbf{C}^n$ is said to be a *tuboid of profile* $\Omega + i\Gamma$ if $W \subset \Omega + i\Gamma$ and if for any $\Omega_1 \Subset \Omega$ and $\Gamma_1 \Subset \Gamma$ there is an $r > 0$ such that $W \supset \Omega_1 + i(\Gamma_1 \cap B_r)$, where B_r is the ball of radius r.

We write $F(z) \in \mathcal{O}(\Omega + i\Gamma 0)$ if F is a holomorphic function on a tuboid of profile $\Omega + i\Gamma$. We define the boundary value operator

$$b_\Gamma : \mathcal{O}(\Omega + i\Gamma 0) \longrightarrow \mathcal{B}(\Omega) \tag{43}$$

as follows. For any $\Omega_0 \Subset \Omega$ we take an Ω_1 such that $\Omega_0 \Subset \Omega_1 \Subset \Omega$, and $\Gamma_1 \Subset \Gamma$ such that $W \supset \Omega_1 + (\Gamma_1 \cap B_r)$. Then we define

$$H_\gamma(z,t) := \int_{\Omega_1 + i\gamma} P_t(z-w)F(w)dw. \tag{44}$$

If $\gamma \in \Gamma_1$ is sufficiently small, $H_\gamma(x,t)$ is continued to a harmonic function on an upper neighborhood V_0 of Ω_0. It is easy to see that $H_\gamma(x,+0) \in \mathcal{B}(\Omega_0)$ does not depend on Ω_1 nor on γ. Thus we define the *boundary value* $F(x + i\Gamma 0)$ by

$$F(x + i\Gamma 0) \,|_{\Omega_0} := H_\gamma(x,+0). \tag{45}$$

Theorem 9. *The boundary value operator $b_\Gamma : \mathcal{O}(\Omega + i\Gamma 0) \longrightarrow \mathcal{B}(\Omega)$ is injective. A hyperfunction $f(x) \in \mathcal{B}(\Omega)$ is in the image of b_Γ if and only if it has the singularity spectrum*

$$SS\, f \subset \Omega \times (\Gamma^\circ \cap S^{n-1}), \tag{46}$$

where $\Gamma^\circ := \{\xi \in \mathbf{R}^n; \langle y, \xi\rangle \geqq 0$ *for all* $y \in \Gamma\}$ *is the polar of* Γ.

If $F(x+i\Gamma 0)$ *vanishes on an open subset of* Ω, *then it vanishes on its connected component.*

Theorem 10. *The following conditions are equivalent for* $F(z) \in \mathcal{O}(\Omega + i\Gamma 0)$:

(a) $$F(x+i\Gamma 0) \in \mathcal{D}^{*\prime}(\Omega) \ \ (\text{resp.}\, \mathcal{E}^*(\Omega));$$

(b) $F(x+iy)$ *converges in* $\mathcal{D}^{*\prime}(\Omega)$ (*resp.* $\mathcal{E}^*(\Omega)$) *as* $y \longrightarrow 0$ *in an (and any)* $\Gamma_1 \Subset \Gamma$;

(c) *For an (and any)* $\Gamma_1 \Subset \Gamma$ *and* $r > 0$ $\{F(x+iy); y \in \Gamma_1, \ |y| < r\}$ *is bounded in* $\mathcal{D}^{*\prime}(\Omega)$ (*resp.* $\mathcal{E}^*(\Omega)$);

(d) *For any compact set* $K \Subset \Omega$ *and* $\Gamma_1 \Subset \Gamma$ *there are constants* $r > 0$ *and* C *and a growth function* $G(t)$ *of class* $*$ *such that*

$$\sup_{x \in K} |F(x+iy)| \leqq C \ \exp\ G(|y|), \quad y \in \Gamma_1, \ |y| < r \tag{47}$$

(*resp. and for any ultradifferential operator* $P(\partial)$ *of class* $*$ $\{P(\partial)F(x+iy); x \in K, y \in \Gamma_1, |y| < r\}$ *is bounded*).

Then $F(x+iy)$ *converges to* $F(x+i\Gamma 0)$ *in* $\mathcal{D}^{*\prime}(\Omega)$ (*resp.* $\mathcal{E}^{*(\Omega)}$) *as* $y \in \Gamma_1$ *tends to zero.*

See [14] for proofs of Theorems 9 and 10. It follows from Theorems 5 and 9 that our sheaf $\mathcal{C}$ of microfunctions is isomorphic to Sato's.

Theorem 10 shows that the boundary value $F(x+i\Gamma 0)$ in the sense of hyperfunction coincides with the topological limit if the boundary value is an ultradistribution.

Let $\Gamma_j, j = 1, \cdots, m$, be proper convex cones in $\mathbf{R}^n$. If a hyperfunction $f(x) \in \mathcal{B}(\Omega)$ (resp. $\mathcal{D}^*(\Omega)$, resp. $\mathcal{E}^*(\Omega)$) satisfies the condition

$$SS\ f \subset \Omega \times \bigcup_{j=1}^{m} \Gamma_j^\circ, \tag{48}$$

then by the flabbiness of $\mathcal{C}$ (or the suppleness of $\mathcal{C}^*$ and $\mathcal{C}_*$) the microfunction sp f is represented as the sum $\sum_{j=1}^m$ sp f_j with $f_j \in \mathcal{B}(\Omega)$ (resp. $\mathcal{D}^{*\prime}(\Omega)$, resp. $\mathcal{E}^*(\Omega)$) such that $SS\ f_j \subset \Omega \times \Gamma_j^\circ$. Hence there are $F_j(z) \in \mathcal{O}(\Omega + i\Gamma_j 0)$ (resp. $\mathcal{O}^*(\Omega + i\Gamma_j 0)$, resp. $\mathcal{O}_*(\Omega + i\Gamma_j 0)$) such that

$$f(x) = \sum_{j=1}^{m} F_j(x + \Gamma_j 0). \tag{49}$$

This is called the boundary value representation of f by holomorphic functions.

As for the relations among different representations we have the following Martineau *edge of the wedge theorem* [17], [18]:

Theorem 11. *Let* $\Gamma_j, j = 1, \cdots, m$, *be proper convex cones in* $\mathbf{R}^n$. *In order that* $F_j(z) \in \mathcal{O}(\Omega + i\Gamma_j 0)$ (*resp.* $\mathcal{O}^*(\Omega + i\Gamma_j 0)$, *resp.* $\mathcal{O}_*(\Omega + i\Gamma_j 0)$) *satisfy*

$$\sum_{j=1}^{m} F_j(x + i\Gamma_j 0) = 0 \ \ \text{on}\ \Omega \tag{50}$$

it is necessary and sufficient that there are $F_{jk}(z) \in \mathcal{O}(\Omega + i(\Gamma_j + \Gamma_k)0)$ *(resp.* $\mathcal{O}^*(\Omega + i(\Gamma_j + \Gamma_k)0)$, *resp.* $\mathcal{O}_*(\Omega + i(\Gamma_j + \Gamma_k)0)$) *such that*

$$F_{jk}(z) + F_{kj}(z) = 0, \tag{51}$$

$$F_j(z) = \sum_{k=1}^{m} F_{jk}(z). \tag{52}$$

If $\{\Gamma_j^\circ\}$ covers $\mathbf{R}^n$, then we have the representaton of all hyperfunctions:

$$\mathcal{B}(\Omega) = \oplus \mathcal{O}(\Omega + i\Gamma_j 0)/\sim, \tag{53}$$

where $(F_j(z)) \sim 0$ if and only if there are $F_{jk}(t) \in \mathcal{O}(\Omega + i(\Gamma_j + \Gamma_k)0)$ satisfying (51) and (52). Kaneko [11] adopted this as the definition of hyperfunctions.

REFERENCES

1. G. Bengel and P. Schapira, Décomposition microlocale analytique des distributions, *Ann. Inst. Fourier, Grenoble* **29**-3 (1979), 101–124.
2. G. Björck, Linear partial differential operators and generalized distributions, *Ark. f. Mat.* **6** (1966), 351–407.
3. J.-M. Bony, Propagation des singularités différentiables pour une classe d'opérateurs différentiels à coefficients analytiques, *Astérisque* **34-35** (1976), 43–91.
4. J.-M. Bony and P. Schapira, Solutions hyperfunctions du problème de Cauchy, *Lecture Notes in Math.* **287** (1973), 82-98.
5. J. W. De Roever, Hyperfunctional singular support of ultradistributions, *J. Fac. Sci. Univ. Tokyo, Sect. IA* **31** (1984), 585–631.
6. A. Eida, On microlocal decomposition of ultradistributions and ultradifferentiable functions, *Master's thesis, Univ. Tokyo* (1989).
7. A. Eida, Harmonic analysis of ultradistributions, *Doctor's thesis, Univ. Tokyo*, (1992).
8. A. Grothendieck, Sur les espaces de solutions d'une classe générale d'équations aux dérivées partielles, *J. Analyse Math.* **2** (1952-53), 243–280.
9. L. Hörmander, "An Introduction to Complex Analysis in Several Variables", Van Nostrand, Princeton (1966).
10. L. Hörmander, "The Analysis of Linear Partial Differential Operators I", Springer, Berlin-Heidelberg-New York-Tokyo (1983).
11. A. Kaneko, "Introduction to Hyperfunctions", Kluwer Academic, Dordrecht-Boston-London (1988).
12. H. Komatsu, On the index of ordinary differential operators, *J. Fac. Sci. Univ. Tokyo, Sect. IA* **18** (1971), 379–398.
13. H. Komatsu, Ultradistributions I, Structure theorems and a characterization, *J. Fac. Sci. Univ. Tokyo, Sect. IA* **20** (1973), 25–105.
14. H. Komatsu, Microlocal Analysis in Gevrey classes and in complex domains, *in*: "Microlocal Analysis and Applications", *Lecture Notes in Math.* **1495** (1991), 161–236.
15. H. Komatsu, An elementary theory of hyperfunctions and microfunctions, *Banach Center Publ.* **27** (1992), to appear.
16. P. Lévy, "Leçon d'Analyse Fonctionnel", Gauthier-Villard, Paris (1922).

17. A. Martineau, Distributions et valeurs au bord des fonctions holomorphes, *in*: "Theory of Distributions", Inst. Gulbenkian de Ciência, Lisboa (1964), 195–326.
18. A. Martineau, Le "edge of the wedge theorem" en théorie des hyperfonctions de Sato, *in*: "Proc. Internat. Conf. on Functional Analysis and Related Topics (1969)", Univ. of Tokyo Press (1970), 95-106.
19. C. Roumieu, Sur quelques extensions de la notion de distribution, *Ann. Ecole Norm. Sup.* **77** (1960), 41–121.
20. M. Sato, Theory of hyperfunctions, I, *J. Fac. Sci. Univ. Tokyo, Sect. I* **8** (1959), 139–193.
21. M. Sato, Theory of hyperfunctions, II, *J. Fac. Sci. Univ. Tokyo, Sect. I* **8** (1960), 387–437.
22. M. Sato, Hyperfunctions and partial differential equations, *in*: "Proc. Internat. Conf. on Functional Analysis and Related Topics, 1969", Univ. of Tokyo Press (1970), 91-94.
23. M. Sato, T. Kawai and M. Kashiwara, Microfunctions and pseudo-differential equations, *Lecture Notes in Math.* **287** (1973), 265–529.
24. L. Schwartz, "Théorie des Distributions ", Hermann, Paris (1950-51).
25. J. Siciak, Holomorphic continuation of harmonic functions, *Ann. Polon. Math.* **29** (1974), 67–73.

GENERALIZED FUNCTIONS FOR THE LAPLACE TRANSFORM AND UNIVERSAL APPROXIMANTS

John Boris Miller

Department of Mathematics
Monash University
Clayton, Victoria 3168
Australia

ABSTRACT

Some classes of spaces of generalized functions with special reference to the Laplace transformation are surveyed, the spaces being norm completions. It is proved using a theorem of G. D. Birkhoff that these spaces contain universal approximants.

Mathematics Subject Classification: primary 44A40, 44A10, secondary 44A35, 46F12.

Key words and phrases: Laplace transform, inflation, seat, universal approximant.

1. INTRODUCTION

In [1] the author proposed several types of generalized-function spaces having special affinity to the Laplace transformation

$$(\mathcal{L}x)(s) = x^{\wedge}(s) := \int_0^\infty e^{-st}\, x(t)\, dt. \tag{1.1}$$

The first of these, which will here be written $\mathfrak{X}_\Omega$, was defined as follows. Let Ω denote a compact non-finite subset of the open right halfplane

$$\Pi_0 := \{s \in \mathbb{C} : \ \mathrm{Re}(s) > 0\},$$

and for any complex-valued function $x \in L^1(\mathbb{R}^+)$ write

Generalized Functions and Their Applications, Edited by
R.S. Pathak, Plenum Press, New York, 1993

$$\|x\|_\Omega := \max_{s\in\Omega} |x^\wedge(s)|. \tag{1.2}$$

(As usual, integrable functions which are equal almost everywhere are equated.) This defines on L^1 not merely a seminorm, but a norm: for $x^\wedge$ is holomorphic on Π_0 (see [2], p. 57), hence $\|x\|_\Omega = 0$ implies that $x^\wedge$ is identically zero, by the properties of zeros of holomorphic functions, whence $x = 0$ in L^1. In fact L^1 is a commutative complex normed algebra without identity, with respect to convolution

$$(x.y)(t) := \int_0^t x(u)\, y(t-u)\, du \tag{1.3}$$

as product and with $\|.\|_\Omega$ as norm instead of the usual L^1 norm $\|\cdot\|$, and we have $\|x\|_\Omega \leq \|x\|$ for all $x \in L^1$. But $(L^1, \|.\|_\Omega)$ is not complete: the sought space of generalized functions $\mathfrak{X}_\Omega$ is defined to be its completion. We refer to $\mathfrak{X}_\Omega$ as an *inflation* of L^1, and Ω as the *seat* of the inflation. L^1 is dense in $\mathfrak{X}_\Omega$, and convergence in L^1 implies convergence in $\mathfrak{X}_\Omega$.

Norm (1.2) uses the usual maximum modulus norm

$$|\phi|_\Omega := \max_{s\in\Omega} |\phi(s)|$$

on $C(\Omega)$, the algebra (under pointwise operations) of all continuous complex–valued functions on Ω. We shall interpret the Laplace transformation (1.1) as the operator

$$\mathcal{L} : x \mapsto x^\wedge|_\Omega, \quad L^1 \to C(\Omega);$$

it is bounded, indeed an isometric monomorphism since $\|x\|_\Omega = |\mathcal{L}(x)|_\Omega$; therefore it has an isometric monomorphic extension to $\mathfrak{X}_\Omega$ which we also write as $\mathcal{L}$.

Let us dispense with some unnecessary refinements by supposing *henceforth* that the seat is always a compact convex set with nonempty convex interior. Then the completion $\mathfrak{X}_\Omega$ is mapped by the extended $\mathcal{L}$ onto the completion of its range, which is the algebra

$$H(\Omega) := \{ f \in C(\Omega) : \ f \text{ is holomorphic on the interior } \Omega^\circ \text{ of } \Omega\}; \tag{1.4}$$

thus every continuous function of Ω which is holomorphic on Ω° is the Laplace transform of a unique generalized function in $\mathfrak{X}_\Omega$. This space of generalized functions is by definition a norm completion, not the dual of a space of test functions, and hence has the advantage of a straightforward topology.

As a space of generalized functions, $\mathfrak{X}_\Omega$ contains a Dirac δ, the identity of the algebra, with $\delta^\wedge(s) = 1$ on Ω. It carries generalized translation operators T_k, defined for all complex k by

$$(T_k\, x)^\wedge(s) = e^{-ks}\, x^\wedge(s) \qquad \text{for all } s \in \Omega, \tag{1.5}$$

so that when $x \in L^1$ and $k > 0$ we have

$$(T_k\, x)(t) = \begin{cases} 0 & (\text{if } 0 < t < k), \\ x(t-k) & (\text{if } t > k). \end{cases} \tag{1.6}$$

It carries generalized differentiation and integration operators D and K respectively, defined by

$$(Dx)^\wedge(s) = s\, x^\wedge(s), \qquad (Kx)^\wedge(s) = (1/s)\, x^\wedge(s); \tag{1.7}$$

and

$$Dx = \lim_{k \to 0} \frac{1}{k}(x - T_k x) ,$$

the limit being in norm (1.2). These are all bounded linear operators mapping $\mathfrak{X}_\Omega$ onto $\mathfrak{X}_\Omega$, with $DK = KD = I$.

Let $\mathfrak{W}_\Omega$ denote the space of all (classes of) locally integrable functions w on $\mathbb{R}^+$ whose Laplace transform integral $w^\wedge(s)$ converges on some open half-plane containing Ω, say on $\Pi_\sigma := \{s \in \mathbb{C}: \operatorname{Re}(s) > \sigma\}$, where σ depends upon w, but $\sigma < \inf\{\operatorname{Re}(s): s \in \Omega\}$. $\mathfrak{W}_\Omega$ is naturally embeddable in $\mathfrak{X}_\Omega$. In particular the Heaviside function $h(t) = 1$ (if $t > 0$), 0 (if $t < 0$) belongs to $\mathfrak{W}_\Omega$, and $D(h) = \delta$. If $x \in \mathfrak{W}_\Omega$ and $x : \mathbb{R}^+ \to \mathbb{C}$ is absolutely continuous and its usual derivative dx/dt also belongs to $\mathfrak{W}_\Omega$ then

$$Dx = dx/dt + x(0+)\delta.$$

These and other elementary formulae famous in theories of generalized functions are discussed in [1].

In a similar way we can embed $L^2(\mathbb{R}^+)$ naturally in $\mathfrak{X}_\Omega$; for if $x \in L^2$ then its transform integral converges in mean square to a function which is holomorphic on Π_0 ([2], p. 80).

Functions which are locally integrable on all of $\mathbb{R}$, which vanish on a neighbourhood of $-\infty$, and whose transform integral converges on a open half-plane containing Ω can also be embedded in $\mathfrak{X}_\Omega$. The theory for convenience uses the bilateral Laplace tranform with a modified definition of convolution, or alternatively uses translations to the left, T_k with $k < 0$.

The reader will recognise the construction of $\mathfrak{X}_\Omega$ as an instance of the following general procedure. We start with a commutative Banach algebra $\mathfrak{A}$ without identity (here L^1) whose locally compact carrier space (here identifiable with $\{s: \operatorname{Re}(s) \geq 0\}$) contains an analytic disc P, i.e. a subset analytically isomorphic to the unit disc in $\mathbb{C}$ (here Π_0), on which the Gelfand transform $x^\wedge$ (here the Laplace transform) of every element x in $\mathfrak{A}$ is holomorphic. Choosing a compact subset Ω of the disc, we renorm $\mathfrak{A}$ by the definition (1.2), and then complete it with respect to this norm. This procedure for constructing spaces of generalized functions, and variants of it, are elaborated in [3].

For example, we could start with the convolution subalgebra of $L^1(\mathbb{R})$ consisting of all (classes of) functions with compact support: for every such function its Fourier transform is holomorphic, by the Paley-Wiener Theorem, so we can proceed in the same way, although the starting subalgebra is not complete. Another variant of the outlined method uses the Mellin transformation in place of the Laplace transformation, and the usual Mellin convolution; here it is necessary to arrange first that we start with a space of functions whose Mellin transforms are all holomorphic on a common strip, and the space of generalized functions which eventuates has a novel definition of derivative. We do not pursue these byways here.

Some spaces of generalized functions obtained by more general types of norm completion of incomplete normed spaces are discussed in [4], [5], [6]. These include constructions appropriate to Fourier sine and cosine transformations, and Watson transforms. In [7] such spaces are used in discussing a type of generalization of Fourier series.

2. SOME SERIES EXPANSIONS IN $\mathfrak{X}_\Omega$

As with other theories, such as Mikusinski's [8] and Weston's [9], certain holomorphic functions of s correspond under $\mathcal{L}^{-1}$ to generalized functions in $\mathfrak{X}_\Omega$, while others are best seen as corresponding to operators (although of course they also correspond at the same time to generalized functions). Thus a product x^(s) y^(s) is sometimes to be interpreted as the Laplace transform of a convolution, since

$$x\hat{}(s)y\hat{}(s) = \mathcal{L}\left(\int_0^t x(u)y(t-u)du\right)(s) = (x.y)\hat{}(s) \tag{2.1}$$

(for example when $x, y \in \mathfrak{W}_\Omega$), and sometimes as the Laplace transform of Yx for some operator Y (for example, when y^(s) is e^{-ks}, s or s^{-1}, as in (1.5), (1.7)). Consider the case

$$y\hat{}(s) := \sum_{n=0}^{\infty} \alpha_n s^n , \tag{2.2}$$

where the righthand side is any power series whose open disc of convergence contains the seat Ω. Then $\mathfrak{X}_\Omega$ contains a generalized function $y := \mathcal{L}^{-1}(y\hat{})$, and if $y \in \mathfrak{W}_\Omega$ then

$$\int_0^t x(u)\, y(t-u)\, du = \sum_{n=0}^{\infty} \alpha_n x^{(n)}(t) \tag{2.3}$$

provided x belongs to the space

$$C^\infty_{0,\Omega} := \{x \in C(\mathbb{R}^+) : \text{ x is infinitely differentiable on } \mathbb{R}^+ \text{ with } x^{(k)}(0) = 0 \text{ and } x^{(k)} \in \mathfrak{W}_\Omega \text{ for } k = 0, 1, 2, \ldots \}. \tag{2.4}$$

This formula and variants of it produce some generalizations of Taylor's series, see [10]. For example, the choice of function

$$y\hat{}(s) = \frac{e^{\omega s}}{e^{\omega s} - 1} \qquad (\omega \text{ given, } > 0) \tag{2.5}$$

corresponds to the standard summation operator S_ω, defined on $\mathfrak{W}_\Omega$ by

$$(S_\omega x)(t) = \sum_{r=0}^{[t/\omega]} x(t - r\omega), \qquad (S_\omega x)\hat{}(s) = e^{\omega s}(e^{\omega s} - 1)^{-1} x\hat{}(s); \tag{2.6}$$

when y^(s) is expanded as a power series

$$\frac{e^{\omega s}}{e^{\omega s}-1} = \frac{1}{\omega s} + \frac{1}{2} + \sum_{k=1}^{\infty}(-1)^{k-1}\frac{B_K\omega^{2k-1}}{(2k)!}s^{2k-1},$$

as in (2.2) (with a term in s^{-1}), in place of (2.3) we get the Euler-Maclaurin sum formula, in the form

$$\sum_{r=0}^{[t/\omega]} x(t-r\omega) = \frac{1}{\omega}\int_0^t x(u)\,du + \frac{1}{2}x(t) + \sum_{k=1}^{\infty}(-1)^{k-1}\frac{B_k\omega^{2k-1}}{(2k)!}x^{(2k-1)}(t), \tag{2.7}$$

provided $x \in C^\infty{}_{0,\Omega}$ and $\Omega \subset \Pi_0 \cap \{s: |s| < 2\pi\omega\}$. This is an application of the operator equation

$$S_\omega = \frac{1}{\omega}K + \frac{1}{2}I + \sum_{k=1}^{\infty}(-1)^{k-1}\frac{B_k\,\omega^{2k-1}}{(2k)!}D^{2k-1}; \tag{2.8}$$

here B_k is the kth Bernoulli number, in the notation of [11], §7.2.

There is an elegant formula for the partial sums of the series. Let $S_{\omega,N}$ denote the operator which is the righthand side of (2.8) with the infinite sum replaced by the sum over k from 1 to N; we have

$$(S_{\omega,N}\,x)^\wedge(s) = \left(1 + \frac{(\omega s)^{2N}}{(2N)!}\hat{\phi}_{2N}(\omega s)\right)x^\wedge(s) \tag{2.9}$$

where ϕ_{2N} is the $2N^{th}$ Bernoulli polynomial. There is also a formula for the remainder, as a convolution with the periodic extension of ϕ_{2N}.

The integral of $x \in C^\infty{}_{0,\Omega}$ is displayed as a limit in norm (1.2) of Riemann sums by the formula

$$\lim_{\omega \searrow 0} \omega S_\omega x = Kx. \tag{2.10}$$

Some variants of the Euler-Maclaurin formula are provable by these methods: for example, the periodic Euler-Maclaurin summation formula of Berndt and Schoenfeld, which includes Boole's summation formula. See [12] and [13].

3. DEPENDENCE UPON THE SEAT

Let Φ and Ψ be seats. If $\Phi \subset \Psi$, there exists an obvious natural embedding

$$\kappa_{\Psi,\Phi} : \mathfrak{X}_\Psi \to \mathfrak{X}_\Phi,\quad x \mapsto \mathcal{L}^{-1}(x^\wedge|_\Phi), \qquad (\kappa_{\Psi,\Phi}(x))^\wedge = x^\wedge|_\Phi; \tag{3.1}$$

κ is an algebra monomorphism, and continuous since

$$|x|_\Phi < |x|_\Psi \quad \text{if} \quad x \neq 0.$$

Let $R_{\Psi,\Phi}$ denote the range of $\kappa_{\Psi,\Phi}$; it is a subalgebra of $\mathfrak{X}_\Omega$ containing the identity δ.

If more generally Φ and Ψ are arbitrary seats, we can identify certain subsets of $\mathfrak{X}_\Phi$ and $\mathfrak{X}_\Psi$ as follows. Let $\Omega = \text{conv}(\Phi \cup \Psi)$, the closed convex hull of the union; Ω is a seat, and

$$\theta_{\Psi,\Phi} := \kappa_{\Omega,\Phi} \circ \kappa_{\Omega,\Psi}^{-1} : \quad R_{\Omega,\Psi} \to R_{\Omega,\Phi} \tag{3.2}$$

is an isomorphism between algebras, with $\theta_{\Psi,\Phi} = \kappa_{\Psi,\Phi}$ when $\Phi \subset \Psi$.

The family of spaces $\mathfrak{X}_\Omega$ and maps $\kappa_{\Psi,\Phi}$, indexed by the set of all seats contained in Π_0 right directed under $\subseteq$, constitute a projective spectrum, with projective limit PL_0, defined to be

$$PL_0 = \{ f \in \Pi_\Omega \mathfrak{X}_\Omega : \kappa_{\Psi,\Phi}(f_\Psi) = f_\Phi \}; \tag{3.3}$$

the Laplace transforms of the elements of PL_0 can be identified with all functions which are holomorphic on Π_0. Every element of PL_0 has its natural projection in every $\mathfrak{X}_\Omega$; this we may loosely think of as the same generalized function appearing in different inflations: we call these generalized functions *ubiquitous*. All functions in $L^1 \cup L^2$ are ubiquitous, for instance, as are all locally integrable functions with transform integral converging on Π_0.

Of course in the construction of PL_0 we lose the simplicity of the norm topology, and must use instead the family of all possible norms (1.2), or at least a sufficiency of them. PL_0 is then topologically isomorphic under $\mathcal{L}$ to the space of all holomorphic functions on Π_0 with the compact-open topology.

The definition of PL_0 can be generalized to spaces PL_ω, $\omega > 0$, where the Laplace transforms are holomorphic on the half-plane $\Pi_\omega = \{s : \text{Re}(s) > \omega\}$; then the inductive limit ILPL of these spaces over the ordered index set $\mathbb{R}^+\backslash\{0\}$ gives a space of generalized functions which is characterized by the fact that their Laplace transforms constitute all functions holomorphic on open right halfplanes of their own choosing. These spaces are described more fully in [1].

The dependence of the space of generalized functions $\mathfrak{X}_\Omega$ upon the seat Ω gives a certain arbitrariness in the theory, but it also offers a distinct advantage, namely: given any nonzero x in L^1, we can always find a seat Ω such that $\mathfrak{X}_\Omega$ contains a (convolution) inverse x^{-1} of x. For x is invertible in $\mathfrak{X}_\Omega$ if and only if $x^\wedge$ does not vanish on Ω. The properties of zeros of holomorphic functions ensure that such a seat can always be found. More generally, if x is locally integrable and the transform integral $x^\wedge(s)$ converges on Π_σ say, we can always find an Ω in this half-plane for which x^{-1} exists in $\mathfrak{X}_\Omega$. Again, given Ω and $0 \neq x \in \mathfrak{X}_\Omega$ we can always find an extension $\mathfrak{X}_\Phi$, $\Phi \subset \Omega$, in which x (more precisely $\kappa_{\Omega,\Phi}(x)$) has an inverse.

For $x \in \mathfrak{X}_\Omega$, the spectrum of x as an element of this Banach algebra is

$$\begin{aligned} Sp_\Omega(x) &= \{\lambda \in \mathbb{C} : \lambda\delta - x \text{ is singular in } \mathfrak{X}_\Omega\} \\ &= \text{range of } x^\wedge \text{ on } \Omega, \end{aligned}$$

and clearly $Sp_\Phi(x) \subseteq Sp_\Omega(x)$ when $\Phi \subset \Omega$.

As an application, consider the Wiener-Hopf equation of the form

$$\lambda x(t) = y(t) + \int_0^t z(t-u)x(u)\,du, \tag{3.4}$$

where y and z are given in $L^1(\mathbb{R}^+)$ and a solution x is sought. Though its solvability in L^1 depends upon the eigenparameter λ, we can (whatever the value of λ) always choose a seat Ω such that λ does not belong to $Sp_\Omega(z)$, and then (3.4) has a unique solution in $\mathfrak{X}_\Omega$ for that Ω, obtained as

$$x = \mathcal{L}^{-1}(x\wedge), \qquad \text{where } x\wedge(s) = \frac{y\wedge(s)}{\lambda - z\wedge(s)} .$$

The more general case where y and z are locally integrable on $\mathbb{R}$ and zero on a neighbourhood of $-\infty$ with convergent transforms can be handled in the same way.

These advantages do not exist in any of the projective limits PL_ω, since there exist L^1 functions whose Laplace transforms have real zeros accumulating at $+\infty$, and which therefore have inverse in no PL_ω, nor in ILPL. An example is

$$x(t) = t^{-\frac{1}{2}} \sin(\tfrac{1}{2}t^{-1}), \qquad x\wedge(s) = \pi^{\frac{1}{2}} s^{-\frac{1}{2}} \exp(-s^{\frac{1}{2}}) \sin(s^{\frac{1}{2}}). \tag{3.5}$$

For another illustration of the effect of the choice of seat in an application, consider again the analysis leading to (2.7), but now suppose that Ω is chosen to lie in the annulus

$$A = \{s: \ 2\pi q < |\omega s| < 2\pi(q+1)\}$$

for some $q \in \mathbb{N}$ ((2.7) assumes $q = 0$). The series in (2.7) no longer converges in norm $\|.\|_\Omega$; but the Laurent series for $y\wedge(s)$ of (2.5) on A is

$$\frac{e^{\omega s}}{e^{\omega s}-1} = \sum_{k=1}^{\infty}(-1)^k \frac{2(2\pi)^{2k}\,\phi_{2k+1}(q+1)}{\omega^{2k+1}(2k+1)}\frac{1}{s^{2k+1}} + \frac{2q+1}{\omega s} + \frac{1}{2} + $$
$$+ \sum_{k=1}^{\infty}(-1)^{k-1}\frac{2\omega^{2k-1}\,\zeta(2k, q+1)}{(2\pi)^{2k}} s^{2k-1}, \tag{3.6}$$

where $\zeta(m,p) := \sum_{j=p}^{\infty} j^{-m}$; and consequently in place of (2.7) we have, for $x \in C^{\infty}_{0,\Omega}$,

$$\sum_{r=0}^{[t/\omega]} x(t-r\omega) = \sum_{k=1}^{\infty}(-1)^k\frac{2(2\pi)^{2k}\,\phi_{2k+1}(q+1)}{\omega^{2k+1}\,(2k+1)!}\int_0^t (t-u)^{2k}x(u)du +$$
$$+ \frac{2q+1}{\omega}\int_0^t x(u)du + \frac{1}{2}x(t) +$$
$$+ \sum_{k=1}^{\infty}(-1)^{k-1}\frac{2\omega^{2k-1}\zeta(2k, q+1)}{(2\pi)^{2k}} x^{(2k-1)}(t). \tag{3.7}$$

4. UNIVERSAL APPROXIMANTS IN $\mathfrak{X}_\Omega$

With $\zeta \in \mathbb{C}$ let $E_\zeta : \mathfrak{X}_\Omega \rightsquigarrow \mathfrak{X}_\Omega$ denote the linear operator defined by

$$(E_\zeta x)\hat{}(s) = x\hat{}(s + \zeta); \tag{4.1}$$

its domain consists of all elements x for which there exists an analytic continuation of the function

$$y_\zeta\hat{} : s \mapsto x\hat{}(s + \zeta),$$

(initially defined on $\Omega - \zeta$) to an element of $H(\Phi)$, where

$$\Phi = \mathrm{conv}(\Omega \cup (\Omega - \zeta)),$$

thereby defining a generalized function y_ζ in $\mathfrak{X}_\Omega$. In particular if $x \in \mathfrak{W}_\Omega$ and the Laplace integral $x\hat{}(s)$ converges on Π_σ say, then

$$y_\zeta\hat{}(s) = \int_0^\infty e^{-st} . e^{-\zeta t} x(t) dt \tag{4.2}$$

so long as

$$\mathrm{Re}(\zeta) > \sigma - \min\{\mathrm{Re}(s) : s \in \Omega\}, \tag{4.3}$$

in which case

$$(E_\zeta x)(t) = e^{-\zeta t} x(t). \tag{4.4}$$

In the spirit of theories of generalized functions we regard E_ζ as the operator of "pointwise multiplication by $e^{-\zeta t}$". By assumption, the righthand side of (4.3) is negative. The operator E_ζ is not bounded. We shall prove the following surprising result.

Theorem 1

There exists a ubiquitous generalized function b with the following property: For every choice of disc seat Ω in Π_0,

$$b \in \bigcap\{\mathrm{dom}\,(E_\zeta) : \zeta \in \mathbb{C}\}, \textit{ and the set } \{E_\zeta b : \zeta \in \mathbb{C}\} \textit{ is dense in } \mathfrak{X}_\Omega . \tag{4.5}$$

In fact there are infinitely many such elements b; we call them *universal approximants*. The proof is an easy consequence of the Weierstrass-Stone Theorem and the following theorem of Birkhoff's, asserting the existence of universal entire functions.

Theorem (G. D. Birkhoff, [14]). There exists an entire function ϕ with the following property: For every entire function f, every compact subset Ω of $\mathbb{C}$ and every $\varepsilon > 0$ there exists a complex number μ such that

$$\max_{\lambda \in \Omega} | f(\lambda) - \phi(\lambda + \mu) | < \varepsilon.$$

Proof of Theorem 1. Let ϕ be any universal entire function as in Birkhoff's theorem. Given a seat Ω, we have

$$\phi|_{\Omega} = b^{\wedge}$$

for a uniquely determined b of $\mathfrak{X}_{\Omega}$ which is clearly ubiquitous. Since every translate $\phi(\cdot + \zeta)$ restricts to a function in $H(\Omega)$, $E_{\zeta}b$ is defined in $\mathfrak{X}_{\Omega}$ with

$$(E_{\zeta}b)^{\wedge}(s) = \phi(s + \zeta)$$

for all $\zeta \in \Omega$, so the first statement in (4.5) holds.

Let x be any element of $\mathfrak{X}_{\Omega}$, and $\varepsilon > 0$. Now when Ω is the closed unit disc about 0 the algebra $C(\Omega)$ carries the involution $f^*(\lambda) = f(\lambda^-)^-$ (where $^-$ denotes complex conjugation), and the polynomials restricted to Ω form a strongly point-separating selfadjoint subalgebra of $H(\Omega)$, itself a selfadjoint subalgebra of $C(\Omega)$; so the polynomials are dense in $H(\Omega)$ by the Weierstrass-Stone Theorem. This conclusion easily extends to any seat which is a disc. Thus there exists an entire function p (for example, a polynomial) such that

$$|x^{\wedge} - p|_{\Omega} < \tfrac{1}{2}\varepsilon .$$

Then by Birkhoff's theorem there exists μ such that $| p - \phi(\cdot + \mu) |_{\Omega} < \tfrac{1}{2}\varepsilon$, and we have

$$\| x - E_{\mu}b \|_{\Omega} = |x^{\wedge} - \phi(\cdot + \mu)|_{\Omega} < \varepsilon,$$

completing the proof.

Note that every translate of ϕ is also a universal entire function, so $E_{\zeta}b$ is also a universal approximant, for every ζ.

The theorem shows that every generalized function is the limit of some sequence loosely written

$$(e^{-\mu_n t} b(t))_{n=1}^{\infty} ;$$

in general the sequence (μ_n) does not converge, instead $\mu_n \to \infty$.

REFERENCES

1. J. B. Miller, Generalized-function calculi for the Laplace transformation, *Arch. for Rat. Mech. and Analysis,* 12:409 (1963).
2. D. V. Widder, "The Laplace transform", Princeton U. P., Princeton (1946).
3. J. B. Miller, Generalized Gâteaux and Fréchet derivatives in convolution algebras, *Proc. Camb. Phil. Soc.* 59:707 (1963).
4. J. B. Miller, Hilbert spaces of generalized functions extending L^2, I and II, *J. Austral. Math. Soc.* 1: 281 (1960) and 3: 267 (1963).
5. J. B. Miller, On the solution of certain integral equations by generalized functions, *Proc. Camb. Phil. Soc.* 57:767 (1961).

6. J. B. Miller, Normed spaces of generalized functions, *Compositio Mathematica* 15:127 (1963).
7. J. B. Miller, A class of series involving integral kernels, I and II. *Math. Zeitsch.* 76:257 (1961) and 79:227 (1962).
8. J. Mikusinski, "Operational Calculus", Pergamon, New York (1959).
9. J. D. Weston, Positive perfect operators, *Proc. London Math. Soc.* 10:545 (1960).
10. J. B. Miller, Series like Taylor's series, *Aequationes Math.* 26:208 (1983).
11. E. T. Whittaker and G. N. Watson, "A course in modern analysis", Camb. U. P., Cambridge (1946).
12. B. C. Berndt, Periodic Bernoulli numbers, summation formulae and applications, *in:* "Theory and application of special functions", Academic Press, New York, 1975.
13. J. B. Miller, The standard summation operator, the Euler-Maclaurin sum formula, and the Laplace transformation, *J. Austral. Math. Soc. (Series A)* 39:367 (1985).
14. G. D. Birkhoff, Démonstration d'un théorème élémentaire sur les fonctions entiéres, *Comptes rendus, Acad. des Sc. Paris* 189: 473 (1929).

AN EXTENSION OF THE RADON TRANSFORM

Piotr Mikusiński and Ahmed Zayed

Department of Mathematics
University of Central Florida
Orlando, FL 32816

1. INTRODUCTION

Since the publication of Schwartz's work on the theory of distributions (Schwartz, 1950-1951), in particular his work on the Fourier transform of distributions, there has been a tremendous influx of research papers on integral transforms of generalized functions. One of the main sources on this topic is the book by Zemanian (1968).

In recent years, the Radon transform, which was defined and studied by Radon (1917), has emerged as one of the most important mathematical tools used in various applications in physics, engineering, and medicine, in particular in what is known as the reconstruction problem. This is the problem of determining the internal structure of an object, such as a human brain, without having to cut it. The role that the Radon transform plays in the reconstruction problem is closely related to its main property of providing an algorithm to reconstruct a function in N variables from the values of its integrals over hyperplanes of dimension $N-1$.

The Radon transform is not classified as an integral transform; in fact, it consists of infinitely many integral transforms with the δ-function as their kernels. Nevertheless, it has also been extended to spaces of generalized functions by many people (Helgason, 1980; Ludwig, 1966), including Gelfand, Graev and Vilenkin (1966), who used the method of the adjoint transform to define the Radon transform of a tempered distribution F, i.e., $F \in \mathcal{S}'(\mathbb{R}^N)$, where $\mathcal{S}(\mathbb{R}^N)$ is the space of rapidly decreasing functions on $\mathbb{R}^N$.

Denote by S^{N-1} the unit sphere in $\mathbb{R}^N$ and by Σ^N the generalized cylinder $\mathbb{R} \times S^{N-1}$. We recall that if $f \in \mathcal{S}(\mathbb{R}^N)$, then its Radon transform $\widehat{f}(p,\underline{\xi})$, where $(p,\underline{\xi})$ is a point on Σ^N, has the following properties:

(i) $\widehat{f}(p,\underline{\xi})$ is an even homogeneous function of degree -1, i.e., $\widehat{f}(\alpha p,\alpha\underline{\xi}) = \frac{1}{|\alpha|}\widehat{f}(p,\underline{\xi}), \quad \alpha \neq 0;$

(ii) $\widehat{f}(p,\underline{\xi})$ is infinitely differentiable with respect to $\underline{\xi}$ and with respect to p for $\underline{\xi} \neq 0$;

(iii) $\widehat{f}(p,\underline{\xi})$ and all its derivatives with respect to p and $\underline{\xi}$ are of order $o\left(\frac{1}{|p|^k}\right)$ as $|p| \to \infty$ for all k and uniformly in $\underline{\xi}$;

Generalized Functions and Their Applications, Edited by
R.S. Pathak, Plenum Press, New York, 1993

(iv) For any nonnegative integer k, the integral $\int_{\mathbb{R}} \widehat{f}(p,\underline{\xi})\, p^k\, dp$ is a polynomial in $\underline{\xi}$, homogeneous of degree k.

Let

$$\psi(p,\underline{\xi}) = \frac{\partial^{N-1}\widehat{f}(p,\underline{\xi})}{\partial p^{N-1}},$$

where $\widehat{f}$ satisfies (1)-(4). Provided with an appropriate topology, the linear space of all such ψ's becomes a testing-function space and will be denoted by $\widehat{\mathcal{S}}_\Sigma$; see Gelfand et al (1966) and Zemanian (1968). Thus, according to Gelfand et al, the Radon transform $\widehat{F}$ of a tempered distribution F is defined as a continuous linear functional on $\widehat{\mathcal{S}}_\Sigma$ given by

$$< F, f > = C_N < \widehat{F}, \psi > ,$$

where C_N is a constant depending only on the dimension N. The space of all continuous linear functionals on $\widehat{\mathcal{S}}_\Sigma$ is denoted $\widehat{\mathcal{S}}'_\Sigma$.

The theory of distributions and generalized functions as developed by Schwartz (1950-1951), Gelfand and Shilov (1964), and later by Zemanian (1968), regards generalized functions as elements of the dual space of some space of testing functions. Following a different approach to generalized functions, Mikusiński (1959) developed his theory of the Mikusiński operators in which a generalized function is regarded as an equivalence class of convolution quotients. Later, extending an idea of regular operators introduced by Boehme (1973), J. Mikusiński and P. Mikusiński (1981) constructed a larger class of mathematical objects, commonly known as Boehmians, which embraces regular operators, Schwartz distributions, and other spaces of generalized functions (Mikusiński, 1988). A Boehmian is defined as an equivalence class of "quotients of sequences". In Mikusiński and Zayed (1992) the Radon transform is extended to the space of the so-called rapidly decreasing Boehmians. The extension is based on some algebraic properties of the Radon transform.

In this note we show that an equivalent extension of the Radon transform can be defined by using the continuity of the transform with respect to the Δ-convergence in $\mathcal{S}(\mathbb{R}^N)$. The extension includes some generalized functions which are not embraced by spaces of generalized functions considered by other authors.

2. THE RADON TRANSFORM

The Radon transform $\check{f}$ or $\mathcal{R}f$ of a function of two variables $f(x,y)$ is defined by

$$\check{f} = \mathcal{R}f = \int_L f(x,y)\, ds, \tag{2.1}$$

whenever the integral exists, where L runs over all possible straight lines in the xy-plane and ds is an increment of length along L. To be more precise, for a fixed straight line L, the integral (2.1) is called a "slice" or a "sample" of the Radon transform and the totality of all these samples constitutes the Radon transform of f. Since the equation of any straight line in the xy-plane can be written in the form $p = x\cos\phi + y\sin\phi = \underline{x}\cdot\underline{\xi}$, where $\underline{x} = (x,y)$, $\underline{\xi} = (\cos\phi, \sin\phi)$, and $0 \le \phi \le 2\pi$, it is not difficult to see (Deans, 1983) that (2.1) can be put in the form

$$\check{f}(p,\underline{\xi}) = \int_{\mathbb{R}^2} f(\underline{x})\,\delta(p - \underline{x}\cdot\underline{\xi})\, d\underline{x}, \tag{2.2}$$

where $d\underline{x} = dxdy$ and δ is the Dirac delta distribution.

It is easy to see that if f is integrable over $\mathbb{R}^2$ then the function $\check{f}(p,\underline{\xi})$ exists for almost all p and $\underline{\xi}$.

The last form of the Radon transform enables us to extend it to higher dimensions as follows. Let $\underline{\xi}$ denote a unit vector in $\mathbb{R}^N$. Then, since the equation of any hyperplane in $\mathbb{R}^N$ can be written in the form $p = \underline{x} \cdot \underline{\xi}$, we define the N-dimensional Radon transform of a function of N variables $f(\underline{x})$ as

$$\check{f}(p,\underline{\xi}) = (\mathcal{R}f)(p,\underline{\xi}) = \int_{\mathbb{R}^N} f(\underline{x})\delta(p - \underline{x}\cdot\underline{\xi})\,d\underline{x}. \tag{2.3}$$

Since $\underline{\xi}$ can be regarded as a point on the generalized sphere S^{N-1}, $\check{f}$ can be viewed as a function defined on the generalized cylinder $\Sigma^N = \mathbb{R}\times S^{N-1}$. For more details on the Radon transform we refer the reader to Deans (1983), Gelfand et al (1966), and Helgason (1980).

The following properties of the Radon transform are easy to establish:

$$\check{f}(\alpha p, \alpha\underline{\xi}) = \frac{1}{|\alpha|}\check{f}(p,\underline{\xi}) \quad \text{for any real } \alpha \neq 0; \tag{2.4}$$

$$\mathcal{R}(f(A\underline{x})) = |\det B|\,\check{f}(p, B^{\mathrm{T}}\underline{\xi}) \quad \text{where } B = A^{-1}; \tag{2.5}$$

$$\mathcal{R}(f(\underline{x}+\underline{a})) = \check{f}(p+\underline{\xi}\cdot\underline{a},\underline{\xi}); \tag{2.6}$$

$$\check{f}(p,\underline{\xi}) = \int_{\mathbb{R}} \tilde{f}(\omega\underline{\xi})e^{2\pi i\omega p}d\omega, \tag{2.7}$$

where $\tilde{f}$ denotes the Fourier transform of f:

$$\tilde{f}(\underline{y}) = \int_{\mathbb{R}^N} f(\underline{x})e^{-2\pi i\underline{x}\cdot\underline{y}}\,d\underline{x}.$$

It is proven in Deans (1983), Ch. 5, that if f is a nice function then it can be recovered from its Radon transform according to the following formulas

$$f(\underline{x}) = \frac{1}{(2\pi i)^N}\int_{|\underline{\xi}|=1} d\underline{\xi}\int_{-\infty}^{\infty} dp\,\frac{\left(\frac{\partial}{\partial p}\right)^{N-1}\check{f}(p,\underline{\xi})}{p-\underline{\xi}\cdot\underline{x}} \tag{2.8}$$

when n is even, and

$$f(\underline{x}) = \frac{1}{2(2\pi i)^{N-1}}\int_{|\underline{\xi}|=1}\left(\frac{\partial}{\partial p}\right)^{N-1}\check{f}(\underline{\xi}\cdot\underline{x},\underline{\xi})\,d\underline{\xi} \tag{2.9}$$

when n is odd, where

$$\frac{\partial}{\partial p}\check{f}(\underline{\xi}\cdot\underline{x},\underline{\xi}) = \left[\frac{\partial}{\partial p}\check{f}(p,\underline{\xi})\right]_{p=\underline{\xi}\cdot\underline{x}}.$$

Therefore, we can talk about the inverse Radon transform $\mathcal{R}^{-1}$. Thus $\mathcal{R}^{-1}\check{f} = f$.

3. Δ-CONVERGENCE

By $\mathcal{S}(\mathbb{R}^N)$ we denote the space of complex-valued infinitely differentiable rapidly decreasing functions on $\mathbb{R}^N$. More precisely, an infinitely differentiable function f on $\mathbb{R}^N$ is in $\mathcal{S}(\mathbb{R}^N)$ if and only if

$$\|f\|_m = \sup_{|\alpha|\le m}\ \sup_{\underline{x}\in\mathbb{R}^N}\left(1+\|\underline{x}\|^2\right)^m\left|\frac{\partial^{|\alpha|}f}{\partial\underline{x}^{\alpha}}\right| < \infty, \tag{3.1}$$

where m is a non-negative integer, $\alpha = (\alpha_1, \ldots, \alpha_N)$, $|\alpha| = \alpha_1 + \ldots + \alpha_N$, and

$$\frac{\partial^{|\alpha|}}{\partial \underline{x}^{\alpha}} = \frac{\partial^{|\alpha|}}{\partial x_1^{\alpha_1} \ldots \partial x_N^{\alpha_N}}.$$

If $f, g \in \mathcal{S}(\mathbb{R}^N)$, then by $f*g$ we mean the convolution product of f and g, i.e.,

$$(f*g)(\underline{y}) = \int_{\mathbb{R}^N} f(\underline{x})\, g(\underline{y} - \underline{x})\, d\underline{x},$$

where $\underline{x} = (x_1, \ldots, x_N) \in \mathbb{R}^N$ and $d\underline{x} = dx_1 \ldots dx_N$.

For $\phi \in \mathfrak{D}(\mathbb{R}^N)$, define $s(\phi)$ to be the radius of the smallest closed ball at the origin that contains the support of ϕ, i.e.,

$$s(\phi) = \inf \{ \varepsilon > 0 \colon \operatorname{supp} \phi \subset B_\varepsilon \}$$

where $\| \underline{x} \| = \| (x_1, \ldots, x_N) \| = \sqrt{x_1^2 + \ldots + x_N^2}$ and $B_\varepsilon = \{ \underline{x} \in \mathbb{R}^N \colon \| \underline{x} \| < \varepsilon \}$.

A sequence of real-valued functions $\delta_1, \delta_2, \ldots \in \mathfrak{D}(\mathbb{R}^N)$ is called a *delta sequence* if

$$\int_{\mathbb{R}^N} \delta_n(\underline{x})\, d\underline{x} = 1 \quad \text{for every } n \in \mathbb{N}, \tag{3.2}$$

$$\int_{\mathbb{R}^N} | \delta_n(\underline{x}) |\, d\underline{x} < M \quad \text{for some } M > 0 \text{ and all } n \in \mathbb{N}, \tag{3.3}$$

$$s(\delta_n) \to 0 \quad \text{as} \quad n \to \infty. \tag{3.4}$$

The collection of all delta sequences will be denoted by Δ.

Let $f, f_n \in \mathcal{S}(\mathbb{R}^N)$. We say that the sequence $\{f_n\}$ is Δ-*convergent* to f, and write $\Delta\text{-}\lim_{n \to \infty} f_n = f$, if there exists a delta sequence $\{\delta_n\} \in \Delta$ such that the sequence $(f - f_n)*\delta_n \to 0$ in $\mathcal{S}(\mathbb{R}^N)$, i.e., $\| (f - f_n)*\delta_n \|_m \to 0$ for every non-negative integer m.

It can be shown that the topology of Δ-convergence is metrizable (Mikusiński, 1983). Clearly, if $f_n \to f$ in $\mathcal{S}(\mathbb{R}^N)$, then $\Delta\text{-}\lim_{n \to \infty} f_n = f$. On the other hand, there are Δ-convergent sequences which are not convergent in $\mathcal{S}(\mathbb{R}^N)$.

Let $\mathcal{S}(\Sigma^N)$ denote the space of all complex-valued infinitely differentiable functions f on Σ^N such that

$$\| f \|_{\Sigma, m} = \sup_{\underline{\xi} \in S^{N-1}} \ \sup_{k \le m} \ \sup_{p \in \mathbb{R}} \left(1 + p^2\right)^m \left| \frac{\partial^k}{\partial p^k} f(p, \underline{\xi}) \right| < \infty. \tag{3.5}$$

For $f, g \in \mathcal{S}(\Sigma^N)$, by $f \odot g$ we denote the convolution of f and g with respect to p, i.e.,

$$(f \odot g)(p, \underline{\xi}) = \int_{\mathbb{R}} f(t, \underline{\xi})\, g(p - t, \underline{\xi})\, dt.$$

Thus, if $f, g \in \mathcal{S}(\Sigma^N)$, then $f \odot g \in \mathcal{S}(\Sigma^N)$.

Let $\mathfrak{D}(\Sigma^N)$ be the space of all infinitely differentiable functions on Σ^N with compact support. Let Δ_Σ denote the family of all sequences $\phi_1, \phi_2, \ldots \in \mathfrak{D}(\Sigma^N)$ such that

$$\int_{\mathbb{R}} \phi_n(p, \underline{\xi})\, dp = 1 \quad \text{for every } \underline{\xi} \in S^{N-1} \text{ and } n \in \mathbb{N}, \tag{3.6}$$

$$\sup_{\underline{\xi} \in S^{N-1}} \int_{\mathbb{R}} |\, \phi_n(p, \underline{\xi}) \,|\, dp < M \quad \text{for some } M > 0 \text{ and all } n \in \mathbb{N}, \tag{3.7}$$

$$\sup_{\underline{\xi} \in S^{N-1}} s(\phi_n(\,\cdot\,, \underline{\xi})) \to 0 \quad \text{as} \quad n \to \infty. \tag{3.8}$$

Δ-convergence in $\mathcal{S}(\Sigma^N)$ is defined as in $\mathcal{S}(\mathbb{R}^N)$:

A sequence $f_n \in \mathcal{S}(\Sigma^N)$ is called Δ-*convergent* to $f \in \mathcal{S}(\Sigma^N)$, $\Delta\text{-}\lim_{n\to\infty} f_n = f$, if there exists a delta sequence $\{\phi_n\} \in \Delta_\Sigma$ such that $\| (f - f_n) \odot \delta_n \|_{\Sigma, m} \to 0$ for every non-negative integer m.

The following theorem is crucial for the extension of the Radon transform presented in this paper.

Theorem 1. If $\Delta\text{-}\lim_{n\to\infty} f_n = f$ in $\mathcal{S}(\mathbb{R}^N)$, then $\Delta\text{-}\lim_{n\to\infty} \mathcal{R} f_n = \mathcal{R} f$ in $\mathcal{S}(\Sigma^N)$.

The proof of this theorem is based on the following two lemmas; the proof of the first lemma can be found in Mikusiński and Zayed (1992).

Lemma 1. If $\{\phi_n(\underline{x})\} \in \Delta$, then $\{\check{\phi}_n(p, \underline{\xi})\} \in \Delta_\Sigma$.

Lemma 2. If $f_n \to f$ in $\mathcal{S}(\mathbb{R}^N)$, then $\mathcal{R} f_n \to \mathcal{R} f$ in $\mathcal{S}(\Sigma^N)$.

Proof. It suffices to show that if $f_n \to 0$ in $\mathcal{S}(\mathbb{R}^N)$ then $\mathcal{R} f_n \to 0$ in $\mathcal{S}(\Sigma^N)$. Let $f_n \to 0$ in $\mathcal{S}(\mathbb{R}^N)$. Since the Fourier transform is a continuous mapping from $\mathcal{S}(\mathbb{R}^N)$ into itself, we have $\tilde{f}_n \to 0$ in $\mathcal{S}(\mathbb{R}^N)$, where $\tilde{f}_n$ is the Fourier transform of f_n. From (2.7) it follows, by integration by parts, that

$$\begin{aligned} p^l \frac{\partial^k}{\partial p^k} (\mathcal{R} f_n)(p, \underline{\xi}) &= \int_{\mathbb{R}} p^l (2\pi i \omega)^k \tilde{f}_n(\omega \underline{\xi})\, e^{2\pi i \omega p}\, d\omega \\ &= (-1)^l (2\pi i)^{k-l} \int_{\mathbb{R}} \frac{d^l}{d\omega^l} \Big(\omega^k \tilde{f}_n(\omega \underline{\xi}) \Big) e^{2\pi i \omega p}\, d\omega, \end{aligned}$$

for a fixed $\underline{\xi} \in S^{N-1}$.

Since the function $\tilde{f}_n$ is in $\mathcal{S}(\mathbb{R}^N)$, $g_n(\omega \underline{\xi}) = \omega^k \tilde{f}_n(\omega \underline{\xi})$ is also in $\mathcal{S}(\mathbb{R}^N)$. Moreover $g_n \to 0$ in $\mathcal{S}(\mathbb{R}^N)$. Therefore, for any given α and any $\varepsilon_1 > 0$, we can chose N large enough so that

$$\sup_{\underline{x} \in \mathbb{R}^N} \left| \frac{\partial^{|\alpha|} g_n(\underline{x})}{\partial \underline{x}^\alpha} \right| \le \frac{\varepsilon_1}{(1 + \| \underline{x} \|^2)^m},$$

for all $n \ge N$ and any m. On the other hand,

$$\frac{d^l g_n(\omega \underline{\xi})}{d\omega^l} = \sum_{|\alpha| = l} A_\alpha \frac{\partial^{|\alpha|} g_n(\underline{x})}{\partial \underline{x}^\alpha},$$

for some constants A_α that depend on $\underline{\xi}$ with

$$|\, A_\alpha \,| \le \begin{cases} \dbinom{l}{l/2} & \text{if } l \text{ is even} \\[2ex] \dbinom{l}{(l+1)/2} & \text{if } l \text{ is odd} \end{cases}.$$

Hence,

$$\sup_{\omega\in\mathbb{R}}\left|\frac{d^l g_n(\omega\underline{\xi})}{d\omega^l}\right| \le \frac{C_1\varepsilon_1}{(1+\|\underline{x}\|^2)^m} = \frac{C_2\varepsilon_1}{(1+|\omega|^2)^m},$$

for some constants C_1 and C_2 that depend on l. Then

$$\sup_{p\in\mathbb{R}}\left|p^l\frac{\partial^k(\mathcal{R}f_n)(p,\underline{\xi})}{\partial p^k}\right| \le C\varepsilon_1\int\frac{d\omega}{(1+|\omega|^2)^m}.$$

By choosing $\varepsilon_1 < \varepsilon\left(C\int\frac{d\omega}{(1+|\omega|^2)^m}\right)^{-1}$, we obtain

$$\sup_{p\in\mathbb{R}}\left|p^l\frac{\partial^k(\mathcal{R}f_n)(p,\underline{\xi})}{\partial p^k}\right| < \varepsilon,$$

which implies $\|\mathcal{R}f_n\|_{\Sigma,m}\to 0$.

Proof of Theorem 1. Let $f, f_n \in \mathcal{S}(\mathbb{R}^N)$ and $\Delta\text{-}\lim_{n\to\infty} f_n = f$. Then there exists a delta sequence $\{\delta_n\}\in\Delta$ such that $\|(f-f_n)*\delta_n\|_m\to 0$ for every non-negative integer m. Let $\phi_n = \mathcal{R}\delta_n$, $n\in\mathbb{N}$. Then, by Lemma 1, $\{\phi_n\}\in\Delta_\Sigma$. Moreover, since $\mathcal{R}(f*g) = \mathcal{R}f\odot\mathcal{R}g$, see Deans (1983), p. 95, we have

$$\|(\mathcal{R}f-\mathcal{R}f_n)\odot\phi_n\|_{\Sigma,m} = \|\mathcal{R}(f-f_n)\odot\mathcal{R}\delta_n\|_{\Sigma,m} = \|\mathcal{R}((f-f_n)*\delta_n)\|_{\Sigma,m}\to 0,$$

by Lemma 2.

The following property of Δ-convergence will be used in the next section.

Lemma 3. (Mikusiński, 1983) Let $f_n\in\mathcal{S}(\mathbb{R}^N)$, $n = 1,2,3,\ldots$. If there exists a delta sequence $\{\delta_n\}\in\Delta$ such that $\|f_n*\delta_k\|_m\to 0$ for every $k\in\mathbb{N}$, then $\Delta\text{-}\lim_{n\to\infty} f_n = f$.

4. THE EXTENSION OF THE RADON TRANSFORM

The space $\mathcal{S}(\mathbb{R}^N)$ is not complete with respect to Δ-convergence. For example, if $f_n(\underline{x}) = n^N\phi(n\underline{x})$, where $\phi\in\mathcal{S}(\mathbb{R}^N)$, then $\{f_n\}$ is a Δ-Cauchy sequence in $\mathcal{S}(\mathbb{R}^N)$, but it does not converge to an element of $\mathcal{S}(\mathbb{R}^N)$, unless $\phi = 0$. In this example, however, the sequence $\{f_n\}$ converges in the distributional sense. In the next example we are going to construct a sequence of functions in $\mathcal{S}(\mathbb{R}^N)$ which is Δ-Cauchy, but is not convergent in $\mathcal{S}'(\mathbb{R}^N)$. We use an idea of Boehme (1973).

Example 1. From the Denjoy-Carleman theorem it follows that there exists $\psi\in\mathcal{D}(\mathbb{R})$ such that $\int\psi = 1$ and

$$\left|\psi^{(n)}(t)\right| \le AB^n(2n)! \quad \text{for all } t\in\mathbb{R}; \tag{4.1}$$

see Rudin (1974), p. 412. Define

$$\phi(\underline{x}) = \psi(x_1)\cdots\psi(x_N) \quad\text{and}\quad \phi_n(\underline{x}) = n^N\phi(n\underline{x})$$

and

$$f_n(\underline{x}) = \sum_{j=0}^{\infty}\frac{1}{(3j)!}\frac{\partial^j}{\partial x_1^j}\phi_n(\underline{x}) \tag{4.2}$$

In view of (4.1), the series in (4.2) converges and $f_n\in\mathcal{S}(\mathbb{R}^N)$ for all $n\in\mathbb{N}$. We will show that $\{f_n\}$ is a Δ-Cauchy sequence. According to Lemma 3, it suffices to show that there exists a delta sequence $\{\delta_n\}$ such that, for every $k\in\mathbb{N}$, $(f_n - f_m)*\delta_k\to 0$

in $\mathcal{S}(\mathbb{R}^N)$ as $n, m \to \infty$. In fact, for $\delta_k = \phi_k$, we have

$$(f_n - f_m)*\delta_k = \left(\sum_{j=0}^{\infty} \frac{1}{(3j)!} \frac{\partial^j}{\partial x_1^j}(\phi_n - \phi_m)\right)*\phi_k = \left(\sum_{j=0}^{\infty} \frac{1}{(3j)!} \frac{\partial^j}{\partial x_1^j}\big((\phi_n - \phi_m)*\phi_k\big)\right)$$

$$= \left(\sum_{j=0}^{\infty} \frac{1}{(3j)!} \frac{\partial^j}{\partial x_1^j}\phi_k\right)*(\phi_n - \phi_m) = f_k*\phi_n - f_k*\phi_m \to f_k - f_k = 0$$

as $n, m \to \infty$.

The sequence $\{f_n\}$ cannot converge in $\mathcal{S}'(\mathbb{R}^N)$ because its limit would be a distribution of infinite order with compact support. This example shows that the completion $\mathcal{S}(\mathbb{R}^N)$ with respect to Δ-convergence contains some object which cannot be identified with distributions.

Denote by $\mathfrak{B}_{\mathcal{S}}$ and $\mathfrak{B}_{\Sigma}$ the completions with respect to Δ-convergence of $\mathcal{S}(\mathbb{R}^N)$ and $\mathcal{S}(\Sigma^N)$, respectively. Since $\mathcal{S}(\mathbb{R}^N)$ is dense in $\mathfrak{B}_{\mathcal{S}}$, the Radon transform has a unique extension to a continuous linear transform form $\mathfrak{B}_{\mathcal{S}}$ into $\mathfrak{B}_{\Sigma}$.

Definition 1. Let $F \in \mathfrak{B}_{\mathcal{S}}$ and $F = \Delta\text{-}\lim_{n\to\infty} f_n$, where $f_n \in \mathcal{S}(\mathbb{R}^N)$. Then the Radon transform of F is defined as

$$\mathcal{R}F = \Delta\text{-}\lim_{n\to\infty} \mathcal{R}f_n.$$

The spaces $\mathfrak{B}_{\mathcal{S}}$ and $\mathfrak{B}_{\Sigma}$ can be also defined algebraically. Elements of those spaces are defined as "quotients of sequences" and called Boehmians. The extended Radon transform can be then defined in terms of these quotients. This approach is used in Mikusiński and Zayed (1992).

REFERENCES

Boehme, T.K., 1973, The support of Mikusiński operators, *Trans. Amer. Math. Soc.* 176, 319:334.

Deans, S., 1983, The Radon Transform and Some of its Applications, A Wiley-Interscience Publ., John Wiley & Sons, New York.

Gelfand, I.M., and Shilov, G.E., 1964, Generalized Functions, Vols. 1 & 2, Academic Press, New York.

Gelfand, I.M., Graev, M.I., and Vilenkin, N.Ya., 1966, Generalized Functions, Vol. 5, Academic Press, New York.

Helgason, S., 1980, The Radon Transform, Birkhäuser, Boston-Basel-Stuttgart.

Ludwig, D., 1966, The Radon transform on Euclidean spaces, *Comm. Pure and Appl. Math.* 19, 49:81.

Mikusiński, J., 1959, Operational Calculus, Pergamon Press, London.

Mikusiński, J., and Mikusiński, P., 1981, Quotients de suites et leurs applications dans l'analyse fonctionnelle, *Comptes Rendus* 293 Serie I, 463:464.

Mikusiński, P., 1983, Convergence of Boehmians, *Japan. J. Math.* 9, 159:179.

Mikusiński, P., 1988, Boehmians and generalized functions, *Acta Math. Hungarica* 51, 271:281.

Mikusiński, P., and Zayed, A., 1992, The Radon transform of Boehmians, *Proc. Amer. Math. Soc.*, (to appear).

Radon, J., 1917, Über die Bestimmung von Funktionen durch ihre Integralwärte längs gewisser Manningfaltigkeiten, *Ber. Verh. Sächs. Akad.* 69, 262:277.

Rudin, W., 1974, Real and Complex Analysis, 2nd Edition, McGraw-Hill, New York.

Schwartz, L., 1950-1951, Théorie des Distributions, Hermann, Paris.

Zemanian, A.H., 1968, Generalized Integral Transformations, Interscience Publishers, New York.

ON THE EXPANSION OF ZONAL HOLOMORPHIC FUNCTIONS ON THE COMPLEX SPHERE

Mitsuo Morimoto[1], Ryoko Wada[2], and Keiko Fujita[1]

[1]Faculty of Science and Technology, Sophia University
Chiyoda-ku Tokyo 102 Japan
[2]Faculty of Integrated Arts and Sciences, Hiroshima University, Hiroshima City, Hiroshima 730 Japan

We start this note with two motivations. The first motivation is the Fourier series. Suppose S^1 is the circle (i.e. 1 dimensional sphere). If we want to study functions or generalized functions on it, we study the Fourier series. If we are working in the analytic category, our objects are real analytic functions and hyperfunctions on S^1 and they can be characterized by the growth conditions of their Fourier coefficients. For example,

$$f(t) = \sum_{n=-\infty}^{\infty} a_n t^n, \quad |t| = 1 \tag{1}$$

is real analytic if and only if

$$\limsup_{|n|\to\infty} \sqrt[|n|]{|a_n|} < 1.$$

If we consider the condition

$$\limsup_{|n|\to\infty} \sqrt[|n|]{|a_n|} \leq \frac{1}{r} \quad (1 < r), \tag{2}$$

then the function (1) can be extended to a holomorphic function in the complex neighborhood

$$\{w \in \mathbb{C}; \ \frac{1}{r} < |w| < r\} \tag{3}$$

and the converse is also true. Thus the Fourier series and the Laurent series correspond to each other in this way and the annular neighborhood (3) is a good neighborhood of S^1 in the sense that it reflects the convergence of (1).

How can we generalize (3) in higher dimension? One possible generalization is the "multiple Fourier series" and the d dimensional torus $S^1 \times \cdots \times S^1$. But the first and second named authors established a generalization to the d dimensional sphere S^d considering the d (complex) dimensional complex sphere $\tilde{S}^d$ (Morimoto [4], Wada [11], [12] and Morimoto-Wada [5]).

Generalized Functions and Their Applications, Edited by
R.S. Pathak, Plenum Press, New York, 1993

We write the well known 1 dimensional case in the form which may be extended to S^d and $\tilde{S}^d$. First we introduce 1 dimensional complex sphere

$$\tilde{S}^1 = \{(z_1, z_2) \in \mathbb{C}^2;\ z_1^2 + z_2^2 = 1\}.$$

It turns out that $\tilde{S}^1 \cong \mathbb{C}^*$ and our good neighborhood (3) corresponds biholomorphically to

$$\tilde{S}^1(r) = \{(z_1, z_2) \in \tilde{S}^1;\ \max |z_1 \pm i z_2| < \frac{1}{r}\}.$$

Replacing the norm $\max\{|z_1 \pm i z_2|\}$ on $\mathbb{C}^2$ by the Lie norm, we can generalize the condition (2) on the Fourier-Laurent series to the higher dimensional sphere case (Morimoto [3], [4]).

The second motivation is a relation of the power series and the Legendre series expansion. Consider the following power series and Legendre series:

$$\varphi(z) = \sum_{n=0}^{\infty} a_n z^n, \tag{4}$$

$$g(t) = \sum_{n=0}^{\infty} a_n P_{n,d}(t), \tag{5}$$

where $P_{n,d}(t)$ denotes the Legendre polynomial of order n and of dimension d. Suppose $r > 1$. It is known that (4) converges in

$$B_r = \{z \in \mathbb{C};\ |z| < r\}$$

if and only if (5) converges in

$$E_r = \{t \in \mathbb{C};\ |t - 1| + |t + 1| < r + \frac{1}{r}\}.$$

(We refer the reader to Szegö [9] or Geronimus [2].)

During her master's course, the third named author studied A. I. Zayed [15] and earlier papers by G. Szegö [10], Z. Nehari [7] and G. Walter [14] which treat this fact and related problem of analytic continuation for $d = 2$. We found that this phenomenon is related to the complex sphere of the first motivation.

Furthermore if we take a sequence of non-zero numbers $\{c_n\}$ with

$$\limsup_{n\to\infty} \sqrt[n]{|c_n|} = 1$$

and form

$$\varphi_c(z) = \sum_{n=0}^{\infty} a_n c_n z^n, \tag{6}$$

then we can replace (4) by (6) in the above statement. The following questions arise; What is a natural choice of $\{c_n\}$? How can we explain this relation between the power series expansion and the Legendre polynomial expansion?

In this note, we explain this relation and propose one meaningful choice of $\{c_n\}$ from the point of view of the expansion of holomorphic functions on the complex sphere. Our interpretation is given in Theorem 4 below.

1. SPHERICAL HARMONIC EXPANSION

Let $S^d = \{(x_1, x_2, \cdots, x_{d+1}) \in \mathbb{R}^{d+1};\ \|x\| = 1\}$ be the d dimensional real sphere, where $\|x\|^2 = x_1^2 + x_2^2 + \cdots + x_{d+1}^2$.

We denote the volume of S^d by

$$\Omega_d = |S^d| = \frac{2\pi^{(d+1)/2}}{\Gamma((d+1)/2)}.$$

Let $\tilde{H}_{n,d}$ be the linear space of n-homogeneous polynomials of $d+1$ variables with complex coefficients and $\tilde{H}^{\Delta}_{n,d}$ the subspace of $\tilde{H}_{n,d}$ consisting of harmonic polynomials:

$$\tilde{H}^{\Delta}_{n,d} = \{\tilde{P} \in \tilde{H}_{n,d};\ \Delta\tilde{P} = 0\},$$

where Δ is the Laplacian:

$$\Delta = \frac{\partial^2}{\partial x_1^2} + \cdots + \frac{\partial^2}{\partial x_{d+1}^2}.$$

$H_{n,d} = \tilde{H}^{\Delta}_{n,d}|_{S^d}$ is, by definition, the linear space of *spherical harmonics* of order n and of dimension d. It is known that

$$\begin{aligned} N(n,d) &= \dim H_{n,d} = \dim \tilde{H}^{\Delta}_{n,d} \\ &= \frac{(2n+d-1)(n+d-2)!}{n!(d-1)!} = O(n^{d-1}). \end{aligned}$$

Among general references on spherical harmonics is C. Müller [6].

Let $\varepsilon_0 = (1, 0, \cdots, 0)$ be the north pole in S^d. A function f on S^d is said to be *zonal*, if $f(A\omega) = f(\omega)$ for any rotation A with $A\varepsilon_0 = \varepsilon_0$. It is classical that there exists one and only one zonal polynomial $\tilde{P} \in \tilde{H}^{\Delta}_{n,d}$ such that $\tilde{P}(\varepsilon_0) = 1$. The *Legendre polynomial* $P_{n,d}$ of order n and of dimension defined by

$$\tilde{P}(z) = (\sqrt{z^2})^n P_{n,d}(\langle \varepsilon_0, \frac{z}{\sqrt{z^2}} \rangle),$$

where we denote $z^2 = z_1^2 + z_2^2 + \cdots + z_{d+1}^2$ and $\langle z, w\rangle = z_1 w_1 + z_2 w_2 + \cdots + z_{d+1} w_{d+1}$.

The n-th spherical harmonic component of $f \in L^2(S^d)$ is given by

$$S_n(f,\omega) = \frac{N(n,d)}{\Omega_d} \int_{S^d} f(\tau) P_{n,d}(\langle \omega, \tau\rangle) d\tau,$$

where $d\tau$ is the invariant measure on S^d induced from the Euclidean measure on $\mathbb{R}^{d+1}$. We know that $S_n(f,\cdot)$ belongs to $H_{n,d}$ and f can be expanded as follows:

$$f(\omega) = \sum_{n=0}^{\infty} S_n(f,\omega) \text{ in } L^2(S^d).$$

Further we have the direct sum decomposition of the Hilbert space:

$$L^2(S^d) = \bigoplus_{n=0}^{\infty} H_{n,d}.$$

The space $\mathcal{A}(S^d)$ of real analytic functions on S^d is a subspace of $L^2(S^d)$.

2. LIE NORM

We define the *Lie norm* on $\mathbb{C}^{d+1}$ as

$$L(z) = \left\{ \sum_{j=1}^{d+1} |z_j|^2 + \left\{ (\sum_{j=1}^{d+1} |z_j|^2)^2 - |\sum_{j=1}^{d+1} z_j^2|^2 \right\}^{1/2} \right\}^{1/2}.$$

Drużkowsky [1] proved that the Lie norm is a cross norm of $\|x\|$ in the following sense:

$$L(z) = \inf\{\sum_{j=1}^{m} |\lambda_j|\|x_j\|;\ z = \sum_{j=1}^{m} \lambda_j x_j, \lambda \in \mathbb{C}, x_j \in \mathbb{R}^{d+1}, m \in \mathbb{N}\}.$$

The ball with respect to the Lie norm

$$\tilde{B}(r) = \{z \in \mathbb{C}^{d+1};\ L(z) < r\}$$

is called the *Lie ball* of radius r. It is known that any harmonic function in the real ball $\{x \in \mathbb{R}^{d+1};\ \|x\| < r\}$ can be extended analytically to the Lie ball $\tilde{B}(r)$ (Siciak [8]). The Bergman-Šilov boundary of $\tilde{B}(r)$ is given by

$$\Sigma(r) = \{e^{i\theta}x;\ \|x\| = r, x \in \mathbb{R}^{d+1}, \theta \in \mathbb{R}\}.$$

For an n-homogeneous polynomial $\tilde{P} \in \tilde{H}_{n,d}$, we have

$$\|\tilde{P}\|_{L^\infty(S^d)} = \sup\{|\tilde{P}(z)|;\ z \in \tilde{B}(1)\}.$$

Hence we have

$$|\tilde{P}(z)| \leq L(z)^n \|\tilde{P}\|_{L^\infty(S^d)} \text{ for } z \in \mathbb{C}^{d+1}.$$

As the Legendre polynomial satisfies

$$\sup_{-1\leq t\leq 1} |P_{n,d}(t)| = 1,$$

we have the following proposition:

Proposition 1

$$|P_{n,d}(\langle z, \tau\rangle)| \leq L(z)^n$$

for $\tau \in S^d$ *and* $z \in \mathbb{C}^{n+1}$ *with* $z^2 = 1$.

3. COMPLEX NEIGHBORHOOD OF THE REAL SPHERE

The *complex sphere*

$$\tilde{S}^d = \{z \in \mathbb{C}^{d+1};\ z^2 = 1\}$$

is the natural complexification of the real sphere S^d. For $r > 1$, we put

$$\tilde{S}^d(r) = \{z \in \tilde{S}^d;\ L(z) < r\}.$$

Let $\mathcal{O}(\tilde{S}^d(r))$ be the space of holomorphic functions on $\tilde{S}^d(r)$.

Then the family $\{\tilde{S}^d(r);\ r > 1\}$ forms a base of complex neighborhoods of S^d and we have

$$\mathcal{A}(S^d) = \operatorname{ind}\lim_{r\downarrow 1} \mathcal{O}(\tilde{S}^d(r)).$$

As the restriction mapping of $\mathcal{O}(\tilde{S}^d(r))$ to $\mathcal{A}(S^d)$ is injective, we identify $\mathcal{O}(\tilde{S}^d(r))$ with a subspace of $\mathcal{A}(S^d)$. We have the following theorem:

Theorem 1 (Morimoto [4]).

A function f on S^d belongs to $\mathcal{O}(\tilde{S}^d(r))$, if and only if

$$\limsup_{n\to\infty} \left(\|S_n(f, \cdot)\|_{L^\infty(S^d)}\right)^{1/n} \leq \frac{1}{r}.$$

Corollary 1

A function f on S^d belongs to $\mathcal{A}(S^d)$, if and only if

$$\limsup_{n\to\infty}\left(\|S_n(f,\cdot)\|_{L^\infty(S^d)}\right)^{1/n} < 1.$$

Let $\mathcal{O}(\tilde{B}(r))$ be the linear space of holomorphic functions on the Lie ball. We denote by $\mathcal{O}_\lambda(\tilde{B}(r))$ the subspace of $\mathcal{O}(\tilde{B}(r))$ defined by $(\Delta_z + \lambda^2)f = 0$, where

$$\Delta_z = \frac{\partial^2}{\partial z_1^2} + \cdots + \frac{\partial^2}{\partial z_{d+1}^2}.$$

Theorem 2 (R.Wada[12])

The restriction mapping

$$\alpha_\lambda : \mathcal{O}_\lambda(\tilde{B}(r)) \longrightarrow \mathcal{O}(\tilde{S}^d(r))$$

is bijective provided

$$(\lambda/2)^{-n-(d-1)/2} J_{n+(d-1)/2}(\lambda) \neq 0 \text{ for } n = 0, 1, 2, \cdots. \tag{7}$$

It is clear that $\lambda = 0$ satisfies the condition (7).

4. COMPLEX LIGHT CONE

Now we consider the *complex light cone* $M = \{z \in \mathbb{C}^{d+1};\ z^2 = 0\}$. For $r > 0$, $M(r) = \{z \in M;\ L(z) < r\}$ denotes the truncated complex light cone.

Theorem 3 (R.Wada-Morimoto[13])

For any $\lambda \in \mathbb{C}$, the restriction mapping

$$\beta_\lambda : \mathcal{O}_\lambda(\tilde{B}(r)) \to \mathcal{O}(M(r))$$

is bijective.

Corollary 2

The mapping $\beta_0 \circ \alpha_0^{-1}$ is a linear isomorphism of $\mathcal{O}(\tilde{S}^d(r))$ onto $\mathcal{O}(M(r))$.

We will show in the following section that the isomorphism $\beta_0 \circ \alpha_0^{-1}$ applied to zonal functions gives rise to a result similar to the theorem of the second motivation.

5. ZONAL FUNCTIONS

The following proposition is well known (See Müller [6]).

Proposition 2

Suppose $f \in L^2(S^d)$ is zonal.

(i) $\hat{f}(t) = f((t, \pm\sqrt{1-t^2}, 0, \cdots, 0))$ *is a function on the interval* $[-1, 1]$ *and we have* $f(z) = \hat{f}(\langle\varepsilon_0, z\rangle)$.

(ii) *The L^2 norm of f is given as follows :*

$$\|f\|^2_{L^2(S^d)} = \Omega_{d-1} \int_{-1}^{1} |\hat{f}(t)|^2 (1-t^2)^{(d-2)/2} dt.$$

(iii) *$S_n(f,\cdot)$ is also zonal. By the uniqueness of zonal spherical harmonics, we have*

$$S_n(f,\omega) = a_n P_{n,d}(\langle \varepsilon_0, \omega \rangle)$$

and

$$\hat{f}(t) = \sum_{n=0}^{\infty} a_n P_{n,d}(t),$$

where the coefficients a_n is given by

$$\begin{aligned} a_n &= S_n(f,\varepsilon_0) \\ &= \frac{N(n,d)}{\Omega_d} \int_{S^d} f(\tau) P_{n,d}(\langle \varepsilon_0, \tau \rangle) d\tau \\ &= \frac{N(n,d)}{\Omega_d} \Omega_{d-1} \int_{-1}^{1} \hat{f}(t) P_{n,d}(t) (1-t^2)^{(d-2)/2} dt. \end{aligned}$$

By direct calculation, we have the following lemma:

Lemma 1

For $z = (z_1, z_2, 0, \cdots, 0)$, we have $L(z) = \max\{|z_1 + iz_2|, |z_1 - iz_2|\}$.

Proposition 3

We have

$$|P_{n,d}(t)| \leq \max |t \pm \sqrt{t^2-1}| \ \textit{for}\ t \in \mathbb{C}. \tag{8}$$

Proof

Take $z = (t, \sqrt{1-t^2}, 0, \cdots, 0)$ in Proposition 1. The preceding lemma implies (8).

q.e.d.

As usual, we denote $\max |t \pm \sqrt{t^2-1}|$ by $|t + \sqrt{t^2-1}|$. We have

$$\begin{aligned} E_r &= \{t \in \mathbb{C};\ |t + \sqrt{t^2-1}| < r\} \\ &= \{t \in \mathbb{C};\ |t+1| + |t-1| < r + \frac{1}{r}\}. \end{aligned}$$

A result of R. Wada [12] implies the following proposition.

Proposition 4

If a zonal function f is holomorphic in $\tilde{S}^d(r)$, we can expand it as follows :

$$f(z) = \sum_{n=0}^{\infty} a_n P_{n,d}(\langle \varepsilon_0, z \rangle), \quad z \in \tilde{S}^d(r).$$

Then

$$\alpha_0^{-1} f(z) = \tilde{f}(z) = \sum_{n=0}^{\infty} a_n (\sqrt{z^2})^n P_{n,d}(\langle \varepsilon_0, \frac{z}{\sqrt{z^2}} \rangle)$$

is the complex harmonic extension of f, where α_0 is the restriction mapping defined in Theorem 2.

Proposition 5

Supose $f \in L^2(S^d)$ *is zonal and* $f(z) = \hat{f}(\langle \varepsilon_0, z \rangle)$, *where* $\hat{f}(t) = \sum a_n P_{n,d}(t)$. *Let* $r > 1$. *Then the following conditions are equivalent :*

(i) $f \in \mathcal{O}(\tilde{S}^d(r))$.

(ii) $\hat{f}$ *is holomorphic in* E_r.

(iii) $\limsup_{n\to\infty} \sqrt[n]{|a_n|} \le \frac{1}{r}$.

Proof

(iii) $\Rightarrow$ (ii). For $1 < r_1 < r$ there exists $M \ge 0$ such that

$$|a_n| \le M \left(\frac{1}{r_1}\right)^n.$$

Therefore by (8) we have

$$|a_n P_{n,d}(t)| \le M \left(\frac{1}{r_1}\right)^n (t + \sqrt{t^2 - 1})^n,$$

which implies that $\hat{f}(t)$ converges uniformly in E_{r_2} for every $r_2 < r_1$.

(ii) $\Rightarrow$ (i). If we put $f(z) = g(\langle \varepsilon_0, z \rangle)$, $f(z)$ is a zonal function on S^d and holomorphic in $\tilde{S}^d(r)$. The harmonic extension $\tilde{f}$ of f is expanded as in Proposition 4.

(i) $\Rightarrow$ (iii). Because of Proposition 2 (ii), this is a consequence of Theorem 1.

q.e.d.

Theorem 4

Let $g(t) = \sum_{n=0}^{\infty} a_n P_{n,d}(t)$ *is holomorphic in* E_r. *Then*

$$\varphi_0(\zeta) = \sum_{n=0}^{\infty} a_n \frac{1}{N(n,d)} \frac{\Gamma(n + (d+1)/2)}{\Gamma((d+1)/2)} \frac{2^n}{n!} \zeta^n$$

is holomorphic in $B_{r/2} = \{\zeta \in \mathbb{C};\ |\zeta| < \frac{r}{2}\}$. *The correspondence* $g \mapsto \varphi_0$ *is the mapping* $\beta_0 \circ \alpha_0^{-1}$ *applied to the zonal functions and establishes a linear topological isomorphism of* $\mathcal{O}(E_r)$ *onto* $\mathcal{O}(B_{r/2})$.

Proof

By the assumption

$$f(z) = \sum_{n=0}^{\infty} a_n P_{n,d}(\langle \varepsilon_0, z \rangle)$$

is holomorphic in $\tilde{S}^d(r)$. We know

$$\alpha_0^{-1} f(z) = \tilde{f}(z) = \sum_{n=0}^{\infty} a_n (\sqrt{z^2})^n P_{n,d}(\langle \varepsilon_0, \frac{z}{\sqrt{z^2}} \rangle).$$

On the other hand, it is known that the highest term of the Legendre polynomial is given as follows: (Müller [6])

$$P_{n,d}(t) = \frac{1}{N(n,d)} \frac{\Gamma(n + (d+1)/2)}{\Gamma((d+1)/2)} \frac{2^n}{n!} t^n + \cdots.$$

Therefore if we restrict $\tilde{f}$ to the complex light cone M, we get

$$\tilde{f}(z)|_M = \tilde{f}_M(z) = \sum_{n=0}^{\infty} a_n \frac{1}{N(n,d)} \frac{\Gamma(n+(d+1)/2)}{\Gamma((d+1)/2)} \frac{2^n}{n!} \langle \varepsilon_0, z \rangle^n .$$

On the complex light cone M, we have $z^2 = 0$ and

$$L(z)^2 = 2 \sum_{j=1}^{d+1} |z_j|^2 .$$

If $z = (z_1, iz_1, 0, \cdots, 0)$, we have $L(z)^2 = 4|z_1|^2$.

q.e.d.

Corollary 3

(4) *converges in* B_r *if and only if* (5) *converges in* E_r.

REFERENCES

1. L. Drużkowski, *Effective formula for the crossnorm in the complexified unitary spaces*, Zeszyty Nauk. Uniw. Jagielloń, Prace Mat. **16** (1974), 47–53.
2. Ja. L. Geronimus, *Polynomials Orthogonal on a Circle and Interval*, Pergamon Press, 1960.
3. M. Morimoto, *A generalization of the Fourier-Borel transformation for the analytic functionals with non-convex carrier*, Tokyo J. Math. **2** (1979), 301–322.
4. ———, *Analytic functionals on the sphere and their Fourier-Borel transformations*, Complex Analysis (Warsaw), Banach Center Publications, vol. 11, PWN-Polish Scientific Publishers, 1983, pp. 223–250.
5. M. Morimoto and R. Wada, *Analytic functionals on the complex light cone and their Fourier-Borel transformation*, Algebraic Analysis, vol. 1, Academic Press, 1988, pp. 439–455.
6. C. Müller, *Spherical Harmonics*, Lecture Notes in Mathematics, vol. 17, Springer, 1966.
7. Z. Nehari, *On the singularities of Legendre expansions*, Arch. Rational Mech. Anal. **5** (1956), 987–992.
8. J. Siciak, *Holomorphic continuation of harmonic functions*, Ann. Pol. Math. **29** (1974), 67–73.
9. G. Szegö, *Orthogonal Polynomials*, AMS Colloquium Publications, vol. 23, AMS, 1939, Third edition, 1967.
10. ———, *On the singularities of zonal harmonic expansions*, J. Rational Mech. and Anal. **3** (1954), 561–564.
11. R. Wada, *On the Fourier-Borel transformations of analytic functionals on the complex sphere*, Tôhoku J. Math. **38** (1986), 417–432.
12. ———, *Holomorphic functions on the complex sphere*, Tokyo J. Math. **11** (1988), 205–218.
13. R. Wada and M. Morimoto, *A uniqueness set for the differential operator* $\Delta_z + \lambda^2$, Tokyo J. Math. **10** (1987), 93–105.
14. G. Walter, *On real singularities of Legendre expansions*, Proc. Amer. Math. Soc. **19** (1968), 1407–1412.
15. A. I. Zayed, *On the singularities of Gegenbauer (ultraspherical) expansions*, Trans. Amer. Math. Soc. **262** (1981), 487–503.

A CONVOLUTION SRTUCTURE FOR ALMOST INVARIANT OPERATORS ON A HOMOGENEOUS BANACH SPACE OF DISTRIBUTIONS

G.S. Pandey

School of Studies in Mathematics
Vikram University, Ujjain, India

1. INTRODUCTION

Let G be a compact abelian group and Γ its dual. Let $L_1(G)$ be the Banach space of functions with respect to the norm

$$|| f ||_1 = \int_G |f(x)| \, dx < \infty . \quad (1.1)$$

We suppose that $C^\infty(G)$ is the set of all infinitely differentiable complex-valued functions defined on G and D the space of all continuous linear functionals on it.
The space $L_1(G)$ may be regarded as a linear subspace of D, because for each $f \in L_1(G)$ and $u \in C^\infty(G)$ there exists a regular distribution $f \in D$ defined by

$$f[u] = \int_G f(x) \, u(x)dx \, , \; x \in G. \quad (1.2)$$

Following Braun and Feichtinger [1], we say that $(X(G) : ||.||_X)$ is a Banach space of distributions provided it is isometrically translation invariant and continuously embedded in D. For details see also Sinha and Kant [6].

We denote by R_x the translation operator on X(G) such that

$$R_x \, f(y) = f(y-x), \qquad f \in X(G).$$

If for every $f \in X(G)$ the function $x \longrightarrow R_x f$ is continuous from G to X(G), then we say that X(G) is a homogeneous Banach

Generalized Functions and Their Applications, Edited by
R.S. Pathak, Plenum Press, New York, 1993

space of distributions. The space X(G) is called character invariant if

$$f \in X(G) \longrightarrow (x,\gamma) f \in X(G)$$

for all $\gamma \in \Gamma$ and $x \in G$, where

$$(x,\gamma)\ f[u] = f[(x,\gamma)u], \qquad u \in C^{\infty}(G).$$

Let M(G) be the set of all Radon measures on G. On the lines of Edward [3, p. 73], we define the convolution $\mu * f$ by

$$\mu * f[u] = \mu\ [f * \overset{\vee}{u}\ ]^{\vee},$$

where

$$f * \overset{\vee}{u} = f\ [R_x\ \overset{\vee}{u}\ ]$$

and

$$\overset{\vee}{u}\ (x) = u(-x) \quad \forall\ u \in C^{\infty}\ (G) \ \text{ and } x \in G.$$

It is well known that

$$\mu \in M(G)\ ,\quad f \in X(G) \longrightarrow \mu * f \in X(G).$$

In the present paper we study the Fourier series associated with an almost invariant operator on a homogeneous Banach space of distributions defined on a compact abelian group. Our main result provides a factorization theorem for this operator.

2. HOMOGENEITY OF X(G)

We now prove the following :

Theorem 2.1. If $\{e_\alpha\}_{\alpha\in I}$ is a bounded approximate identity in the convolutable Banach space X(G) of distributions such that $e_\alpha \geq 0$ and $||e_\alpha||_1 = 1$ for all $\alpha \in I$, then X(G) is homogeneous if and only if

$$\lim_{\alpha} ||\ e_\alpha * f - f\ ||_X = 0\ . \tag{2.1}$$

Proof. Let G be a compact abelian group and let X(G) be a homogeneous Banach space of distributions on G. From the homogeneity of the space X(G), it follows that $R_x f$ is continuous X-valued on G $\forall f \in X(G)$ and $x \in G$.

Hence the vector-valued integral

$$\int_G (x,\gamma)\ R_X\ f[u]\ dx$$

is defined for each $\gamma \in \Gamma$.

Thus for each $f \in X(G)$ and $u \in C^{\infty}(G)$, we have

$$(e_\alpha * f)[u] - f[u] = \int_G e_\alpha(y) \{R_x f [\check{u}(y)] - f[u(y)]\} dy$$

which implies that

$$\lim_\alpha \| e_\alpha * f - f \|_X \leq \lim_\alpha \int_G \| R_x f - f \|_X e_\alpha(y)\, dy$$

$$= \lim_\alpha \int_V \| R_x f - f \|_X e_\alpha(y)\, dy$$

$$+ \lim_\alpha \int_{G/V} \| R_x f - f \|_X e_\alpha(y)\, dy$$

$$= I_1 + I_2, \tag{2.2}$$

say, where V is a neighourhood of $\{e_\alpha\}$ in G such that

$$\| R_x f - f \|_X < \varepsilon/2 \tag{2.3}$$

for all $x \in V$. Hence we infer that

$$I_1 < \varepsilon/2 . \tag{2.4}$$

Also, there exists $\alpha_0 \in I$ such that for all $\alpha \geq \alpha_0$, we have

$$I_2 \leq \max_{x \in G} \| R_x f - f \|_X \int_{G/V} e_\alpha(y)\, dy$$

$$\leq 2 \| R_x f - f \|_X \int_{G/V} e_\alpha(y)\, dy$$

$$< 2 \| R_x f - f \|_X \cdot \frac{\varepsilon}{4 \| R_x f \|_X} < \varepsilon/2. \tag{2.5}$$

Combining (2.2), (2.4) and (2.5), we infer that (2.1) holds for each $f \in X(G)$ and $x \in G$.

Conversely, let

$$\lim_\alpha \| e_\alpha * f - f \|_X = 0$$

for all $f \in X$ and $x \in G$.

Hence we have

$$\| R_x f - f \|_X \le \| R_x (f - e_\alpha * f) \|_X + \| R_x (e_\alpha * f) - (e_\alpha * f) \|_X$$

$$+ \| e_\alpha * f - f \|_X$$

$$\le (A + 1) \| f - e_\alpha * f \|_X + \| R_x (e_\alpha * f) - (e_\alpha * f) \|_X$$

$$= J_1 + J_2 , \tag{2.6}$$

say, where A is a positive constant not necessarily the same at each occurrence.

By the hypothesis, we have

$$J_1 = o(1) \quad \text{as } x \longrightarrow 0.$$

Also we see that

$$J_2 = \| f * (R_x e_\alpha - e_\alpha) \|_X$$

$$= \| \int_G f(x-y) \, [R_x e_\alpha(y) - e_\alpha(y)] dy \|_X$$

$$\le \int_G \| f \|_X \, | R_X e_\alpha(y) - e_\alpha (y) | \, dy$$

$$= o(1) \text{ as } x \longrightarrow 0.$$

Hence the theorem holds.

3. ALMOST INVARIANT OPERATORS

We suppose that X(G) is a character invariant convolutable homogeneous Banach space of distributions. Let $\mathscr{L}$ be the Banach algebra of bounded linear operators on X(G) with respect to the norm $\|.\|_{\mathscr{L}}$.

If an operator $T \in \mathscr{L}$ satisfies the condition

$$T R_x = R_x T \qquad \forall \; x \in G,$$

then it is called an invariant operator. We denote by $\mathscr{L}_i$ the set of all invariant operators in $\mathscr{L}$.

An operator $T \in \mathscr{L}$ is called almost invariant provided

$$\lim_{x \longrightarrow 0} \| T R_x - R_x T \|_{\mathscr{L}} = 0 \cdot$$

We denote by $\mathcal{L}_{ai}$ the set of all almost invariant operators in $\mathcal{L}$. On the lines of De Leeuw [2], it can be easily seen that $\mathcal{L}_i$ and $\mathcal{L}_{ai}$ are closed subalgebras of $\mathcal{L}$.

Following Tewari and Madan [7], we define the Fourier transform π_γ, $\gamma \in \Gamma$, of an operator $T \in \mathcal{L}_{ai}$ by the vector-valued integral

$$\pi_\gamma (T) = \int_G (-x,\gamma) \; [R_{-x} \; T \; R_x] \; dx. \tag{3.1}$$

It is easy to check that the map $\phi : G \longrightarrow \mathcal{L}_{ai}$ such that

$$\phi(x) = R_{-x} \; T \; R_x$$

is continuous in the operator norm. Hence the integral on the right-hand side of (3.1) makes sense.

The formal expansion of an opertor T in the form of Fourier series is given by

$$T \sim \sum_{\gamma\in\Gamma} \pi_\gamma (T) \tag{3.2}$$

The convolution of a Radon measure $\mu \in M(G)$ and an operator $T \in \mathcal{L}_{ai}$ is defined by

$$(\mu * T) = \int_G R_x \; T \; R_{-x} \; d\,\mu(x). \tag{3.3}$$

It is easy to verify that $\mu * T \in \mathcal{L}_{\alpha i}$ and

$$|| \mu * T ||_{\mathcal{L}} \leq || \mu || \,.\, || T ||_{\mathcal{L}} \,.$$

On the lines of De Leeuw (loc. cit.), it can be easily seen that

$$\pi_\gamma \; (\mu * T) = \hat{\mu} \; (\gamma) \; \pi_\gamma \; (T) \; ,$$

where

$$\hat{\mu} \; (\gamma) = \int_G \; (- x,\gamma) \; d \; \mu \; (x) \; .$$

4. SUMMABILITY OF FOURIER SERIES (3.2)

De Leeuw [2] was the first to have studied in detail the Fourier transform and Fourier series associated with an operator $T \in \mathcal{L}_{ai}$. Tewari and Madan [7] have generalized the results of De Leeuw for an operator T defined on a Segal algebra S(G), where G is a compact abelian group.

Following Tewari and Madan, we say that the Fourier series (3.2) is e_α - summable to T in the norm topology of $\mathcal{L}$ provided for any given $\varepsilon > 0$, there exists an $\alpha_0 \in I$ such that

$$\| \sum_{\gamma \in \Gamma} \hat{e}_\alpha(\gamma)\, \pi_\gamma(T) - T \|_{\mathcal{L}} < \varepsilon \qquad (4.1)$$

for all $\alpha \geq \alpha_0$, where $\{e_\alpha\}_{\alpha \in I}$ denotes a bounded approximate identity of X(G) consisting of trigonometric polynomials.

We prove the following :

Theorem 4.1 For an operator $T \in \mathcal{L}$ the following statements are equivalent :

i) $T \in \mathcal{L}_{ai}$.

ii) The series (3.2) is e_α-summable to T in the norm topology of $\mathcal{L}$.

iii) $T = g * S$,

where $S \in \mathcal{L}_{ai}$ and $g \in C(G)$.

The proof of the theorem depends on the following :

Lemma 4.2. If G is a compact abelian group and C(G) is the set of all continuous positive-definite functions on G, then every g in C(G) has the form

$$g(x) = \sum_{n=1}^{\infty} \lambda_n\, (-x, \gamma_n),$$

where $\lambda_n \geq 0$, $\gamma_n \in \Gamma$ and

$$\sum_{n=1}^{\infty} \lambda_n < \infty .$$

Proof See [4. p. 332].

Proof of the Theorem. In order to prove that i) $\longrightarrow$ ii), it is enough to show that $e_\alpha * T$ converges to T in the norm topology of $\mathcal{L}$.

From the definition of convolution it follows that

$$e_\alpha * T - T = \int_G [\, R_x\, T\, R_{-x} - T \,]\, e_\alpha(x)\, dx.$$

Hence we have

$$\| e_\alpha * T - T \|_{\mathcal{L}} \le \int_G \| R_x\, T\, R_{-x} - T \|_{\mathcal{L}}\, e_\alpha(x) dx$$

$$= \int_V \| R_x\, T\, R_{-x} - T \|_{\mathcal{L}}\, e_\alpha(x) dx$$

$$+ \int_{G/V} \| R_x\, T\, R_{-x} - T \|_{\mathcal{L}}\, e_\alpha(x) dx$$

$$= I_1 + I_2\,, \qquad (4.2)$$

say, where V is a neighbourhood of the unit element e in G such that

$$\| R_x\, T\, R_{-x} - T \|_{\mathcal{L}} < \varepsilon/2 \qquad \forall\ x \in V.$$

Thus we have

$$I_1 < \tfrac{1}{2}\,\varepsilon \int_V e_\alpha(x)\ dx$$

$$< \tfrac{1}{2}\,\varepsilon\ . \qquad (4.3)$$

Next, since $\{e_\alpha\}_{\alpha\in I}$ is an approximate identity, we choose V such that

$$\lim_\alpha \int_{G/V} e_\alpha(x)\ dx = 0.$$

Hence we may assume that there exists $\alpha_0 \in I$ such that

$$\int_{G/V} e_\alpha(x)\ dx < \frac{\varepsilon}{4\|T\|_{\mathcal{L}}}$$

for all $\alpha \ge \alpha_0$. Thus we see that

$$I_2 \le 2\, \| T \|_{\mathcal{L}} \int_{G/V} e_\alpha(x)\ dx$$

$$< \tfrac{1}{2}\,\varepsilon \qquad (4.4)$$

for all $\alpha \ge \alpha_0$.

Combining (4.2), (4.3) and (4.4), we obtain

$$\lim_{\alpha} \| e_{\alpha} * T - T \|_{\mathscr{L}} = 0 \qquad \forall\, T \in \mathscr{L}_{ai} \;.$$

Next we shall show that ii) $\longrightarrow$ iii). We suppose that the series

$$\sum_{\gamma\in\Gamma} \pi_{\gamma}\ (S)$$

is e_{α} - summable to S in the norm topology of $\mathscr{L}$ and

$$T \sim \sum_{\gamma\in\Gamma} \lambda_{\gamma}\ \pi_{\gamma}\ (S),$$

where λ_{γ} satisfies the conditions of the above lemma and $S \in \mathscr{L}_{ai}$. Then we have

$$\pi_{\gamma}\ (T) = \int_G (-t,\gamma)\ R_{-t}\ T\ R_t\ dt\ , \qquad t \in G.$$

$$= \int_G (-t,\gamma)\ [R_{-t}\ T\ R_t - \sum_{\gamma\in\Gamma} \hat{e}_{\alpha}(\gamma)\ \pi_{\gamma}\ (T)\ ]\ dt$$

$$+ \int_G (-t,\gamma)\ [\ \sum_{\gamma\in\Gamma} \hat{e}_{\alpha}(\gamma)\ \pi_{\gamma}\ (T)\ ]\ dt$$

$$= I_1 + I_2 \quad \text{(say)}.$$

Since the Fourier series of S is e_{α} - summable to S in the norm topology of $\mathscr{L}$, I_1 approaches to zero as $\alpha \longrightarrow \infty$.

Now we consider I_2.

$$I_2 = \int_G (-t,\gamma)\ [\sum_{\gamma\in\Gamma} \hat{e}_{\alpha}(\gamma) .\ \pi_{\gamma}\ (T)\ ]dt$$

$$= \int_G (-t,\gamma)\ [\sum_{\gamma\in\Gamma} \hat{e}_{\alpha}(\gamma)\ .\lambda_{\gamma}\ \pi_{\gamma}\ (S)\ ]dt$$

$$= \int_G (-t,\gamma)\ [\sum_{\gamma\in\Gamma} \lambda_{\gamma}\ \hat{e}_{\alpha}(\gamma)\ .\lambda_{\gamma} (S)]\ dt$$

$$= \int_G (-t,\gamma)\ [\sum_{\gamma\in\Gamma} \lambda_{\gamma}\ \pi_{\gamma}\ (e_{\alpha} * S)]\ dt$$

$$= \int_G (-t,\gamma)\ [g * (e_\alpha * S)]\ dt$$

(see [5, p. 66])

$$= \int_G e_\alpha(t) * [(-t,\gamma)\ (g * S)]\ dt$$

$$= \int_G e_\alpha(t)\ .\ \pi_\gamma\ (g * S)\ ]dt$$

$$= \pi_\gamma\ (g * S)$$

Hence we obtain

$$\pi_\gamma\ (T) = \pi_\gamma\ (g * S)$$

which implies that

$$T = g * S. \tag{4.5}$$

Finally, it remains to show that iii) $\longrightarrow$ i). We have

$$\| R_x\ (g * S)\ R_{-x} - (\ g * S\ )\|_{\mathcal{L}}$$

$$\leq \int_G \| R_x\ (g * S)\ R_{-x} - (g * S)\ \|_{\mathcal{L}}\ dx$$

$$\leq A \int_G \| R_x\ S\ R_{-x} - S\ \|_{\mathcal{L}}$$

$$= o(1) \text{ as } x \longrightarrow 0, \text{ for } S \in \mathcal{L}_{ai}\ ,$$

which implies that

$$g * S \in \mathcal{L}_{ai} \Rightarrow T \in \mathcal{L}_{ai}.$$

This completes the proof of the theorem.

REFERENCES

1. W.Braun and H.G. Feichtinger, Banach spaces of distributions having two moudle structures, *J. Functional Analysis,* 51 : 174-212 (1983).
2. K.DeLeeuw, An harmonic analysis for operators I, *Illinois J. Math.* 19 : 593-606(1975).
3. R.E. Edwards, "Fourier Series", Vol. II, Springer-Verlag, New York (1979).
4. E.Hewitt and K.A. Ross, "Abstract Harmonic Analysis", Vol. II, Springer-Verlag, New York (1970).
5. S.Madan, "Some Problems on Segal Algebras and their Multipliers", Ph.D. Thesis, *Indian Institute of Technology,* Kanpur (1979).

6. R.P. Sinha and V. Kant, Homogeneous Banach spaces of distributions, *Commet. Math.*, 13 : 286-298 (1983).
7. U.B. Tewari and S. Madan, An harmonic analysis for operators : F. and M. Ries theorem, *Illinois Journ. Maths.*, 28 : 286-298 (1984).

CHARACTERIZATION OF FUNCTIONS WITH FOURIER TRANSFORMS SUPPORTED ON ORTHANTS

J. N. Pandey[1], O.P.Singh[2]

[1]Department of Mathematics and Statistics
Carleton University, Ottawa, Canada K1S 5B6
[2]Department of Applied Mathematics
Banaras Hindu University, Varanasi-221005, India

ABSTRACT

We characterize the functions in $L^p(\mathbf{R}^n)$ and generalized functions in $D'_{L^p}(\mathbf{R}^n)$, $1 < p < \infty$, whose Fourier transform vanishes on one or more orthants of $\mathbf{R}^n$. We present here a brief summary of the main results and related matters. Detailed proofs will appear in the Journal of Mathematical Analysis and Applications.

1. INTRODUCTION

It is a fairly difficult problem to characterize the functions in $L^p(\mathbf{R}^n)$ whose Fourier transform vanishes in some orthants of $\mathbf{R}^n$. Very little is known concerning this problem except the classical Paley-Wiener Theorem in one dimension which characterizes the functions in $L^2(\mathbf{R})$ having their Fourier transforms vanish for negative values of the variable [14, p. 175]. Later some results for the space $L^2(\mathbf{R}^n)$ were obtained by Stein and Weiss [28, p. 112].

Concerning the Fourier transform of a distribution with compact support, it was shown that the Fourier transform of a distribution f with bounded support is a function $F(z) = f(\exp(-2\pi iz \cdot x)$, which may be continued to all complex numbers z as an entire function of exponential growth. The converse is also true [32, p. 15]. For further references see [1, 3, 4, 5, 7, 8]. But none of those give the explicit characterization of functions in $L^p(\mathbf{R}^n)$ and distributions in the Schwartz space $D'_{L^p}(\mathbf{R}^n)$, whose Fourier transforms are supported on a given number of orthants in $\mathbf{R}^n$. The aim of the present paper is to give a complete answer to the problem for functions in $L^p(\mathbf{R}^n)$ and distributions in $D'_{L^p}(\mathbf{R}^n)$, $1 < p < \infty$.

For $f \in L^p(\mathbf{R}^n)$, $1 < p < \infty$, we construct the following holomorphic function $F(z)$, $z \in \mathbf{C}^n$, as

$$F(z) = \frac{1}{(2\pi i)^n} \int_{\mathbf{R}^n} f(t) \frac{1}{\prod_{j=1}^{n} (t_j - z_j)} \, dt, \tag{1}$$

where $z_j = x_j + iy_j$ and $y_j \neq 0 \ \forall \ j = 1, 2, \ldots, n$. For a distribution $f \in D'_{L^p}(\mathbf{R}^n)$ the corresponding function $F(z)$ is defined as

$$F(z) = \frac{1}{(2\pi i)^n} \left\langle f(t), \frac{1}{\prod_{j=1}^{n} (t_j - z_j)} \right\rangle, \quad y_j \neq 0 \ \forall \ j. \tag{2}$$

There are 2^n different ways in which $y \to 0$ depending upon the way the various components y_j of y tend to either 0_+ or 0_-. Thus we get 2^n different boundary values of $F(z)$ as $y \to 0$. To denote them we adopt the following notation:

Let $\sigma_k = \{\sigma_k(1), \sigma_k(2), \ldots, \sigma_k(n)\}$ be a sequence of length n whose elements are $+$ and $-$ for $1 \leq k \leq 2^n$. Then 2^n orthants of $\mathbf{R}^n$ are denoted by S_{σ_k}, $1 \leq k \leq 2^n$, where

$$S_{\sigma_k} = \left\{ x \in \mathbf{R}^n \middle| x_j > 0 \text{ if } \sigma_k(j) = + \text{ and } x_j < 0 \text{ if } \sigma_k(j) = -, \ j = 1, 2, \ldots, n \right\}. \tag{3}$$

For example, when $n = 2$ the various quadrants of $\mathbf{R}^2$ are denoted by S_{++}, S_{-+}, S_{+-} and S_{--}, where

$$S_{+-} = \left\{ x \in \mathbf{R}^2 \middle| x_1 > 0 \text{ and } x_2 < 0 \right\}, \text{ etc.}$$

Similarly the various limits of $F(z)$ as $y \to 0$ are denoted by

$$F_{\sigma_k}(x) = \lim_{y_1 \to 0_{\sigma_k(1)}, \ldots, y_n \to 0_{\sigma_k(n)}} F(z), \tag{4}$$

where $0_{\sigma_k(j)} = 0_+$ if $\sigma_k(j) = +$ otherwise it is 0_-.

For $f \in D'_{L^p}(\mathbf{R}^n)$ or $L^p(\mathbf{R}^n)$, with the limits taken in the respective spaces, we have proved that

$$f = \sum_{k=1}^{2^n} (-1)^{m_k} F_{\sigma_k} \quad \text{in } D'_{L^p}(\mathbf{R}^n) \text{ (or } L^p(\mathbf{R}^n)), \quad 1 < p < \infty \tag{5}$$

and

$$(-1)^{m_k} \widehat{F}_{\sigma_k}(\xi) = \begin{cases} \hat{f}(\xi) & \text{for } \xi \in S_{\sigma_k} \\ 0 & \text{elsewhere,} \end{cases} \tag{6}$$

where

m_k is the number of minus signs in the sequence σ_k,

and $\hat{f}$ is the Fourier transform of f in the following sense:

$$\begin{aligned} \langle \hat{f}, \varphi \rangle &= \langle f, \hat{\varphi} \rangle \\ &= \left(\int_{\mathbf{R}^n} f \hat{\varphi} \text{ if } f \in L^p(\mathbf{R}^n) \right), \ \forall \ \varphi \in S(\mathbf{R}^n) \end{aligned} \tag{7}$$

where $\hat{\varphi}$ is the classical Fourier transform of φ defined as

$$\hat{\varphi} = \int_{\mathbf{R}^n} \varphi(t) e^{it \cdot x} \, dt.$$

[20, 29, 34]

The space $S(\mathbf{R}^n)$ is the testing function space of rapid descent [20, 26].

From (1.6), we are able to prove the following Paley-Wiener Theorem For $D'_{L^p}(\mathbf{R}^n)$ (or $L^p(\mathbf{R}^n)$ $(1 < p < \infty)$:

THEOREM: For $f \in D'_{L^p}(\mathbf{R}^n)$,

$$\hat{f}(\xi) = 0 \text{ for } \xi \in \bigcup_{k=1}^{\ell} S_{\sigma_k}, \quad 1 \le \ell \le 2^n$$

$$\text{iff } \sum_{k=1}^{\ell}(-1)^{m_k} F_{\sigma_k}(\xi) = 0 \text{ for } \xi \in \bigcup_{k=1}^{\ell} S_{\sigma_k}, \text{ in some space } S_0'(\mathbf{R}^n).$$

The space $S_0(\mathbf{R}^n)$ is a subspace of $S(\mathbf{R}^n)$ which is closed with respect to the multiplication by the function $sgn\ x$. The sgn is defined as

$$sgn(x) = \prod_{i=1}^{n} sgn(x_i). \tag{8}$$

In the process we proved the M. Riesz and Titchmarsh Inequality and many related classical results for $L^p(\mathbf{R}^n)$.

As an application of our theory, we characterize the solution space of the following Dirichlet boundary value problem:

$$\Delta u = 0, \text{ where } \Delta = \prod_{j=1}^{n}\left(\frac{\partial^2}{\partial x_j^2} + \frac{\partial^2}{\partial y_j^2}\right) \tag{9}$$

with boundary conditions

$$\lim_{y \to 0_{\sigma_k}} u = F_{\sigma_k} \text{ in } D'_{L^p}(\mathbf{R}^n) \text{ (or } L^p(\mathbf{R}^n)) \ (1 \le k \le 2^n). \tag{10}$$

Here F_{σ_k}, $1 \le k \le 2^n$, are arbitrary elements of $D'_{L^p}(\mathbf{R}^n)$. Incidently for fixed F_{σ_k} $(1 \le k \le 2^n)$ in $D'_{L^p}(\mathbf{R}^n)$ (or $L^p(\mathbf{R}^n)$), the system (9) and (10) has

$$F(z) = \frac{1}{(2\pi i)^n}\left\langle \sum_{k=1}^{2^n}(-1)^{m_k} F_{\sigma_k}(t), \frac{1}{\prod_{j=1}^{n}(t_j - z_j)}\right\rangle, \tag{11}$$

$$(I_m z_j \neq 0\ \forall\ j)$$

as a unique solution.

2. MAIN THEOREMS

We now summarize our main results as follows:

THEOREM 1. For $f \in D'_{L^p}(\mathbf{R}^n)$, $1 < p < \infty$, define

$$F(z) = \frac{1}{(2\pi i)^n}\left\langle f(t), \frac{1}{\prod_{j=1}^{n}(t_j - x_j)}\right\rangle, y_j \neq 0,\ j = 1, 2, \ldots, n \tag{12}$$

where y_i is the imaginary part of z_j. Then

$$f = \sum_{k=1}^{2^n}(-1)^{m_k} F_{\sigma_k} \text{ in } D'_{L^p}(\mathbf{R}^n),$$

and

$$(-1)^{m_k}\widehat{F}_{\sigma_k}(\xi) = \begin{cases} \hat{f}(\xi), & \text{for } \xi \in S_{\sigma_k} \\ 0, & \text{elsewhere} \end{cases}$$

on $S_0(\mathbf{R}^n)$, where

$$F_{\sigma_k} = \lim_{y_1 \to 0_{\sigma_k}(1), \ldots, y_n \to 0_{\sigma_k}(n)} F(z).$$

Here m_k stands for the number of negative signs in the sequence σ_k.

Lemma. For $f \in D'_{L^p}(\mathbf{R}^n)$, $1 < p < \infty$, and $F(z)$ defined as in (12) we have

$$\sum_{k=1}^{\ell}(-1)^{m_k}\widehat{F}_{\sigma_k}(\xi) = \begin{cases} \hat{f}(\xi), & \text{if } \xi \in \bigcup_{k=1}^{\ell} S_{\sigma_k} \\ 0, & \text{elsewhere} \end{cases}$$

for $1 \leq \ell \leq 2^n$, equality is in the sense of $S'_0(\mathbf{R}^n)$.

THEOREM 2. (Paley-Wiener Theorem or $D'_{L^p}(\mathbf{R}^n)$)

Let $f \in D'_{L^p}(\mathbf{R}^n)$, $1 < p < \infty$. Define $F(z)$ by

$$F(z) = \frac{1}{2\pi i)^n}\left\langle f(t), \frac{1}{\prod_{j=1}^{n}(t_j - x_j)}\right\rangle,\ Imz_j \neq 0, j = 1, 2 \ldots, n.$$

Then we have

$$\hat{f}(\xi) = 0 \text{ for } \xi \in \bigcup_{k=1}^{\ell} S_{\sigma_k} \text{ in } S'_0(\mathbf{R}^n)$$

$$\text{iff} \sum_{k=1}^{\ell}(-1)^{m_k} F_{\sigma_k} = 0 \,\forall x \in \bigcup_{k=1}^{\ell} \text{ in } (F(S_0(\mathbf{R}^n)))' \ i.e$$

$$\left\langle \sum_{k=1}^{\ell}(-1)^{m_k} F_{\sigma_k}, \hat{\phi}\right\rangle = 0,\ \forall \phi \in S_0(\mathbf{R}^n)$$

with the support of ϕ contained in $\bigcup_{k=1}^{\ell} S_{\sigma_k}$.

DEFINITION: We define the space M of holomorphic functions as follows: A function $F(z_1, z_2, \ldots, z_n) \in M$ iff

$$\sup_{\substack{x_j \in \mathbf{R}, |y_j| \geq \delta > 0 \\ 1 \leq j \leq n}} \left|F(z_1, z_2, \ldots, z_n)\right| < A_\delta < \infty$$

and

$$\left|F(z_1, z_2, \ldots, z_n)\right| = o(1), \text{ as } |y_1|, |y_2|, |y_3|, \ldots, |y_n| \to \infty$$

independently of each other and the asymptotic order is valid uniformly $\forall x \in \mathbf{R}^n$ and

$$\lim_{y \to 0_{\sigma_k}} F(z) = F_{\sigma_k}(x) \text{ in } D'_{L^p}(\mathbf{R}^n), k = 1, 2, \ldots, 2^n.$$

where $y \to 0_{\sigma_k}$ mean $y_j \to 0_{\sigma_k(j)}, 1 \leq j \leq n$. Now we have the following

THEOREM 3. For $F(z) \in M$ we have

$$F(z) = \frac{1}{(2\pi i)^n} \left\langle \sum_{k=1}^{2^n} (-1)^{m_k} F_{\sigma_k}(t), \frac{1}{\prod_{j=1}^{n} (t_j - z_j)} \right\rangle.$$

For example, when $n = 2$

$$F(z) = \frac{1}{(2\pi i)^n} \left\langle F_{++} + (-1)F_{-+} - (-1)F_{+-} + (-1)^2 F_{--}, \frac{1}{(t_1 - z_1)(t_2 - z_2)} \right\rangle.$$

THEOREM 4. (The Dirichlet boundary value problem)
Let

$$\triangle = \prod_{j=1}^{n} \left(\frac{\partial^2}{\partial x_j^2} + \frac{\partial}{\partial y_j^2} \right)$$

Then the solution to the boundary value problem

$$\triangle F(z) = 0, \; F(z) \in M$$

$$\lim_{y \to 0_{\sigma_k}} F(z) = F_{\sigma_k} \text{in } D'_{L^p}(\mathbf{R}^n), \; 1 \leq k \leq 2^n \text{ is}$$

$$F(z) = \frac{1}{(2\pi i)^n} \left\langle \sum_{k=1}^{2^n} (-1)^{m_k} F_{\sigma_k}(t), \frac{1}{\prod_{j=1}^{n} (t_j - z_j)} \right\rangle$$

REFERENCES

1. K. Andersen, Weighted norm inequalities for Hilbert transforms and conjugate functions of even and odd functions, Proc. Amer. Math. Soc. 56 (1976), 99-107.
2. Hans Bremmerman, Distributions, complex variables and Fourier transforms, Addison-Wesley Publishing Company Inc., 1965.
3. S.Y.A. Cheng and R. Fefferman, Some recent developments in Fourier analysis on product domains, Bull. Amer Math Soc. (Ns) 12 (1985), 1-44.

4. R.R. Coifman and C. Fefferman, Weighted norm inequalities for maximal functions and singular integrals, Studia Math. 51 (1974), 241-250.

5. M. Cotlar, Some generalizations of Hardy-Littlewood maximal theorem, Rev. Math., Cuyana, 1 (1955), 85-104.

6. L. Ehrenpreis, Analytic functions and the Fourier transform of distributions I, Ann. Math. 63 (Jan. 1956), 129-159.

7. E.B. Fabes and N.M. Riviere, Singular integrals with mixed homogeneity, Studia Math. 27 (1966), 19-38 MR 35 #683.

8. C. Fefferman, Estimates for double Hilbert transform, Studia Math., TXLIV (1972) 1-15.

9. ——, Recent progress on classical Fourier analysis, Internat. Congress Math., Vancouver, 1974, 95-118.

10. Avner Freidman, Generalized functions and partial differential equations, Prentice Hall Inc., 1963.

11. I.M. Gel'fand and G.E. Shilov, Generalized functions, Vol. II, Academic Press, 1968.

12. E. Hewit and K. Stromberg, Real and abstract analysis, Springer Verlag, 1969.

13. R. Hunt, B. Muckenhaupt, R.Wheeden, Weighted norm inequalities for the conjugate functions and Hilbert transform, Trans. Amer. Math. Soc., 176 (1973), 227-251.

14. Y. Katznelson, An introduction to harmonic analysis, John Wiley and Sons, Inc. 1968.

15. V.M. Kokilashvili, Singular integral operators in weighted spaces, Colloquia Mathematica Societatis in Janos Bolyai Vol. 35, Functions, Series, Operators Budapest (Hungary) 1980.

16. H.A. Lauwerier, The Hilbert problem for generalized functions, Arch. Rational Mech. Anal. 13 (1963), 157-166.

17. D. Mitrovic, A distributional representation of analytic functions, Mathematics. Balkanica, 79 (1974), 437-440.

18. B. Muckenhoupt, Weighted norm inequalities for the Hardy maximal function, Trans. Amer. Math. Soc. 165 (1972) 207-226.

19. N.I. Muckhelishvili, Singular integral equations, Moscow, 1946.

20. J. Barros-Neto, An introduction to the theory of distributions, Marcel Dekkar, Inc., NY 1973.

21. M. Orton, Hilbert transforms, Plemelj relations and Fourier transform of distributions, SIAM J. Math. Anal. 4 (Nov. 1973), 656-667.

22. J.N. Pandey and M.A. Chaudhry, The Hilbert transform of generalized functions and applications, Can. J. Math., Vol. XXXV, No. 3 (1983), 478-495.

23. J.N. Pandey, The Hilbert transform of Schwartz distributions, Proc. Amer. Math. Soc. Vol. 89, No. 1 (Sept. 1983), 86-90.

24. J.N. Pandey, Action of the Fourier Transform on the Hilbert transform and vice-versa, (submitted).

25. J.N. Pandey and O.P. Singh, On the p-norm of the truncated n-dimensional Hilbert transform, Bulletin of the Australian Mathematical Society (to appear).

26. Laurent Schwartz, Theorie des distributions, Herman Paris, 1966.

27. O.P. Singh and J.N. Pandey, The n-dimensional Hilbert transform of distributions, its inversion and applications, to appear in Can. J. Math.

28. K. Sokel-Soholowski, On trignometric series conjugate to Fourier series in two variables, Fund. Math., 33 (1945) 166-182.

29. E. Stein and G. Weiss, Introduction to Fourier analysis on euclidean spaces, Princeton University Press, Princeton, NJ 1971.

30. E.C. Titchmarsh, Introduction to the theory of fourier integrals, Oxford Univ. Press, 1967.

31. ——-, The Theory of functions, second edition, Oxford University Press, 1964.

32. A. Terras, Harmonic analysis on symmetric spaces and applications I, Springer-Verlag, 1985.

33. Heinz Gunther Tillman, Distributionen als Randverteilungen analytischer Functionen, II, Math. Zeitschr. 76, 5-21 (1961).

34. F. Treves, Topological Vector spaces distributions and kernels, Academic Press, 1967, New York, London.

35. V.S. Vladimirov, Methods of the theory of funcions of many complex variables, The M.I.T. Press, 1966.

36. J.H. Williams, Lebesgue integration, Hilt Rinehart and Winston, NY, 1962.

37. A.H. Zemanian, Distribution theory and transform analysis, McGraw-Hill Book Company, 1965.

38. ——, Generalized integral transforms, Interscience Publishers, NY 1968.

39. A. Zigmund, On the boundary values of functions of several complex variables, Fund. Math. 36 (1949), 201-235.

PSEUDO-ASYMPTOTIC EXPANSION OF GENERALIZED FUNCTIONS

R.S.Pathak

Department of Mathematics
Banaras Hindu University
Varanasi 221005, India

1. INTRODUCTION

Drozzinov and Zavjalov [2] gave the idea of quasi-asymptotic behaviour (q.a.b.) and quasi-asymptotic expansions (q.a.e.) of generalized functions and established tauberian theorems for Fourier-Laplace transforms of generalized functions in the complex domain. In the present work Pseudo-asymptotic expansion (p.a.e.) of generalized functions is defined as an improvement over its quasi-asymptotic expansion (q.a.e.) providing an error term in q.a.e.

We shall use the following functions. According to Bojanic and Karmata [1] function L(x) is slowing varying if L is real valued, positive for $x > 0$ and measurable on every finite interval $[a,b]$, $0 < a < b < \infty$, and if

$$\lim_{t\longrightarrow\infty} L(ut)/L(t) = 1 \qquad \text{for each } u > 0. \tag{1.1}$$

Moreover, for any $\eta > 0$,

$$x^{\eta} L(x) \longrightarrow \infty \text{ and } x^{-\eta} L(x) \longrightarrow 0 \quad (x \longrightarrow \infty) . \tag{1.2}$$

The generalized function f_α defined by

Generalized Functions and Their Applications, Edited by
R.S. Pathak, Plenum Press, New York, 1993

$$f_\alpha = 1_+(t)\ t^\alpha \in S'_+ \quad , \qquad \alpha \geq -1 \tag{1.3}$$

is also of considerable importance to us. Here $1_+(t)$ means the following :

$$1_+(t) = 1 \qquad (t > 0) \tag{1.4}$$
$$= 0 \qquad (t < 0) \ ;$$

and S'_+ denotes the space of tempered distributions having support in $[0,\infty)$.

2. PSEUDO-ASYMPTOTIC EXPANSIONS IN S'_+

The following definition can easily be extended to any Schwartz distributions or its generalizations the so called ultradistributions. To simplify the analysis we consider the case of S' .

Definition 2.1. A tempered distribution $T \in S'_+$ possesses pseudo-asymptotic expansion (p.a.e.) at infinity of order $\alpha \in \mathbb{R}$ with respect to $(k^\alpha\ L(k))$ if there exist $T_i \in S'_+$ such that

$$\lim_{k \longrightarrow \infty} (k^\alpha\ L(k))^{-1} \{ T(kt) - \sum_{i=1}^{N} C_i\ T_i\ (kt) \} : = \gamma(t) \tag{2.1}$$

in S' ; and T has p.a.e. at zero of order α with respect to $(k^\alpha\ L(1/k))$ if

$$\lim_{k \longrightarrow 0+} (k^\alpha\ L(1/k))^{-1} \{ T(kt) - \sum_{i=1}^{N} C_i\ T_i\ (kt) \} : = \gamma(t) \tag{2.2}$$

in S' provided $\gamma(t) \neq 0$. Then we write

$$T(t) \sim \sum_{i=1}^{N} C_i\ T_i + \gamma(t) \tag{2.3}$$

in S' when $t \longrightarrow \infty$ (0+ respectively).

If for any α, $\gamma(t) = 0$, the p.a.e. (2.1) and (2.2) are called closed quasi-asymptotic expansions of T of order α and length 0 in the terminology of [2].

Since for $\phi \in S$ and $b > 0$,

$$<\gamma(bt),\ \phi(t)> = \lim_{k \longrightarrow \infty} <(k^\alpha\ L(k))^{-1}\{T(k(bt)) - \sum C_i T_i\ (k(bt))\}, \phi(t)>$$

$$= \lim_{bk \to \infty} <((kb)^{\alpha} L(kb))^{-1} \{T(k(bt)) - \sum C_i T_i (k(bt))\}.$$

$$b^{\alpha} (L(kb)/L(k)), \phi(t) >$$

$$= b^{\alpha} < \gamma(t), \phi(t) >$$

on using (1.1), it follows that $\gamma(t)$, if it exists in S'_+, is a homogeneous generalized function of order α. Hence by a result of Gel'fand and Shilov [3 , p.80],

$$\gamma(t) = C f_{\alpha}(t) \tag{2.4}$$

for some constant $C \neq 0$ and $t > 0$.

The following theorem gives a characterization of elements of S'_+ having p.a.e. in the special case when

$$T_i = f_{\alpha_i}(t) = C_i t^{\alpha_i} 1_+(t).$$

Theorem 2.2. Let $T \in S'_+$ be such that

$$T = D_t^m G(t) \tag{2.5}$$

where $G(t)$ is a continuous function growing at infinity slower than some polynomial and having support in $[0,\infty)$. If the T has p.a.e. at infinity of order $\alpha > -1$ with respect to $k^{\alpha} L(k)$, then there exists a constant C such that for $\alpha_i > -1$, i=1,2,...,N,

$$\lim_{k \to \infty} (k^{\alpha+m+1} L(k))^{-1} [\overline{G}(s/k) - \sum_{i=1}^{N} C_i \Gamma(\alpha_i+1) (s/k)^{-(\alpha_i+m+1)}]$$

$$= C \Gamma(\alpha+1) s^{-(\alpha+m+1)}, \tag{2.6}$$

where $\overline{G}(s) = \mathscr{L} [G(t)$.

Also, if the $T \in S'_+$ has p.a.e. at zero of order $\alpha > -1$ with respect to $k^{\alpha} L(k)$ then there exists a constant C such that for $\alpha_i > -1$, i = 1,2,...,N,

$$\lim_{k \to 0+} (k^{\alpha+m+1} L(1/k))^{-1} [\overline{G}(s/k) - \sum_{i=1}^{N} C_i \Gamma(\alpha_i+1) (s/k)^{-(\alpha_i+m+1)}]$$

$$= C\,\Gamma\,(\alpha+1)\ s^{-(\alpha+m+1)}. \qquad (2.7)$$

Proof. Let $T \in S'_+$ be given by (2.5) [4, p.272] and have p.a.e. at infinity of order α. Then by Definition 2.1 there is an element $\gamma \in S'_+$ such that

$$\lim_{k\to\infty} (k^{\alpha} L(k))^{-1} < \{ T(kt) - \sum_{i=1}^{N} C_i\ f_{\alpha_i}(kt) \}\ ,\ \phi(t) >$$

$$= < \gamma\ ,\ \phi >$$

for each $\phi \in S$, where $f_{\alpha_i} = C_i\ t^{\alpha_i}\ 1_+(t) \in S'_+$, $i=1,2,..,N$.

Now, let $\phi(t) = e^{-st}$, Re $s > 0$. Since $\gamma(t) = C\ t^{\alpha}$ we have

$$\lim_{k\to\infty} k^{-\alpha-1}\ (L(k))^{-1} < T(u) - \sum_{i=1}^{N} C_i\ T_i(u)\ ,\ e^{-su/k} >$$

$$= C < 1_+(t)\ t^{\alpha},\ e^{-st} >.$$

Hence, in view of (2.5),

$$\lim_{k\to\infty} k^{-\alpha-1}\ (L(k))^{-1}\ < D^m\ G(u) - \sum_{i=1}^{N} C_i\ 1_+(u)\ u^{\alpha_i},\ \ e^{-su/k}\ >$$

$$= C\,\Gamma\,(\alpha+1)\ s^{-\alpha-1}\ .$$

So that

$$\lim_{k\to\infty} k^{-\alpha-m-1}\ (L(k))^{-1}\ \ [\ \bar{G}(s/k) - \sum_{i=1}^{N} C_i\Gamma(\alpha_i+1)\ (s/k)^{-\alpha_i-m-1}\ ]$$

$$= C\,\Gamma\,(\alpha+1)\ s^{-\alpha-m-1}\ .$$

where $\bar{G}(s) = \mathcal{L}\ [G(t)]$. The proof of (2.7) is similar.

A special case of (2.7) is of some interest.

THEOREM 2.3. Let $T \in S'_+$ be as in (2.5) where $G(t)$ is a continuous function and $e^{-ct}G(t) \in L(0,\infty)$, $C > 0$. Let T have p.a.e. at zero of order $\alpha > -1$ with respect to k^{α}. Then there exists a constant C such that for $\alpha_i > -1$, $i=1,2,...,N$,

$$G(t) = \sum_{i=1}^{N} C_i \, \Gamma(\alpha_i+1)/\,\Gamma(\alpha_i+m+1)\; t^{\alpha_i+m} \;+\; C\,\Gamma(\alpha+1)/\Gamma(\alpha+m+1)\,t^{\alpha+m} \tag{2.8}$$

as $t \longrightarrow 0+$.

Proof. From the proof of (2.7) we have

$$\overline{G}(s/k) \sim \sum_{i=0}^{N} C_i \, \Gamma(\alpha_i+1) \, (s/k)^{-(\alpha_i+m+1)} \qquad \text{as } k \longrightarrow 0+$$

where $C_0 = C$ and $\alpha_0 = \alpha$. Putting $s/k = S$ we have

$$\overline{G}(S) \sim \sum_{i=0}^{N} D_i \, \Gamma(\alpha_i+m+1) \, (S)^{-(\alpha_i+m+1)} \qquad \text{as } S \longrightarrow \infty,$$

where $D_i = C_i \, \Gamma(\alpha_i+1)/\,\Gamma(\alpha_i+m+1)$.

Now, using converse to Watson's lemma [5 , p 31] we get

$$G(t) \sim \sum_{i=0}^{N} C_i \, \Gamma(\alpha_i+1)/\,\Gamma(\alpha_i+m+1) \; t^{\alpha_i+m} \qquad \text{as } t \longrightarrow 0+.$$

Next, we obtain a converse result.

THEOREM 2.4. Let T be the same as in (2.5) where $G(t)$ is now assumed to be a continuous function such that $t^{-q}\,G(t) \in L(0,\infty)$ for $q > \alpha+m+1$. Assume that (2.8) holds for $\alpha > -1$ and $\alpha_i > \alpha$, $i=1,2,\dots,N$. Then T has p.a.e. at zero of order α with respect to k^{α}.

Proof. For each $\phi \in S$ we have

$$S_k^N := k^{-\alpha} < T(kt) - \sum_{i=1}^{N} C_i \, f_{\alpha_i}(kt), \, \phi(t) >$$

$$= k^{-\alpha-m} < D_t^m \, G(kt) - \sum_{i=1}^{N} C_i \, \Gamma(\alpha_i+1)/\Gamma(\alpha_i+m+1) \; k^{\alpha_i+m} \; D_t^m \, t^{\alpha_i+m}, \phi(t) >$$

$$= (-1)^m \, k^{-\alpha-m} \int_0^{\infty} \{ G(kt) - \sum_{i=1}^{N} C_i \Gamma(\alpha_i+1)/\Gamma(\alpha_i+m+1)\,(kt)^{\alpha_i+m} \} \; \phi^{(m)}(t)\,dt$$

$$= (-1)^m k^{-\alpha-m} \left[\int_0^1 C\, \Gamma(\alpha+1) / \Gamma(\alpha+m+1)\ (kt)^{\alpha+m}\ \phi^{(m)}(t)\,dt \right.$$

$$\left. + \int_1^\infty G(kt)\phi^{(m)}(t)\,dt - \int_1^\infty \sum C_i \Gamma(\alpha_i+1)/\Gamma(\alpha_i+m+1)(kt)^{\alpha_i+m}\ \phi^{(m)}(t)\,dt \right]$$

$$= I_1 + I_2 + I_3 \quad \text{(say)},$$

on using (2.8).

Now,

$$I_1 = (-1)^m C\, \Gamma(\alpha+1) / \Gamma(\alpha+m+1) \int_0^1 t^{\alpha+m}\ \phi^{(m)}(t)\ dt < \infty$$

and

$$|I_2| = |\ (-1)^m k^{-\alpha-m} \int_1^\infty G(kt)\ \phi^{(m)}(t)\ dt\ |$$

$$\le \sup_{0<t<\infty} |\ t^q\ \phi^{(m)}(t)\ |\ k^{-\alpha-m} \int_1^\infty t^{-q}\ |\ G(kt)|dt$$

$$\le \sup_t |\ t^q\ \phi^{(m)}(t)\ |\ k^{-\alpha-m+q-1} \int_0^\infty u^{-q}\ |\ G(u)|du$$

$\longrightarrow 0$ as $k \longrightarrow 0+$ provided that $q > \alpha+m+1$.

Next,

$$|I_3| = \sum_{i=1}^N C_i k^{-\alpha}\ \Gamma(\alpha_i+1)/\ \Gamma(\alpha_i+m+1) \int_1^\infty (kt)^{\alpha_i} t^m\ |\ \phi^{(m)}\ (t)|\ dt$$

$\longrightarrow 0$ as $k \longrightarrow \infty$ provided that $\alpha_i > \alpha$.

Thus we have

$$\lim_{k\longrightarrow 0+} S_k^N = \lim_{k\longrightarrow 0+} [I_1 + I_2 + I_3]$$

$$= C\, \Gamma(\alpha+1)/\ \Gamma(\alpha+m+1)\ < D_t^m\ [\ (1_+(1-t)t^{\alpha+m}],\ \phi(t)>.$$

Hence T possesses p.a.e. at zero of order $\alpha > -1$.

ACKNOWLEDGEMENT

The work is supported by C.S.I.R. under Research Project No. P 25/118.

REFERENCES

1. R.Bojanic and J. Karamata, On slowly varying functions and asymptotic relations, M R C Technical Summary Report 432, Madisin , Wisconsin, October (1973).
2. Ju.N.Drozzinov and B.I. Zov'jalov, The quasi-asymptotics of generalized functions and Tauberian theorems in the complex domain, *Math. USSR Sbornik* , 31 (3) : 329-345(1977).
3. I.M.Gel'fand and G.E. Shilov, "Generalized functions, vol.1", Academic Press, London (1964).
4. F.Treves, "Topological Vector Spaces, Distributions and Kernels", Academic Press, London (1967).
5. R.Wong, "Asymptotic Approximations of Integrals", Academic Press, London (1989).

MULTIPLIERS, CONVOLUTORS AND HYPOELLIPTIC CONVOLUTORS FOR TEMPERED ULTRADISTRIBUTIONS

Stevan Pilipović

Instiute of Mathematics,
University of Novi Sad
Trg Dositeja Obradovića 4
21000 Novi Sad, Yugoslavia

1. INTRODUCTION

Malgrange's, Ehrenpreis's and Hörmander's results on the solvability and the hypoellipticity of convolution equations in Schwartz's spaces of distributions stimulated many mathematicians to study such problems in various subspaces of distributions. We cite here only the results of Zielezny ([15], [16]) and Swartz ([12], [13]), since they are connected with the results of this paper. Convolution equations for ultradistribution spaces were studied by Meise, Taylor, Voigt and their cooperators; (see [1], [7] and references there.)

In this paper we start the investigations of convolution equations in subspaces of ultradistribution spaces similarly as it was done for distributions. Some facts from distribution theory can be simply transferred but many problems are specific for subspaces of ultradistribution spaces and need to be solved.

By following Komatsu's approach [3] we introduced and studied spaces of tempered ultradistributions ([5], [6]) of Beurling and Roumieu type (see also [9]). The space of multipliers was studied in [5] and [9]. In this paper we prove that these spaces are nuclear, introduce the spaces of convolutors and give the structural theorems for them, and in the last part we give a necessary condition for a convolutor to be hypoelliptic. The sufficient condition which is also given, is very strong; it is given here only as a starting point for further investigations.

2. NOTATION

The sets of nonnegative integers, natural, real and complex numbers are denoted by $\mathbf{N}_0, \mathbf{N}, \mathbf{R}, \mathbf{C}$. We use the symbols $< x >= (1 + x^2)^{1/2}$ and
$D = i^{-1}d/dx$.

Generalized Functions and Their Applications, Edited by
R.S. Pathak, Plenum Press, New York, 1993

By M_p is denoted a sequence of positive numbers. The following conditions on this sequences will be assumed (see [3]): $M_0 = 1$;
(M.1) $M_p^2 \leq M_{p-1}M_{p+1},\ \ p \in \mathbf{N}$;

$$(M.2) \quad M_p \leq AH^p \min_{0\leq q\leq p} \{M_{p-q}M_q\}, \ \ p,q \in \mathbf{N}_0, \quad \text{for some} \quad A,H \geq 0;$$

(M.3) $\sum_{p=q+1}^{\infty} \frac{M_{p-1}}{M_p} \leq Aq\frac{M_q}{M_p},\ \ q \in \mathbf{N}$.
The so-called associated function for the sequence M_p is defined by

$$M(\rho) = \sup_{p\in\mathbf{N}_0} \{\log \frac{\rho^p}{M_p}\}, \ \ \rho > 0.$$

For the definitions and properties of the spaces $\mathcal{D}_{K,r}^{M_p}, \mathcal{D}_K^{(M_p)}, \mathcal{D}_K^{\{M_p\}}, \mathcal{D}^{(M_p)}$, $\mathcal{D}^{\{M_p\}}, \mathcal{E}^{(M_p)}, \mathcal{E}^{\{M_p\}}$ we refer to [3]. Recall only that for a compact set $K \subset \mathbf{R}$ and $r > 0$, $\mathcal{D}_{K,r}^{M_p}$ is the space of smooth functions ϕ on $\mathbf{R}$ ($\phi \in C^\infty$) supported by K such that

$$\|\phi\|_{K,r} := \sup\{\frac{|\ \phi^{(p)}(x)\ |}{r^p M_p};\ p \in \mathbf{N}_0,\ x \in K\} < \infty.$$

The norm in $L^s = L^s(\mathbf{R})$, $s \in [1,\infty]$, is denoted by $\|\ \|_r$. If $f \in L^1$ then its Fourier transform is defined by

$$(\mathcal{F}f)(\xi) = \hat{f}(\xi) = \int_{\mathbf{R}} e^{ix\xi} f(x)dx, \ \ \xi \in \mathbf{R}.$$

By $\Re$ is denoted a set of positive sequences which increases to infinity. If $r_p \in \Re$ and K is a compact set in $\mathbf{R}$ then $\mathcal{D}_{K,r_p}^{\{M_p\}}$ is the space of smooth functions φ on $\mathbf{R}$ supported by K such that

$$\|\varphi\|_{K,r_p} := \sup\{\frac{|\ \varphi^{(p)}(x)\ |}{N_p};\ \ p \in \mathbf{N}_0,\ \ x \in K\} < \infty,$$

where $N_p = M_p \prod_{i=1}^{p} r_i$, $\ p \in \mathbf{N}_0$. Clearly, this a Banach space. It is proved in [3] that

$$\mathcal{D}_K^{\{M_p\}} = \text{proj} \lim_{r_p\in\Re} \mathcal{D}_{K,r_p}^{M_p}.$$

If $\Omega \subset \mathbf{R}$ is a bounded open set and $r > 0$, resp. $r_p \in \Re$, we put

$$\mathcal{D}_{\Omega,r}^{(M_p)} = \text{ind} \lim_{K\subset\subset\Omega} \mathcal{D}_{K,r}^{M_p}, \mathcal{D}_{\Omega,r_p} = \text{ind} \lim_{K\subset\subset\Omega} \mathcal{D}_{K,r_p}^{\{M_p\}}.$$

The associated function for the sequence N_p $(= M_p \prod_{i=1}^{p} r_i)$ is

$$N_{r_p}(\rho) = \sup\{\log \frac{\rho^p}{N_p};\ \ p \in \mathbf{N}_0\}, \ \ \rho > 0.$$

Note ([3]), for given r_p and every $k > 0$ there is $\rho_k > 0$ such that

$$N_{r_p}(\rho) \leq M(k\rho), \ \ \rho > \rho_0.$$

The inequalities

$$M(k\ |\ x - t\ |) \leq M(2k\ |\ x\ |) + M(2k\ |\ t\ |), \ \ x \in \mathbf{R}, \ \ t \in \mathbf{R}, \ \ k > 0, \tag{1}$$

$$N_{r_p}(|x-t|) \le N_{r_p/2}(|x|) + N_{r_p/2}(|t|),\ x \in \mathbf{R},\ \ t \in \mathbf{R},\ \ r_p \in \Re, \tag{2}$$

simply follow from the definitions.

It is said that

$$P(\xi) = \sum_{\alpha \in \mathbf{N}_0} a_\alpha \xi^\alpha, \quad \xi \in \mathbf{R},$$

is an ultrapolynomial of the class (M_p) resp. $\{M_p\}$ whenever the coefficients a_α satisfy the estimate

$$|a_\alpha| \le CL^\alpha / M_\alpha, \quad \alpha \in \mathbf{N}_0, \tag{3}$$

for some $L > 0$ and $C > 0$ resp. for every $L > 0$ and some $C_L > 0$. The corresponding operator $P(D) = \sum_\alpha a_\alpha D^\alpha$ is an ultradifferential operator of the class (M_p) resp $\{M_p\}$.

Put

$$P_r(\xi) = (1+\xi^2)\prod_{p \in \mathbf{N}}(1 + \tfrac{\xi^2}{r^2 m_p^2}), \quad resp.$$
$$P_{r_p}(\xi) = (1+\xi^2)\prod_{p \in \mathbf{N}}(1 + \tfrac{\xi^2}{r_p^2 m_p^2}),\ \xi \in \mathbf{C},$$

where $m_p = M_p/M_{p-1}$ and $r > 0$ resp. $r_p \in \Re$. Conditions (M.1), (M.2) and (M.3) imply that P_r resp. P_{r_p} is an ultradifferential operator of the class (M_p) resp. of the class $\{M_p\}$ (see [3]). The following estimations

$$|P_r(\xi)| \ge e^{M(r|\xi|)}, \quad \xi \in \mathbf{R},$$
$$|P_{r_p}(\xi)| \ge e^{N_{r_p}(|\xi|)}, \quad \xi \in \mathbf{R}, \tag{4}$$

will be used.

We denote by $\mathcal{S}_2^{M_p,m}(\mathbf{R})$, $m > 0$, the space of smooth functions φ which satisfy

$$\sigma_{m,2}(\varphi) := (\sum_{p,q \in \mathbf{N}_0} \int_{\mathbf{R}} |\frac{<x>^p\ \varphi^{(q)}(x)}{m^{p+q} M_p M_q}|^2\, dx)^{1/2} < \infty. \tag{5}$$

It is equipped with the topology induced by the norm $\sigma_{m,2}$. If in (5) "2" is substituted by "p" $p \in [1,\infty]$, the corresponding sequence of norms is denoted by $\sigma_{m,p}$, $m > 0$. We shall use the notion $\gamma_m = \sigma_{m,\infty}$, $m > 0$.

The spaces $\mathcal{S}'^{(M_p)}$ and $\mathcal{S}'^{\{M_p\}}$ of tempered ultradistributions of Beurling and Roumieu type respectively, are defined as the strong duals of the spaces

$$\mathcal{S}^{(M_p)} = \lim\text{proj}_{m\to 0}\mathcal{S}^{M_p,m}(\mathbf{R}) \quad \text{and} \quad \mathcal{S}^{\{M_p\}} = \lim\text{ind}_{m\to\infty}\mathcal{S}_2^{M_p,m}(\mathbf{R}),$$

respectively. The common notation for symbols (M_p) and $\{M_p\}$ will be $*$.

All the "good" properties of $\mathcal{S}^*$ and its strong dual follow from the equivalence of the sequence of norms $\sigma_{m,2}$, $m > 0$, with the each of the following sequences of norms ([6]) .

(a) $\sigma_{m,p}$, $m > 0$, $p \in [1,\infty]$ is fixed ;

(b) $s_{m,p}$, $m > 0$, $p \in [1,\infty]$ is fixed, where $s_{m,p}(\varphi) := \sum_{\alpha,\beta \in \mathbf{N}_0} \frac{\|x^\beta \varphi^{(\alpha)}\|_p}{m^{\alpha+\beta} M_\alpha M_\beta}$;

(c) s_m, $m > 0$, where $s_m(\varphi) := \sup_{\alpha \in \mathbf{N}_0}\{\frac{\|\varphi^{(\alpha)} e^{M(m\cdot)}\|}{m^\alpha M_\alpha}\}$;

(d) θ_δ, $\delta > 0$, where

$$\theta_\delta(\varphi) := \sum_{n \in \mathbf{N}_0} |a_n|^2 e^{2M(\delta\sqrt{2n+1})},\ \varphi \overset{L^2}{=} \sum_{n \in \mathbf{N}_0} a_n h_n,$$

and h_n, $n \in \mathbf{N}_0$, is the Hermite orthonormal basis of $L^2(\mathbf{R})$;

(e) $\bar{s}_{m,2}, m > 0$, where $\bar{s}_{m,2}(\varphi) := \sum_{\alpha,\beta\in\mathbf{N}_0} \frac{\|(x^\beta\varphi)^{(\alpha)}\|_2}{m^{\alpha+\beta}M_\alpha M_\beta}$.

It is proved in [10] that

$$\mathcal{S}^{\{M_p\}} = \text{proj} \lim_{r_i,s_j\in\Re} S^{M_p}_{r_i,s_j},$$

where

$$S^{M_p}_{r_i,s_j} = \{\phi \in C^\infty(\mathbf{R});\ \gamma_{r_i,s_j}(\phi) < \infty\},$$

and

$$\gamma_{r_i,s_j}(\phi) := \sum_{p,q\in\mathbf{N}_0} \{\frac{\| < x >^p \phi^{(q)}\|_{L^\infty}}{(\prod_{i=1}^p r_i)M_p(\prod_{j=1}^q s_j)M_q}\}.$$

Note, $\mathcal{F} : \mathcal{S}^* \to \mathcal{S}^*$ is a topological isomorphism and the Fourier transformation on $\mathcal{S}'^*$ is defined as usual.

3. THE SPACE OF MULTIPLIERS

As in [5] and [9] , O^*_M is defined as the space of smooth functions α from $\mathcal{E}^*$ such that $\alpha \in O^*_M$ iff

$$\begin{array}{ll} \text{for every } \psi \in \mathcal{S}^*,\ \alpha\psi \in \mathcal{S}^* & \text{and the mapping} \\ \psi \to \alpha\psi,\ \ \mathcal{S}^*,\ \ \text{into } \mathcal{S}^* & \text{is continuous .}\end{array}$$

The proof of the next theorem in (M_p)-case is given in [5] and [9]. For $\{M_p\}$ -case the proof is similar and we omit it.

Proposition 1. The following statements are equivalent.

(i) $\alpha \in O^*_M$.

(ii) For every $r > 0$, resp. there is $r > 0$, such that for some $m > 0$, resp. for every $m > 0$,

$$\sup_{p\in\mathbf{N}_0} \{\frac{\|e^{-M(m|\cdot|)}\phi^{(p)}\|_{L^\infty}}{r^p M_p}\} < \infty.$$

(iii) For every $\psi \in \mathcal{S}^{(M_p)}$, resp. $\psi \in \mathcal{S}^{\{M_p\}}$ and every $r > 0$, resp. for some $r > 0$.

$$\gamma_{r,\psi}(\alpha) := \gamma_r(\psi\alpha) < \infty.$$

(iv) $\alpha \in O^{\{M_p\}}_M$ iff for every $r_i, s_j \in \Re$

$$\gamma_{r_i,s_j\psi}(\alpha) := \gamma_{r_i,s_j}(\psi\alpha) < \infty.$$

Denote by $L(\mathcal{S}^*,\mathcal{S}^*)$ the space of continuous linear mappings from $\mathcal{S}^*$ into $\mathcal{S}^*$; O^*_M is its subspace. With $L_w(\mathcal{S}^*,\mathcal{S}^*)$ we denote the weak topology in $L(\mathcal{S}^*,\mathcal{S}^*)$. This topology induces the topology in O^*_M which is equivalent with:

$O^{(M_p)}_M$ equipped with the topology induced by the family of seminorms $\gamma_{r,\psi}$, $r > 0$, $\psi \in \mathcal{S}^{(M_p)}$,

resp.

$O^{\{M_p\}}_M$, equipped with the topology induced by the family of seminorms$\gamma_{r_i,s_j,\psi}$, $r_i, s_j \in \Re$, $\psi \in \mathcal{S}^{\{M_p\}}$.

The space O^*_M, equipped with the topology is denoted by $O^*_{M,w}$. The weak topology in $L(\mathcal{S}'^*,\mathcal{S}'^*)$, induces the same topology on O^*_M. The next proposition is obvious.

Proposition 2. The bilinear mappings

$$O^*_{M,\mathrm{w}} \times \mathcal{S}^* \to \mathcal{S}^*, \ \ (\alpha, \psi) \to \alpha\psi,$$

$$O^*_{M,\mathrm{w}} \times \mathcal{S}'^* \to \mathcal{S}'^*, \ \ (\alpha, f) \to \alpha f,$$

are separately continuous.

Proposition 3. The space $O^*_{M,\mathrm{w}}$ is nuclear.

Proof . The proof is similar to the corresponding one for the space of multipliers for $\mathcal{S}$, given in [14]. Let $\psi \in \mathcal{S}^*$. Put for $r > 0$, resp. $r_i, s_j \in \Re$,

$$U_{r,\psi} = \{\alpha \in O_M^{(M_p)};\ \gamma_{r,\psi}(\alpha) \leq 1\},$$

resp.

$$U_{r_i,s_j,\psi} = \{\alpha \in O_M^{\{M_p\}};\ \gamma_{r_i,s_j,\psi}(\alpha) \leq 1\}.$$

The family

$$\mathcal{U}^{(M_p)} = \{U_{r,\psi}; r > 0, \psi \in \mathcal{S}^{(M_p)}\},$$

resp.

$$\mathcal{U}^{\{M_p\}} = \{U_{r_i,s_j,\psi},\ r_i, s_j \in \Re,\ \psi \in \mathcal{S}^{\{M_p\}}),$$

is a basis of neighborhoods of zero in $O^*_{M,\mathrm{w}}$. We shall show that for every neighborhood $U \in \mathcal{U}^*$ there is $V \in \mathcal{U}^*$ and a positive Radon measure μ on

$$V^o = \{\alpha' \in (O^*_{M,\mathrm{w}})';\ |< \alpha, \alpha' >| \leq 1,\ \ \alpha \in V\},$$

such that

$$p_U(\alpha) \leq \int_{V^o} |< \alpha, \alpha' >|\, d\mu(\alpha'),\ \ \alpha \in O^*_M, \tag{6}$$

where p_U is the Minkowski functional.

According to [8], this will imply the assertion.

One can show that for given $r > 0$, resp. for given $r_i, s_j \in \Re$, there are $C > 0$ and $\tilde{r} > 0$, resp. there are $C > 0$ and $\tilde{r}_i, \tilde{s}_j \in \Re$ such that in $\{M_p\}-$ case:

$$\gamma_{r_i,s_j,\psi}(\alpha) \leq$$

$$C \sum_{p,q \in \mathrm{N}_0} \frac{1}{\prod_{i=1}^p \tilde{r} \prod_{j=1}^q \tilde{s}_j} \int_{-\infty}^{\infty} \frac{1}{(1+t^2)} [\frac{< t >^{q+1} (\psi\alpha)^{(p)}(t)}{M_{q+1}M_p} +$$

$$+ \frac{< t >^{q+2} | (\alpha\psi)^{(p+1)}(t) |}{M_{q+2}M_{p+1}}] dt,$$

and in $(M_p)-$ case the appropriate inequality holds.

As well , for given $\tilde{r} > 0$, resp. for given $\tilde{r}_i, \tilde{s}_j \in \Re$, there is $\tilde{\tilde{r}} > 0$, resp. there are $\tilde{\tilde{r}}_i, \tilde{\tilde{s}}_j \in \Re$ such that in $\{M_p\}-$ case the following implication holds

$$\sum_{p,q \in \mathrm{N}_0} \frac{< x >^q (\alpha\psi)^{(p)}(x)}{\prod_{i=1}^p \tilde{\tilde{r}}_i M_p \prod_{j=1}^q \tilde{\tilde{s}}_j M_q} \leq 1 \Longrightarrow$$

$$\frac{2^{p+q}}{\prod_{i=1}^p \tilde{r}_i \prod_{j=1}^q \tilde{s}_j} [\frac{< t >^{q+1} | (\alpha\psi)^{(p)}(t) |}{M_{q+1}M_p} + \tag{7}$$

$$+\frac{<t>^{q+1}|(\alpha\psi)^{(p+1)}(t)|}{M_{q+2}M_{p+1}}] \leq 1,\ p,q \in \mathbf{N}_0,$$

and in (M_p)–case the appropriate implication holds.

Let in (M_p)– case, $V = \{\alpha \in O_M^{(M_p)};\ \gamma_{\tilde{r},\psi}(\alpha) \leq 1\}$ and in $\{M_p\}$– case, $V = \{\alpha \in O_M^{\{M_p\}};\ \gamma_{\tilde{r}_i,\tilde{s}_j,\psi}(\alpha) \leq 1\}$. We shall show that (6) holds only in $\{M_p\}$– case since it is more complicated.

Define a functional μ on $C(V^o)$ by

$$<\mu, f> = C \sum_{p,q\in\mathbf{N}_0} \frac{1}{2^{p+q}} \int_{-\infty}^{\infty} \frac{1}{1+t^2} f(\alpha'_{t,p,q})dt, \tag{8}$$

where $\alpha'_{t,p,q} \in O_{M,\mathbf{w}}^{\{M_p\}}$, $t \in \mathbf{R}$, $p,q \in \mathbf{N}_0$, is defined on $O_M^{\{M_p\}}$ by

$$<\alpha, \alpha'_{t,p,q}> = \frac{2^{p+q}}{\prod_{i=1}^p \tilde{r}_i \prod_{j=1}^q \tilde{s}_j}[\frac{<t>^{q+1}|(\alpha\psi)^{(p)}(t)|}{M_{q+1}M_p} + \frac{<t>^{q+2}|(\alpha\psi)^{(p+1)}(t)|}{M_{q+2}M_{p+1}}]. \tag{9}$$

If $\alpha \in V$, then (7) implies

$$|<\alpha, \alpha'_{t,p,q}>| \leq 1.$$

This implies that for every $t \in \mathbf{R}$, $p,q \in \mathbf{N}_0$, $\alpha'_{t,p,q} \in V^o$.

One can easily prove that μ is a positive Radon measure. (8) and (9) imply

$$<\mu, \alpha> = C \sum_{p,q} \frac{1}{2^{p+q}} \int_{-\infty}^{\infty} \frac{2^{p+q}}{\prod_{i=1}^p \tilde{r}_i \prod_{j=1}^q \tilde{s}_j}$$

$$\cdot \frac{1}{1+t^2}[\frac{<t>^{q+1}|(\alpha\psi)^{(p)}(t)|}{M_{q+1}M_p} + \frac{<t>^{q+2}|(\alpha\psi)^{(p+1)}(t)|}{M_{q+2}M_{p+1}}]dt.$$

By the definition

$$<\mu, \alpha> = \int_{V^o} <\alpha, \alpha'> d\mu(\alpha'),$$

and this implies

$$p_U(\alpha) = \gamma_{r_i,s_j,\psi}(\alpha) \leq <\mu, \alpha> \leq \int_{V^o} |<\alpha, \alpha'>| d\mu(\alpha').$$

The assertion is proved.

4. THE SPACE OF CONVOLUTORS

Proposition 4. If $\phi \in \mathcal{S}^*$ and $S \in \mathcal{S}'^*$ then,

$$(S * \phi)(x) = <S(t), \phi(x-t)>,\quad x \in \mathbf{R},$$

is a smooth function which satisfies the following condition:

There is $k > 0$, resp. there is $k_p \in \Re$, such that for every operator P of class $*$ and $\phi \in \mathcal{S}^*$

$$P(D)(S*\phi)(x) = O(e^{M(k|x|)}),\ |x| \to \infty,\quad \text{resp.}$$
$$P(D)(S*\phi)(x) = O(e^{N_{k_p}(|x|)}),\quad |x| \to \infty. \tag{10}$$

Proof . It is proved in [10] that there exist an ultradifferential operator $\tilde{P}$ of class $*$ and a continuous function F such that for some $\tilde{k} > 0$, resp. for some $\tilde{k}_p \in \Re$,

$$S = \tilde{P}(D)F \quad \text{where } F(x) = O(e^{M(\tilde{k}|x|)}), \quad \text{resp.}$$

$$F(x) = O(e^{N_{\tilde{k}_p}(|x|)}), \quad |x| \to \infty.$$

This implies that for given $P(D)$,

$$P(D)(S * \phi)(x) = (F * (P(D)\tilde{P}(D)\phi))(x), \quad x \in \mathbf{R}.$$

Since $P(D)\tilde{P}(D)\phi \in \mathcal{S}^*$, by using (1) and (2), we obtain that for some $k > 0$, resp. for some $k_p \in \Re$, (10) holds.

An S from $\mathcal{S}'^*$ is a convolution operator, in short- the convolutor, if for every $\phi \in \mathcal{S}^*$, $S * \phi \in \mathcal{S}^*$ and the mapping

$$\phi \to S * \phi, \quad \mathcal{S}^* \to \mathcal{S}^*,$$

is continuous. The space of convolutors is denoted by $O_C'^*$.

Let $S \in O_C'^*$ and $T \in \mathcal{S}^*$. Then their convolution $S * T$ is defined by

$$< S * T, \varphi > = < T, \check{S} * \varphi >, \quad \varphi \in \mathcal{S}^* \quad (\check{S}(x) = S(-x)).$$

Since for a strong unit sequence η_k ([2]) from $\mathcal{D}^*$

$$< T, \check{S} * \varphi > = \lim_{k \to \infty} < T, \eta_k(\check{S} * \varphi) > =$$

$$= \lim_{k \to \infty} < T(x)(\check{S} * \varphi)(x), \ \eta_k(x) >,$$

[2] implies that $S * T$ exists in the sense of ultratempered convolution of tempered ultradistributions and that both are equal.

Let $S \in O_C'^*$. From the definition easily follows that the mapping

$$T \to S * T, \quad \mathcal{S}'^* \to \mathcal{S}'^*, \quad \text{is continuous.}$$

Proposition 5. Let $S \in \mathcal{S}'^*$. The following statements are equivalent.

a) S is a convolutor.

b) For every $\phi \in \mathcal{D}^*$, $S * \phi \in \mathcal{S}^*$.

c) For every $r > 0$, resp. for every $r_p \in \Re$,

$$\{e^{M(r|x|)} S(\cdot - x); \ x \in \mathbf{R}\} \text{ resp. } \{e^{N_{r_p}(|x|)} S(\cdot - x); \ x \in \mathbf{R}\},$$

is bounded in $\mathcal{D}'^*$.

d) For every $r > 0$, resp. for every $r_p \in \Re$ there is $k > 0$, resp. there is $k_p \in \Re$, and continuous functions F_1 and F_2 such that

$$S = P_k(D)F_1 + F_2, \quad \text{resp.} \quad S = P_{k_p}(D)F_1 + F_2,$$

and

$$\sup_{x \in \mathbf{R}} \{e^{M(r|x|)}(| F_1(x) | + | F_2(x) |)\} < \infty$$

resp.

$$\sup_{x \in \mathbf{R}} \{e^{N_{r_p}|x|)}(| F_1(x) | + | F_2(x) |)\} < \infty.$$

Proof . The appropriate assertion for tempered distribution is given in [11]. We shall prove the assertion only in $\{M_p\}-$ case since it is rather completed. The implications $(a) \Rightarrow (b)$ and $(d) \Rightarrow (a)$ are trivial.

$(b) \Rightarrow (c)$ For given $\phi \in \mathcal{D}^*$ and every $r_p \in \Re$

$$\sup_{x\in\mathbf{R}}\{\frac{1}{(\prod_{i=1}^{p} r_i)M_p} \mid (S*\phi)(x) \mid\} < \infty.$$

This implies that the family $\{e^{N_{k_p}(|x|)}S(\cdot - x),\ \ x \in \mathbf{R}\}$, is pointwisely bounded and since $\mathcal{D}^{\{M_p\}}$ is barrelled it follows that the quoted families are (strongly) bounded in $\mathcal{D}'^{\{M_p\}}$.

$(c) \Rightarrow (d)$ Let r_p be as in the previous part. From [10] it follows that for every open bounded set Ω which contains zero there is an open bounded neighborhood of zero ω and $k_p \in \Re$ such that for every $\phi \in \mathcal{D}^{\{M_p\}}_{\omega,k_p}$,

$$\sup\{e^{N_{k_p}(|x|)}(S(\cdot + x) * \phi)(u); u \in \Omega,\ x \in \mathbf{R}\} < \infty.$$

By putting $u = 0$, this implies that for every $\phi \in \mathcal{D}^{\{M_p\}}_{\omega,k_p}$,

$$x \to e^{N_{r_p}(|x|)}(T * \phi)(x),\ \ x \in \mathbf{R}, \tag{11}$$

is bounded. By [4, Lemma 3.1] there are $\theta \in \mathcal{D}^{\{M_p\}}_{\omega,k_p}$, and $\psi \in \mathcal{D}^{\{M_p\}}_{\omega}$ such that

$$S = P_{k_p}(D)S * \theta + S * \psi.$$

Thus, (11) implies assertion (d)

The ultratempered convolution of $S_1, S_2 \in O'^*_C$ is in O'^*_C. This follows from [2] . As well for any $T \in \mathcal{S}'^*$, and $\psi \in \mathcal{S}^*$

$$< (S_1 * S_2) * T, \psi >=< S_1 * T, \check{S}_2 * \psi >=< T * S_2, \check{S}_1 * \psi >=< T, (S_1 * S_2) * \psi >$$

We equip O'^*_C with the topology from $L_{\mathrm{w}}(\mathcal{S}^*, \mathcal{S}^*)$ and denote it by $O'^*_{C,\mathrm{w}}$. The same topology on this space is induced by $L_{\mathrm{w}}(\mathcal{S}'^*, \mathcal{S}'^*)$.

Proposition 6. The Fourier transformation is a topological isomorphism of $O^*_{M,\mathrm{w}}$ onto $O'^*_{C,\mathrm{w}}$.

Proposition 7. A sequence S_n from O'^*_C converges to O in $O'^*_{C,\mathrm{w}}$ if and only if for every $r > 0$ resp. for every $r_p \in \Re$, there is $k > 0$, resp. there is $k_p \in \Re$ and sequences of continuous functions F_{1n} and F_{2n} such that

$$S_n = P_k(D)F_{1n} + F_{2n}, \quad \text{resp. } S_n = P_{k_p}(D)F_{1n} + F_{2n},$$

and

$$\lim_{n\to\infty} \sup_{x\in\mathbf{R}} e^{M(r|x|)}(\mid F_{1n}(x) \mid + \mid F_{2n}(x) \mid) = 0,$$

resp.

$$\lim_{n\to\infty} \sup_{x\in\mathbf{R}} e^{N_{r_p}(|x|)}(\mid F_{1n}(x) \mid + \mid F_{2n}(x) \mid) = 0.$$

We omit the proof and only note that it follows from the last assertion in [10] similarly as in the proof of Proposition 5. (see also [12]).

5. THE SPACE O_C^*

We denote the space of continuous linear functionals on $O'^*_{C,w}$ by O_C^* and by $\mathcal{ES}'^*$ the space of elements f from $\mathcal{S}'^*$ such that for every $S \in O'^*_C$, $S * f \in \mathcal{E}^*$ and the mapping

$$S \to S * f, \;\; O'^*_{C,\mathrm{w}} \to \mathcal{E}^* \text{ is continuous.}$$

Proposition 8. (i) $\mathcal{ES}'^* \subset \mathcal{E}^* \cap \mathcal{S}'^*$.

(ii) If $f \in \mathcal{ES}'^*$ and $S \in O'^*_C$ then $S * f \in \mathcal{ES}'^*$.

(iii) Let f be a smooth function such that there exists $k > 0$, resp. there is $k_p \in \Re$, such that for every $r > 0$, resp. for every $r_p \in \Re$,

$$P_r(D)f(x) = O(e^{M(k|x|)}), \quad \text{resp. } P_{r_p}(D)f(x) = O(e^{N_{k_p}(|x|)}), \;\; |\, x \,| \to \infty.$$

Then $f \in \mathcal{ES}'^*$.

(iv) $\mathcal{ES}'^* = O_C^*$.

Proof . (*i*) This follows from the fact that the Dirac $\delta-$ distribution is in O'^*_C.

(*ii*) Let $T \in O'^*_C$. Since

$$\mathcal{ES}'^* \ni (T * S) * f = T * (S * f),$$

and the mapping $(T, S) \to T * S, \;\; O'^*_C \times O'^*_C \to O'^*_C$, is separately continuous, mapping

$$T \to T * (S * f), \;\; O'^*_C \to \mathcal{E}^*,$$

is continuous.

(*iii*) We shall prove the assertion only for $\{M_p\}$ - case. Let $S \in O'^{\{M_p\}}_C$ and K be a compact set. We have to show that for every $r_p \in \Re$,

$$\sup_{x \in K}\{|\, P_{r_p}(D)(S * f)(x) \,|\} < \infty. \tag{12}$$

Proposition 5 (*d*) implies that for suitable $\tilde{r}_p \in \Re$ and F_1, F_2

$$P_{r_p}(D)(S * f) = (P_{r_p}(D)P_{\tilde{r}_p}(D)f) * F_1 + (P_{r_p}(D)f) * F_2 \tag{13}$$

and

$$\sup_{x \in \mathbf{R}} e^{N_{s_p}(|x|)}(|\, F_1(x) \,| + |\, F_2(x) \,|) < \infty,$$

where s_p is such that

$$e^{-N_{s_p}(|t|) + N_{r_p}(|x-t|)} \le C(t) e^{N_{k_p}(|x|)} < \;\; t \in \mathbf{R}, \;\; x \in \mathbf{R}, \tag{14}$$

and $C(t)$ is from $L^1(\mathbf{R})$. The existence of such $s_p \in \Re$ follows from (3). Now, (13) and (14) imply (12). Proposition 4, similarly as above, implies that the mapping from $O'^{\{M_p\}}_C \to \mathcal{E}^{\{M_p\}}$, $S \to S * f$, is continuous. Thus, $f \in \mathcal{ES}'^{\{M_p\}}$.

(*iv*) Let $f \in \mathcal{ES}'^*$. By $S \to (f * \check{S})(0)$ an element of $(O'^*_{C,\mathrm{w}})' = O_C^*$ is defined. Let $f \in O_C^*$. [Corollary 4, IV. 4] implies that there is an isomorphism between

$$\mathcal{S}^* \otimes \mathcal{S}'^* \text{ and } (L_{\mathrm{w}}(\mathcal{S}^*, \mathcal{S}'^*))'$$

$$\textstyle\sum_{j=0}^m \varphi_j \otimes T_j \longleftrightarrow J \in (L_{\mathrm{w}}(\mathcal{S}^*, \mathcal{S}'^*))'$$

such that for $S \in L_w(\mathcal{S}^*, \mathcal{S}'^*)$, $J(S) = \sum_{j=0}^{m} < T_j, S(\varphi_j) >$.

Since for $S \in O_C'^*$, $S(\varphi_j) = S * \varphi_j$, $j = 0, ..., m$, it follows

$$J(S) = \sum_{j=0}^{m} < T_j, S * \varphi_j >= \sum_{j=0}^{m} < S, T_j * \varphi_j > .$$

Proposition 4 and part (iii) imply that $T_j * \varphi_j \in \mathcal{ES}'^*$, $j = 0, ..., m$, and thus, every element of O_C^* is the finite sum of elements satisfying the conditions of part (iii). The assertion is proved.

6. THE HYPOELLIPTIC CONVOLUTION EQUATION

The convolution equation

$$S * U = V, \quad S \in O_C'^*, \quad V \in \mathcal{S}'^*,$$

is hypoelliptic if any solution U satisfies the following implication:

$$V \in \mathcal{ES}'^* \quad \Rightarrow \quad U \in \mathcal{ES}'^*.$$

In this case S is called the hypoelliptic convolutor.

Proposition 9. Let $S \in O_C'^*$.

(a) If the convolution equation is hypoelliptic then for every operator $P(D)$ of class (M_p) resp. class $\{M_p\}$, there are $k > 0$, and $a_0 > 0$, resp. for every $k > 0$ there is $a_k > 0$, such that

$$P(D)\hat{S}(\xi) \geq e^{M(k|\xi|)}, |\, \xi \,|> a_0, \quad \text{resp.} \quad |\, \xi \,|> a_k. \tag{15}$$

(b) If for some ψ from $\mathcal{D}^*$ with the property $\psi(\xi) = 1$, $|\, \xi \,| \leq a_0$, the function

$$\tilde{Q}(\xi) = \begin{cases} 0, & |\, \xi \,| \leq a \\ \frac{1-\psi(\xi)}{\hat{S}(\xi)}, & |\, \xi \,|> a, \end{cases}$$

is from O_M^*, then S is hypoelliptic.

Note that the assumption in (b) is very strong and it is given here only as a starting point for further investigations of the problem. The proof of assertion (a) is based on the construction which is similar to one used for tempered distributions ([14]). We shall give the proof for $\{M_p\}-$ case. But in this case the following lemma is needed.

Lemma 1. Let η_j be an increasing sequence of positive numbers which tends to infinite. There exists a sequence d_p from $\Re$ such that

$$e^{N_{d_p}(\eta_j)}/e^{M(\frac{1}{(j+1)}\eta_j)} \geq 1, \quad j \in \mathbf{N}.$$

Proof of Lemma . For every $j \in \mathbf{N}$ there is $p_j \in \mathbf{N}$ such that

$$\sup_{p \in \mathbf{N}}\{\frac{\eta_j^p}{(j+1)^p M_p}\} = \frac{\eta_j^{p_j}}{(j+1)^{p_j} M_{p_j}}.$$

Without losing the generality, assume that p_j is a strictly increasing sequence from $\mathbf{N}$. There are numbers d_i, $i = 1, ..., p_1$ with the properties

$$1 < d_1 < ... < d_{p_1} < 2 \quad \text{and} \prod_{i=1}^{p_1} d_i \leq 2^{p_1}.$$

As well, there are d_i, $i = p_1 + 1, ..., p_2$, such that

$$d_{p_1} < d_{p_1+1} < ... < d_{p_2} < 3 \quad \text{and} \prod_{i=p_1+1}^{p_2} d_i < 3^{p_2-p_1}.$$

Continuing in this way we construct the sequence d_p from $\Re$ such that

$$d_{p_s} < d_{p_s+1} < ... < d_{p_{s+1}} < s+2 \quad \text{and} \prod_{i=p_s+1}^{p_{s+1}} d_i < (s+2)^{p_{s+1}-p_s}, \quad s \in \mathbf{N}.$$

Since

$$\sup_{p\in\mathbf{N}}\{\frac{\xi_j^p}{\prod_{i=1}^p d_i M_p}\} \geq \frac{\xi_j^{p_j}}{\prod_{i=1}^{p_j} d_i M_{p_j}} \geq \frac{\xi_j^{p_j}}{(j+1)^{p_j} M_{p_j}}, \quad j \in \mathbf{N},$$

(6) implies the assertion of the Lemma .

Proof of Proposition 10. $(\{M_p\})-$ case

$a)$ Clearly, it is enough to prove the assertion only for $P(D) = 1$. Let

$$T(x) = \sum_{j=1}^{\infty} a_j e^{ix\xi_j}, \; x \in \mathbf{R}, \quad \text{where} \quad \xi_{j+1} > 2\xi_j, \; j \in \mathbf{N}, \; \xi_1 > 2.$$

Note that formally

$$\hat{T}(\xi) = \sum_{j=1}^{\infty} a_j \delta(\xi - \xi_j), \; \xi \in \Re.$$

One can easily prove that $T \in \mathcal{S}'^{\{M_p\}}$ if for some $C > 0$ and $d_p \in \Re$,

$$| a_j | \leq C e^{N_{d_p}(\xi_j)}, \; j \in \mathbf{N}.$$

We shall prove that $T \in \mathcal{ES}'^{\{M_p\}}$ if and only if for some $d > 0$ and some $C > 0$

$$| a_j | < C/e^{M(d\xi_j)}, \; j \in \mathbf{N}, \tag{16}$$

For every $\tilde{d}_p \in \Re$, (16) implies

$$\sup_{x\in\mathbf{R}}\{| P_{\tilde{d}_p}(D)T(x) |\} < \infty.$$

Thus, Proposition 8 (iii) implies $T \in \mathcal{ES}'^{\{M_p\}}$.

Conversely, let $T \in \mathcal{ES}'^{\{M_p\}}$. Then for every operator $P(D)$ of class $\{M_p\}$ and $\varphi \in \mathcal{S}^{\{M_p\}}$, the Rieman - Lebesgue lemma implies

$$\int e^{ixu} P(D)T(x)\varphi(-x)dx \to 0 \quad \text{as} \quad | u | \to \infty.$$

By using the properties of the Fourier transformation we obtain

$$< \mathcal{F}(e^{ixu}P(D)T(x))(\xi), \; \mathcal{F}(\varphi)(\xi) >=$$

$$=< P(\xi+u)\hat{T}(\xi+u), \hat{\varphi}(\xi) >=< \hat{T}(\xi), P(\xi)\hat{\varphi}(\xi-u) >, \; u \in \mathbf{R}.$$

Hence,

$$\sum_{j=1}^{\infty} a_j P(\xi_j)\hat{\varphi}(\xi_j - u) \to 0, \quad | u | \to \infty.$$

Let $\phi \in \mathcal{S}'^{\{M_p\}}$ such that $\hat{\phi}(0) \geq 1$ and for $| \xi | \geq 1$, $\hat{\phi}(\xi) = 0$. If we assume that (16) does not hold there is a sequence m_j of natural numbers such that

$$| a_{m_j} | > 1/e^{M(\frac{1}{(j+1)}\xi_{m_j})}, \quad j \in \mathbf{N}.$$

With $\xi_{m_j} = \eta_j$ in Lemma 1, (6) implies

$$a_{m_j} P_{d_p}(\xi_{m_j}) \geq 1, \quad j \in \mathbf{N},$$

which is in contradiction with

$$\sum_{j=0}^{\infty} a_j P_{d_p}(\xi_j)\hat{\phi}(\xi_j - u) |_{u=\xi_{m_j}} = a_{m_j} P_{d_p}(\xi_{m_j}) \to 0 \quad \text{as } j \to \infty.$$

Assume that S is a hypoelliptic convolutor but (15) does not hold for $\hat{S}$. This means that there is $d > 0$ and a sequence ξ_j with the property $\xi_{j+1} > \xi_j$, $j \in \mathbf{N}$, $\xi_1 > 2$, such that

$$| \hat{S}(\xi_j) | \leq e^{-M(d\xi_j)}. \tag{17}$$

Put $U(x) = \sum_{j=1}^{\infty} e^{ix\xi_j}$, $x \in \mathbf{R}$. (17) and the previous part of the proof imply that

$$(S * U)(x) = \sum \hat{S}(\xi_j) e^{ix\xi_j}, \quad x \in \mathbf{R},$$

is from $\mathcal{ES}'^{\{M_p\}}$. But U does not belong to $\mathcal{ES}'^{\{M_p\}}$ and this is a contradiction

b) Clearly, $Q = \mathcal{F}^{-1}(\tilde{Q}) \in O_C'^{*}$ and

$$Q * S = \delta - \tilde{\psi} \quad \text{where } \tilde{\psi} = \mathcal{F}^{-1}(\psi) \in \mathcal{S}^{*}.$$

Proposition 4 and 8 imply

$$U = (Q * S * U) + \tilde{\psi} * U = Q * V + \tilde{\psi} * U \in \mathcal{ES}'^{*}.$$

REFERENCES

[1] R.W.Braun, R.Meise, D.Vogt, Existence of fundanemal solutions and surjecivity of convolution operators on class of ultradifferentiable functions, Proc. London Math. Soc. (3)6 (1990), 344-370.

[2] A.Kaminski, D.Kovačević, S.Pilipović, On the Convolution of Ultradistributions, preprint.

[3] H.Komatsu, Ultradistributions, I ; Strucure theorems and a characterizaion, J. Fac. Sci. Univ. Tokyo, Sect. IA Math. 20 (1973), 25-105.

[4] H.Komatsu, Microlocal Analysis in Gevrey Class and in Complex Domains, preprint.

[5] D.Kovačević, Some operations on the space $S'^{(M_p)}$ of tempered ultradistributions, Univ. u Novom Sadu, Zb. Rad. Prirod.- Mat. Fak. Ser. Mat. to appear.

[6] D.Kovačević, S.Pilipović, Structural properties of the space of tempered ultradistributions, Proc. Conf. "Complex Analysis and Applications '91 with Symposium on Generalized Functions, Varna 1991, to appear.

[7] R.Miese, K.Schwerdfeger, B.A.Taylor, On Kernels of Slowly decreasing convolution operators, DOGA Tr. J.Math. 10, 1 (1986).

[8] A.Pietsch, Nukleare lokalkonvexe Raume, Akademie- Verlag, Berlin, 1965.

[9] S.Pilipović, Tempered Ultradistributions, Bul. Un. Math. Ital. (7) 2-B (1988).

[10] S.Pilipović, Characterizations of ultradistributions spaces and bounded sets, preprint.

[11] H.Shaefer, Topological vector spaces, Collier- Mac - Millan, London, 1986.

[12] C.Swartz, Convergence of convolution operators, Studia Math. XLII (1972), 249-257.

[13] C.Swartz, Convolution in $K\{M_p\}$ spaces, Rocky Monntain J. Math. 2 (1972), 259-263.

[14] J.Uryga, On tensor product and convoluion of generalized functions of Gelfand - Shilov type, Generalized Functions and Convergence, Memorial Volume for Prof. J. Mikusinski, World Scientific Singapore, Ed. A. Antosik, A.Kaminski, 251-264.

[15] Z.Zielezny, Hipoelliptic and enire elliptic convolution equations in subspaces of the space of distributions (II), Sudia Math. XXXII (1969), 47-59.

[16] Z.Zielezny, Hipoelliptic and entire elliptic convolution equations in subspaces of the space of disribuions (I), Studia Math. XXVIII (1967), 317-332.

THE HILBERT SPACES OF SZEGÖ TYPE AND FOURIER-LAPLACE TRANSFORMS ON $\mathbb{R}^n$

Saburou Saitoh

Department of Mathematics
Faculty of Engineering
Gunma University
Kiryu 376, Japan

1. INTRODUCTION

Let G be a domain on $\mathbb{R}^n (n \geq 2)$ and $T_G = \mathbb{R}^n + iG \subset \mathbb{C}^n$ be the tube domain over G. Then, following the idea of the general method [9,13] for integral transforms, we will examine the Fourier - Laplace transforms on some subset of $\mathbb{R}^n$ whose images belong to the Hilbert spaces of Szegö type on T_G, in a general situation. The basic idea in this paper is based on the following fact: for any positive matrix $K(p, q)$ on an abstract set E, there exists a uniquely determined Hilbert space H_K admitting the reproducing kernel $K(p, q)$ (Aronszajn [2, p. 344] and Saitoh [13, pp. 7-10]). From the viewpoints of the "Szegö space" on $\mathbb{C}^n$ and the Fourier - Laplace transform on $\mathbb{R}^n$, we shall define a kernel form $K(z, \overline{w})$ on $\mathbb{C}^n$ in a general situation and we shall examine the Hilbert space admitting the reproducing kernel $K(z, \overline{w})$. Some important and typical cases were investigated by Gindikin [6,7] (see Fuks [3]). The situation of Gindikin is perfect in a reasonable sense, but the situation will be restrictive in two senses. The first sense is on the assumption for the boundary values of the members of the Szegö space. Indeed, he assumes that the members of the Szegö space can be extended continuously up to the boundary. Therefore, his space does not form a Hilbert space. In order to define his space as a Hilbert space, we must investigate the properties of boundary values of the members of the space. The second sense is on the assumption for the existence domain of the Szegö space. His existence domain is natural in the sense of holomorphic function theory and perfect in this sense. At the same time, his reproducing kernels will be strongly restrictive for this fact. Compare our explicit examples of reproducing kernels with Gindikin's examples. We will first give a kernel form in a relaxed form and examine the Hilbert space admitting the reproducing kernel. We will state our situation clearly in the next section.

When the images of Fourier-Laplace transforms belong to the Hilbert spaces of Bergman type and the Paley-Wiener space on T_G, the Fourier-Laplace transforms

Generalized Functions and Their Applications, Edited by
R.S. Pathak, Plenum Press, New York, 1993

were investigated by Genchev [4], the author [10,12] and [11], respectively.

The contents are as follows :

§2. Orientation.
§3. Hilbert spaces of Szegö type.
§4. Domains and ranges of Hilbert spaces of Szegö type.
§5. Applications to Schrödinger's equation.
§6. Examples.

The author wishes to thank Professor T. G. Genchev for pleasant and fruitful discussions in Bulgaria.

2. ORIENTATION

In order to find a reasonable and general definition of Hilbert spaces of Szegö type, following the idea of the general method for integral transforms, we first design our situation for the Fourier-Laplace transforms from the viewpoint that the images belong to the Hilbert spaces of Szegö type.

We set

$$\begin{aligned} z &= (z_1, z_2, \cdots, z_n) \in \mathbb{C}^n, & z_j &= x_j + iy_j (x_j, y_j \in R), \\ x &= (x_1, x_2, \cdots, x_n), & y &= (y_1, y_2, \cdots, y_n) \in \mathbb{R}^n, \\ dx &= dx_1 dx_2 \cdots dx_n, & dy &= dy_1 dy_2 \cdots dy_n, \\ t &= (t_1, t_2, \cdots, t_n) \in \mathbb{R}^n, \end{aligned}$$

and

$$(z, t) = \sum_{j=1}^{n} z_j t_j.$$

For some domain $D \subset \mathbb{R}^n$ and for some functions g on D, we consider the Fourier-Laplace transform

$$f(z) = \frac{1}{(2\pi)^{\frac{n}{2}}} \int_D e^{i(z,t)} g(t) dt, \quad z \in \Omega. \tag{2.1}$$

We will discuss the case where the images $f(z)$ belong to a Hilbert space of Szegö type. Here, a Hilbert space of Szegö type on some domain Ω on $\mathbb{C}^n$ will mean that:

(i) $\partial\Omega$ is composed of a finite number of piecewise smooth hypersurfaces on $\mathbb{C}^n$.
(ii) We consider the class $H_2(\Omega)$ of all holomorphic functions $f(z)$ on Ω such that

$$\overline{\lim_{\varepsilon \to 0}} \int_{\partial\Omega_\varepsilon} |f(z)|^2 d\mu_\varepsilon < \infty$$

for $\Omega_\varepsilon = \{z \in \Omega | g(z, w) > \varepsilon\}$ for the Green's function $g(z, w)$ of Ω with pole at any fixed $w \in \Omega$ and for the surface measure $d\mu_\varepsilon$ of $\partial\Omega_\varepsilon$. Then, for any measurable set $S \subset \partial\Omega$, $f(z)$ have nontangential boundary values $f(\xi)$ for a.e. on S and

$$\left\{ \int_S |f(\xi)|^2 \omega(\xi) d\mu(\xi) \right\}^{\frac{1}{2}} < \infty$$

for the surface measure $d\mu$ of S and for any positive continuous function ω on S.

(iii) $H_2(\Omega; S, \omega)$ denotes a Hilbert space equipped with the inner product

$$(f, g)_{H_2(\Omega;S,\omega)} = \int_S f(\xi)\overline{g(\xi)}\omega(\xi)d\mu(\xi),$$

where $\overline{g(\xi)}$ denotes the complex conjugate of $g(\xi)$.

(iv) $H_2(\Omega; S, \omega)$ admits a reproducing kernel $K(z, \overline{w}; \Omega, S, \omega)$ such that for any fixed $w \in \Omega, K(z, \overline{w}; \Omega, S, \omega) \in H_2(\Omega; S, \omega)$ and, for any $f \in H_2(\Omega; S, \omega)$ and for any $w \in \Omega$,

$$f(w) = (f(\cdot), K(\cdot, \overline{w}; S, \omega))_{H_2(\Omega;S,\omega)}.$$

The reproducing kernel $K(z, \overline{w}; \Omega, S, \omega)$ and the Hilbert space $H_2(\Omega; S, \omega)$ are called the Szegö kernel and the Szegö space on Ω, respectively, for both cases of $S = \partial\Omega$ and a subset S of $\partial\Omega$. See, for example, Stein-Weiss [14], Fuks [3] and Aǐzenberg and Yuzhakov [1].

In order to examine the case where the images $f(z)$ by (2.1) belong to $H_2(\Omega; S, \omega)$, we consider the following expression, for a positive and continuous function $\rho(t)$ on D

$$K(z, \overline{w}; \Omega, S, \omega) = \frac{1}{(2\pi)^n} \int_D e^{i(z,t)} e^{-i(\overline{w},t)} \rho(t) dt. \tag{2.2}$$

See Saitoh [9,13] for this idea.

Since the existence domain of the kernel (2.2) is independent of the real parts of z and w, we see immediately that we should take the tube domain T_G as Ω. Since now S is a subset of ∂T_G or ∂T_G itself, we will assume that S is a tube of the form $T_{S_0} = \mathbb{R}^n + iS_0 (S_0 \subset \partial G)$.

We will determine formally the weight $\rho(t)$. From (2.1), we have by Plancherel's theorem

$$\int_{\mathbb{R}^n} |f(z)|^2 dx = \int_D e^{-2(y,t)} |g(t)|^2 dt$$

and so, we have formally, for the surface measure $d\mu$ of S_0

$$\int_{T_{S_0}} |f(z)|^2 \omega(y) dx d\mu(y) = \int_D |g(t)|^2 \left\{ \int_{S_0} e^{-2(y,t)} \omega(y) d\mu(y) \right\} dt. \tag{2.3}$$

Here, we exchanged the function $\omega(y)$ and the measure $d\mu$ as follows: The function $\omega(y)$ is a positive and continuous function on S_0 which is independent of x, and $d\mu$ is the surface measure of S_0 which is not the measure on ∂T_{S_0}. Since S_0 is a subset of ∂G, we must give an interpretation for the "boundary" values $f(z)$ on $\mathbb{R}^n \times S_0$ in (2.3). When the equation (2.3) is valid, for

$$W(t; S_0, \omega) = \int_{S_0} e^{-2(y,t)} \omega(y) d\mu(y) \quad \text{on} \quad D,$$

we obtain the isometrical identity

$$\int_{T_{S_0}} |f(z)|^2 \omega(y) dx d\mu(y) = \int_D |g(t)|^2 W(t; S_0, \omega) dt. \tag{2.4}$$

As we see from the general method for integral transforms, the above arguments will mean that the "Szegö" kernel has the expression

$$K(z, \overline{w}; T_G, S_0, \omega) = \frac{1}{(2\pi)^n} \int_D e^{i(z,t)} e^{-i(\overline{w},t)} W(t; S_0, \omega)^{-1} dt \tag{2.5}$$

for some domains T_G and D, under the condition that the "Szegö" kernel is expressible in terms of the Fourier-Laplace transform (2.1).

In this paper, we will consider conversely the kernel form (2.5), in a general and natural situation and we will examine the Hilbert space admitting the reproducing kernel (2.5). Then, we shall be able to obtain the Hilbert spaces of Szegö type in a general situation. Moreover, the reproducing kernels are explicit and simple, but unusual Hilbert spaces of Szegö type will appear in our situation. For these, we shall give several explicit examples of reproducing kernels which are not covered by Gindikin.

3. HILBERT SPACES OF SZEGÖ TYPE

For a set S_0 of a finite number of piecewise smooth surfaces on $\mathbb{R}^n$ of topological dimension $k < n$ and for a positive and continuous function ω on S_0, we set

$$W(t; S_0, \omega) = \int_{S_0} e^{-2(y,t)} \omega(y) d\mu(y)$$

for the surface measure $d\mu$ of S_0. As we see from the inequality

$$(3.0) \quad u^\alpha v^{1-\alpha} \leq \alpha u + (1-\alpha)v \qquad (u, v \geq 0, \quad 0 < \alpha < 1),$$

note that $W(t; S_0, \omega)$ is a convex function in t and so, it is continuous in t on the interior of a set where $W(t; S_0, \omega)$ is bounded. Let us take any domain $D(S_0, \omega)$ on $\mathbb{R}^n$ contained in the set

$$\{t \in \mathbb{R}^n; 0 < W(t; S_0, \omega) < \infty\}.$$

This mild condition for an arbitrariness of $D(S_0, \omega)$ will enable us to consider a wide class of reproducing kernels containing Gindikin's reproducing kernels.

Next, we consider the following maximal domain

$$G[D(S_0, \omega)] = \left\{ y \epsilon \mathbb{R}^n; \int_{D(S_0,\omega)} e^{-2(y,t)} W(t; S_0, \omega)^{-1} dt < \infty \right\}^\circ,$$

where $\circ$ denotes the interior. Then, we see from (3.0) that $G[D(S_0, \omega)]$ is a convex domain on $\mathbb{R}^n$. Then, the function

$$(3.1) \qquad K(z, \overline{w}; T_{G[D(S_0,\omega)]}) = \frac{1}{(2\pi)^n} \int_{D(S_0,\omega)} e^{i(z,t)} e^{-i(\overline{w},t)} W(t; S_0, \omega)^{-1} dt$$

exists on $T_{G[D(S_0,\omega)]} \times \overline{T_{G[D(S_0,\omega)]}}$ (complex conjugate). Further, note that it is analytic on $T_{G[D(S_0,\omega)]} \times \overline{T_{G[D(S_0,\omega)]}}$ and is a positive matrix on $T_{G[D(S_0,\omega)]}$ in the sense of Aronszajn-Moore. Hence, there exists a *uniquely* determined Hilbert space $H(T_{G[D(S_0,\omega)]})$ comprising holomorphic functions on $T_{G[D(S_0,\omega)]}$ and admitting the reproducing kernel $K(z, \overline{w}; T_{G[D(S_0,\omega)]})$ (see Aronszajn [2]).

On the other hand, as we see from the expression (3.1), this Hilbert space $H(T_{G[D(S_0,\omega)]})$ is composed of all holomorphic functions $f(z)$ on $T_{G[D(S_0,\omega)]}$ which are expressible in the form

$$f(z) = \frac{1}{(2\pi)^n} \int_{D(S_0,\omega)} F(t) e^{i(z,t)} W(t; S_0, \omega)^{-1} dt \tag{3.2}$$

for functions $F \in L_2(D(S_0,\omega), W(t; S_0,\omega)^{-1} dt)$ satisfying

$$\left\{ \frac{1}{(2\pi)^n} \int_{D(S_0,\omega)} |F(t)|^2 W(t; S_0,\omega)^{-1} dt \right\}^{\frac{1}{2}} < \infty. \tag{3.3}$$

Furthermore, we have the isometrical identity

$$\|f\|^2_{H(T_{G[D(S_0,\omega)]})} = \frac{1}{(2\pi)^n} \int_{D(S_0,\omega)} |F(t)|^2 W(t; S_0,\omega)^{-1} dt \tag{3.4}$$

(Saitoh [9,13]).

From the identity

$$\int_{S_0} \left\{ \int_{D(S_0,\omega)} |F(t)|^2 e^{-2(y,t)} W(t; S_0,\omega)^{-2} dt \right\} \omega(y) d\mu(y)$$
$$= \int_{D(S_0,\omega)} |F(t)|^2 W(t; S_0,\omega)^{-1} dt < \infty,$$

we see that for almost all y of S_0 with respect to $d\mu$

$$\int_{D(S_0,\omega)} |F(t)|^2 e^{-2(y,t)} W(t; S_0,\omega)^{-2} dt < \infty. \tag{3.5}$$

We denote the convex hull of S_0 by $\hat{S}_0$. Then, for almost all y of $\hat{S}_0$, (3.5) is valid, as we see from the inequality (3.0). Hence, for almost all y of $\hat{S}_0$, the function $f(x+iy)$ defined by (3.2) can be extended onto $\mathbb{R}^n$ by means of the Fourier integral in the framework of L_2 space and belongs to $L_2(\mathbb{R}^n, dx)$. In this sense, we obtain the isometrical identity

$$\int_{T_{S_0}} |f(z)|^2 \omega(y) dx d\mu(y) = \frac{1}{(2\pi)^n} \int_{D(S_0,\omega)} |F(t)|^2 W(t; S_0,\omega)^{-1} dt \tag{3.6}$$

and so, we obtain a "realization" of the norm in $H(T_{G[D(S_0,\omega)]})$

$$\|f\|^2_{H(T_{G[D(S_0,\omega)]})} = \int_{T_{S_0}} |f(z)|^2 \omega(y) dx d\mu(y), \tag{3.7}$$

by means of the Fourier integral. As to the relationship of S_0 and $G[D(S_0,\omega)]$, we obtain

Theorem 3.1. For the interior $(\hat{S}_0)^\circ$ of $\hat{S}_0$, we have the relation

$$(\hat{S}_0)^\circ \subset G[D(S_0,\omega)].$$

Proof. For any $F \in L_2(D(S_0,\omega), W(t;S_0,\omega)^{-1}dt)$, we have (3.5) for a.e. y of $\hat{S}_0$. For any $y \in (\hat{S}_0)^\circ$, we can easily take a line through the point and connecting two points y_1 and y_2 on $\hat{S}_0$ such that (3.5) is valid for y_1 and y_2. Hence, we see from (3.0) that for the point y, (3.5) is valid. Then, for all $F \in L_2(D(S_0,\omega), W(t;S_0,\omega)^{-1}dt)$,

$$|f(x+iy)| = \frac{1}{(2\pi)^n}\left|\int_{D(S_0,\omega)} F(t)e^{i(x,t)}e^{-(y,t)}W(t;S_0,\omega)^{-1}dt\right| < \infty.$$

Hence, by Landau's theorem for the converse of Hölder's inequality, we have the desired result

$$\int_{D(S_0,\omega)} e^{-2(y,t)}W(t;S_0,\omega)^{-1}dt < \infty.$$

We thus obtain Theorem 3.1.

By Theorem 3.1, if $(\hat{S}_0)^\circ \neq \phi$, a point of S_0 is contained in $G[D(S_0,\omega)]$ or is a boundary point of $G[D(S_0,\omega)]$. When a point y_0 of S_0 is a boundary point of $G[D(S_0,\omega)]$, we want to define $f(z_0)(z_0 = x+iy_0)$ as a boundary value of the analytic function $f(z)$ on $T_{G[D(S_0,\omega)]}$ in (3.7). Now S_0 is composed of a finite number of piecewise smooth surfaces on $\mathbb{R}^n$ of topological dimension $k < n$. Then, for y_0 when we can take an open polyhedron P_{y_0} (i.e. the interior of the convex hull of a finite points) contained in $G[D(S_0,\omega)]$ and having y_0 as a boundary point such that for some constant A

$$\text{(3.8)}\quad \int_{\mathbb{R}^n} |f(z)|^2 dx = \frac{1}{(2\pi)^n}\int_{D(S_0,\omega)} |F(t)|^2 \times e^{-2(y,t)}W(t;S_0,\omega)^{-2}dt \leq A < \infty \quad \text{on} \quad P_{y_0},$$

$f(x+iy)$ tends to $f(x+iy_0)$ in the L_2 norm as y tends to y_0 within P_{y_0}. Here, $f(x+iy_0)$ is defined by the Fourier transform (3.2). See Stein-Weiss [14, Chapter III, Section 2]. The inequality (3.8) is valid when

(i) $(\hat{S}_0)^\circ \neq \phi$ and there exists an open polyhedron $P_{y_0}{}'$ contained in $(\hat{S}_0)^\circ$ and having y_0 as a boundary point such that for $y \in P_{y_0}{}'$,

$$\int_{\mathbb{R}^n} |f(x+iy)|^2 dx = \int_{D(S_0,\omega)} |F(t)|^2 e^{-2(y_0,t)}W(t;S_0,\omega)^{-2}dt < \infty$$

and

$$P_{y_0} \subset P_{y_0}{}',$$

or

(ii) $e^{-2(y,t)}W(t;S_0,\omega)^{-1}$ is bounded for $t \in D(S_0,\omega)$ and $y \in P_{y_0}$.

When (3.8) is valid a.e. on $S_0 \cap \partial G[D(S_0,\omega)]$, we can define the *restricted* boundary values $f(x+iy_0)$ in the above sense. Then, of course, (3.7) is valid. Hence, we can regard (3.7) as a " Szegö " norm. So, we will call this space $H(T_{G[D(S_0,\omega)]})$ and this kernel $K(z,\overline{w};T_{G[D(S_0,\omega)]})$ a Hilbert space and a reproducing kernel of Szegö type, respectively, in our sense.

4. DOMAINS AND RANGES OF HILBERT SPACES OF SZEGÖ TYPE

In order to investigate the interrelationship between $D(S_0,\omega)$ and $G[D(S_0,\omega)]$, we assume that $D(S_0,\omega)$ is a convex domain on $\mathbb{R}^n$, does not contain whole lines, and $\partial D(S_0,\omega)$ is a piecewise smooth hypersurface.

Let O be the origin of coordinates in the y space $\mathbb{R}^n$, and let $\{\Gamma\}$ the hyperplanes pass through it and lie parallel to the limiting positions of the tangent hyperplanes of the hypersurface $\partial D(S_0,\omega)$. We consider the convex cone with vertex at the origin enveloped by these hyperplanes. The nappe of this cone lying on the same side of the hyperplanes $\{\Gamma\}$ as the domain $D(S_0,\omega)$ is called the *asymptotic* cone of the tube domain $T_{D(S_0,\omega)}$. As the asymptotic cone of a domain $T_{D(S_0,\omega)}$ with a bounded base $D(S_0,\omega)$, we take the set $\{O\}$, the origin of coordinates. Further, if V is the asymptotic cone of $T_{D(S_0,\omega)}$, we will say that this domain $T_{D(S_0,\omega)}$ is of *type* V.

Let $\Gamma(\subset \mathbb{R}^n)$ be a convex cone with vertex at the origin. Then, let Γ^* denote the conjugate cone of Γ ; that is, the set defined by $\Gamma^* = \{y^* \in \mathbb{R}^n; (y^*,y) > 0 \text{ for all } y \in \overline{\Gamma}(\text{closure}), y \neq 0\}$.

See Fuks [3] for these definitions. Then, we obtain

Theorem 4.1. We assume that $T_{D(S_0,\omega)}$ is of type V.

(I) If for any constant $L_1 > 0$,

$$W(t;S_0,\omega)^{-1}dt \quad \text{is integrable on} \quad D(S_0,\omega)\cap\{|t|\leq L_1\}$$

and there exist constants m_0 and L_2 such that

$$W(t;S_0,\omega)\geq m_0>0 \quad \text{on} \quad D(S_0,\omega)\cap\{|t|\geq L_2\},$$

then we have

$$G[D(S_0,\omega)]\supset V^*.$$

(II) If there exist constants M_0 and L such that

$$W(t;S_0,\omega)\leq M_0 \quad \text{on} \quad D(S_0,\omega)\cap\{|t|\geq L\},$$

then we have

$$G[D(S_0,\omega)]\subset V^*.$$

Proof. For any $y_0 \in V^*$, we have

$$(y_0,t)=|y_0||t|\cos\theta_0(t)>0 \quad \text{on} \quad V,$$

where, of course, $\theta_0(t)(0\leq\theta_0(t)<\frac{\pi}{2})$ is the angle between the two vectors y_0 and t in the same space $\mathbb{R}^n$. Hence, there exists a constant θ such that

$$\theta_0(t)<\theta<\frac{\pi}{2} \quad \text{on} \quad V.$$

Hence, there exist constants $\varepsilon > 0$ and $L_3 > 0$ such that

$$\theta_0(t) < \theta - \varepsilon < \frac{\pi}{2} \quad \text{on} \quad D(S_0, \omega) \cap \{|t| \geq L_3\}.$$

Note that the identity

$$\int_{|x|<N} f(|x|)dx = \frac{2\sqrt{\pi^n}}{\Gamma(\frac{n}{2})} \int_0^N x^{n-1} f(x)dx \quad [8, p.623].$$

Then, we have:

$$\begin{aligned}
&\int_{D(S_0,\omega)} e^{-2(y_0,t)} W(t; S_0, \omega)^{-1} dt \\
&\quad \leq \int_{D(S_0,\omega)\cap\{|t|\leq \max(L_2,L_3)\}} e^{-2(y_0,t)} W(t; S_0, \omega)^{-1} dt \\
&\quad + \frac{1}{m_0} \int_{D(S_0,\omega)\cap\{|t|\geq \max(L_2,L_3)\}} e^{-2|y_0||t|\cos(\theta-\varepsilon)} dt \\
&\quad \leq \int_{D(S_0,\omega)\cap\{|t|\leq \max(L_2,L_3)\}} e^{-2(y_0,t)} W(t; S_0, \omega)^{-1} dt \\
&\quad + \lim_{N\to\infty} \frac{2\sqrt{\pi^n}}{m_0\Gamma(\frac{n}{2})} \int_0^N t^{n-1} \exp[-t\{2|y_0|\cos(\theta - \varepsilon)\}]dt \\
&\quad < \infty.
\end{aligned}$$

Hence, we have the desired result (I).

Next, for any $y_0 \in \mathbb{R}^n - V^*$, by definition of V^*, there exists a point $t_0 \in V$ such that $(y_0, t_0) < 0$. Then, there exists a narrow nondegenerate (i.e. contains an n-dimensional sphere) convex cone $\Gamma(t_0)$ with vertex at the origin such that

$$(y_0, t) < 0 \quad \text{on} \quad \Gamma(t_0)$$

and

$$D(S_0, \omega) \supset \Gamma(t_0) \cap \{|t| \geq M\} \quad \text{for some constant} \quad M > 0.$$

Then, we have

$$\int_{D(S_0,\omega)} e^{-2(y_0,t)} W(t; S_0, \omega)^{-1} dt \geq \frac{1}{M_0} \int_{\Gamma(t_0)\cap\{|t|\geq \max(L,M)\}} e^{-2(y_0,t)} dt = \infty.$$

Hence, we see that $y_0 \notin G[D(S_0, \omega)]$ and so we have the desired result (II).

The members $f(z)$ of $H(T_{G[D(S_0,\omega)]})$ are analytic on $T_{G[D(S_0,\omega)]}$ and have finite norms (3.7). Conversely, in order to give a characterization of the members of $H(T_{G[D(S_0,\omega)]})$, note that the members $f(z)$ are characterized by the expressions (3.2) with (3.3).

Theorem 4.2. If (3.8) is valid; that is, $f(x + iy)$ are $L_2(\mathbb{R}^n, dx)$ integrable and

(4.1) $$\int_{\mathbb{R}^n} |f(z)|^2 dx \quad \text{are bounded on some} \quad P_{y_0}$$

for a.e. y_0 of $S_0 \cap \partial G[D(S_0, \omega)]$ and $D(S_0, \omega)$ is the maximal domain in the sense of

$$D(S_0, \omega) = D^m(S_0, \omega) = \{t \in \mathbb{R}^n; W(t; S_0, \omega) < \infty\}^\circ,$$

then the analytic functions $f(z)$ on $T_{G[D(S_0,\omega)]}$ having finite norms (3.7) in the sense of the restricted boundary values $f(z)$ belong to the space $H(T_{G[D(S_0,\omega)]})$.

Proof. For any analytic function $f(z)$ on $T_{G[D(S_0,\omega)]}$ having a finite norm (3.7), we set

$$g_y(t) = \frac{1}{(2\pi)^{\frac{n}{2}}} \int_{\mathbb{R}^n} f(z) e^{-i(z,t)} dx. \tag{4.2}$$

From our assumption (4.1), we see that for almost all y of S_0 and on the corresponding polyhedron P_y, $f(x+iy) \in L_2(\mathbb{R}^n)$. Hence, the integrals in (4.2) can be defined by means of the strong convergence and the functions $g_y(t)$ belong to $L_2(\mathbb{R}^n, dt)$. By Theorem 3.1, $(\hat{S}_0)^\circ \subset G[D(S_0, \omega)]$, and so $g_y(t)$ are independent of $y \in G[D(S_0, \omega)]$ as we see from the Cauchy integral formula. See Stein-Weiss [14, pp. 98-101] for this argument. Hence, we set $g_y(t) = g(t)$. Then, we have by Plancherel's theorem

$$\int_{T_{S_0}} |f(z)|^2 \omega(y) dx d\mu(y) = \int_{\mathbb{R}^n} |g(t)|^2 \left\{ \int_{S_0} e^{-2(y,t)} \omega(y) d\mu(y) \right\} dt < \infty.$$

Hence, we see that a.e. on $\mathbb{R}^n - D^m(S_0, \omega)$, $g(t) = 0$. Note that the maximal domain $D^m(S_0, \omega)$ is convex. Therefore, we see that $f(z)$ is expressible in the form (3.2) with (3.3) for

$$g(t) = \begin{cases} \frac{1}{(2\pi)^{\frac{n}{2}}} F(t) W(t; S_0, \omega)^{-1} & \text{on} \quad D^m(S_0, \omega) \\ 0 \quad \text{on} \quad \mathbb{R}^n \setminus D^m(S_0, \omega). \end{cases}$$

We thus have the desired result that $f(z) \in H(T_{G[D(S_0,\omega)]})$.

In Theorem 4.2, when the domain $D(S_0, \omega)$ is not the maximal domain $D^m(S_0, \omega)$, in order to characterize the members of $H(T_{G[D(S_0,\omega)]})$, we need more additional conditions for analytic functions $f(z)$ on $T_{G[D(S_0,\omega)]}$ with finite norms (3.7). See example (1).

5. APPLICATIONS TO SCHRÖDINGER'S EQUATION

As a typical case, we shall determine the case where the solutions of Schrödinger's equation belong to the Hilbert space $H(T_{G[D(S_0,\omega)]})$ of our Szegö type.

The fundamental solution of the Schrödinger equation

$$\frac{\partial u(x,t)}{\partial t} - ib\Delta u(x,t) = 0 \quad (b > 0)$$

is given by

$$E(x,t) = Y(t) \left(\frac{1-i}{2\sqrt{2\pi b t}} \right)^n \exp \frac{ix^2}{4bt}$$

(cf. Szmydt [15, pp. 371-374]). So, we will examine the integral transform

$$u(z,t) = \left(\frac{1-i}{2\sqrt{2\pi b t}} \right)^n \int_D \exp\{\frac{i}{4bt} \sum_{j=1}^n (z_j - \xi_j)^2\} g(\xi) d\xi \tag{5.1}$$

for some functions $g(\xi)$ and for some domain $D \subset \mathbb{R}^n$. For this purpose, let us consider the kernel form

$$(5.2) \qquad K_S(z,\overline{w};\rho) = \int_D \left(\frac{1-i}{2\sqrt{2\pi bt}}\right)^n \exp\{\frac{i}{4bt}\sum_{j=1}^n (z_j - \xi_j)^2\} \cdot \overline{\left(\frac{1-i}{2\sqrt{2\pi bt}}\right)^n \exp\{\frac{i}{4bt}\sum_{j=1}^n (w_j - \xi_j)^2\}}\rho(\xi)d\xi$$

for a positive and continuous function ρ on D. Then, we have

$$\begin{aligned}(5.3) \qquad K_S(z,\overline{w};\rho) &= \left(\frac{1}{4\pi bt}\right)^n \exp\{\frac{i}{4bt}\sum_{j=1}^n z_j^2\}\exp\{\frac{-i}{4bt}\sum_{j=1}^n \overline{w}_j^2\} \\ &\cdot \int_D \exp\{\frac{-i}{2bt}(z,\xi)\}\exp\{\frac{i}{2bt}(\overline{w},\xi)\}\rho(\xi)d\xi \\ &= \exp\{\frac{i}{4bt}\sum_{j=1}^n z_j^2\}\exp\{\frac{-i}{4bt}\sum_{j=1}^n \overline{w}_j^2\} \\ &\cdot \frac{1}{(2\pi)^n}\int_{-\frac{1}{2bt}D} e^{i(z,\xi)}e^{-i(\overline{w},\xi)}\rho(-2bt\xi)d\xi \\ &= \exp\{\frac{i}{4bt}\sum_{j=1}^n z_j^2\}\exp\{\frac{-i}{4bt}\sum_{j=1}^n \overline{w}_j^2\} \\ &\cdot K\left(z,\overline{w};T_{G[-\frac{1}{2\pi t}D(S_0,\omega)]}\right),\end{aligned}$$

for

$$(5.4) \qquad \rho(\xi) = W(-\frac{\xi}{2bt};S_0,\omega)^{-1}.$$

Here, $-\frac{1}{2bt}D$ denotes the domain $\{-\frac{\xi}{2bt};\xi \in D\}$. The above arguments imply that in the integral transform

$$(5.5) \qquad v(z,t) = \left(\frac{1-i}{2\sqrt{2\pi bt}}\right)^n \int_{-\frac{1}{2bt}D} \exp\{\frac{i}{4bt}\sum_{j=1}^n (z_j-\xi_j)^2\}F(\xi)\rho(\xi)d\xi$$

for functions $F(\xi)$ satisfying

$$\int_{-\frac{1}{2bt}D} |F(\xi)|^2\rho(\xi)d\xi < \infty,$$

we obtain the isometrical identity

$$(5.6) \qquad \|v(z,t)\exp\{\frac{-i}{4bt}\sum_{j=1}^n z_j^2\}\|^2_{H(T_{G[D(S_0,\omega)]})} = \int_{-\frac{1}{2bt}D} |F(\xi)|^2\rho(\xi)d\xi.$$

Hence, by setting $g(\xi) = F(\xi)\rho(\xi)$ we obtain

Theorem 5.1. In the situation of Section 3, we construct S_0, ω, a domain $D(S_0,\omega)$ on $\mathbb{R}^n$ contained in the set

$$\{t \in R^n; 0 < W(t;S_0,\omega) < \infty\},$$

and the maximal domain $G[D(S_0, \omega)]$. We assume that

$$O \in G[D(S_0, \omega)].$$

Then, the solution $u(x,t)$ for the Schrödinger equation satisfying

$$u(x,0) = g(x) \quad \text{on} \quad \mathbb{R}^n$$

with

$$\int_{-\frac{1}{2bt}D} |g(\xi)|^2 W(-\frac{\xi}{2bt}; S_0, \omega) d\xi < \infty$$

and

$$g(\xi) = 0 \quad \text{on} \quad \mathbb{R}^n \backslash (-\frac{1}{2bt} D),$$

can be extended analytically onto the tube domain $T_{G[D(S_0,\omega)]}$ and the extension $u(z,t)$ satisfies the identity

$$\|u(z,t) \exp\{\frac{-i}{4bt} \sum_{j=1}^{n} z_j^2\}\|^2_{H(T_{G[D(S_0,\omega)]})} = \int_{-\frac{1}{2bt}D} |g(\xi)|^2 W(-\frac{\xi}{2bt}; S_0, \omega) d\xi.$$

6. EXAMPLES

The typical situation and several explicit examples for reproducing kernels of Szegö type were given by Gindikin [6,7] (see Fuks [3]). So, we will give several explicit examples which are not covered by Gindikin and which support Theorems 3.1, 4.1 and 4.2.

(1)
$$S_0 = \{(y_1, 0); y_1 > 0\}.$$
$$\omega(y) = 1 \quad \text{and} \quad D(S_0, \omega) = \{(t_1, t_2); t_1, t_2 > 0\}.$$

Then,

$$W(t; S_0, \omega) = \frac{1}{2t_1}.$$

Furthermore,

$$G[D(S_0, \omega)] = \{(y_1, y_2); y_1, y_2 > 0\}$$

and

$$K(z, \overline{z}; T_{G[D(S_0,\omega)]}) = \frac{1}{16\pi^2 y_1^2 y_2}.$$

Hence, we have

$$K(z, \overline{w}; T_{G[D(S_0,\omega)]}) = \frac{-1}{\pi(z_1 - \overline{w}_1)^2} \frac{i}{2\pi(z_2 - \overline{w}_2)}.$$

For the set S_0, for all $y_0 \in S_0$, we see that (3.8) is valid for the sake of property (ii). Meanwhile, the maximal domain is $D^m(S_0, \omega) = \{(t_1, t_2); t_1 > 0\}$ and then

$G[D^m(S_0,\omega)] = \phi$. Then, for any fixed $a > 0$ and for the analytic function $f(z_1, z_2)$ on $\{\mathrm{Im} z_1 > 0\} \times \mathbb{C}$ such that

$$f(z_1, z_2) = \frac{1}{(2\pi)^2} \int_{D^m(S_0,\omega)} (e^{-at_1})^{\frac{1}{2}} (e^{-at_2^2})^{\frac{1}{2}} e^{iz_1 t_1} e^{iz_2 t_2} (2t_1) dt_1 dt_2,$$

we have

$$\int_{\mathbb{R}^2} \int_{S_0} |f(z_1, x_2)|^2 dx_1 dx_2 dy_1 = \frac{1}{(2\pi)^2} \int_{D^m(S_0,\omega)} e^{-at_1} e^{-at_2^2} (2t_1) dt_1 dt_2 < \infty,$$

but

$$f(z_1, z_2) \notin H(T_{G[D(S_0,\omega)]}).$$

For this latter fact, recall the characterization of the members of $H(T_{G[D(S_0,\omega)]})$ by expression (3.2) with (3.3).

This example shows that $f(z_1, z_2)$ is analytic on $T_{G[D(S_0,\omega)]}$ and has a finite norm

$$\left\{ \int_{\mathbb{R}^2} \int_{S_0} |f(z_1, x_2)|^2 dx_1 dx_2 dy_1 \right\}^{\frac{1}{2}} < \infty$$

in the sense of the restricted boundary values $f(z_1, x_2)$ on T_{S_0} which was stated below (3.8). However,

$$f(z_1, z_2) \notin H(T_{G[D(S_0,\omega)]}).$$

We will find a characterization of the members $f(z_1, z_2) \in H(T_{G[D(S_0,\omega)]})$.

For any $\varepsilon > 0$, on

$$G_\varepsilon = \{(y_1, y_2); 0 < \arg(y_1 + iy_2) < \frac{\pi}{2} - \varepsilon\},$$

there exists a constant M_ε such that

$$|2t_1 e^{-2y_1 t_1} e^{-2y_2 t_2}| \le M_\varepsilon \quad \text{on} \quad D(S_0, \omega).$$

Hence, for $f(z) \in H(T_{G[D(S_0,\omega)]})$ in the expression (3.2) with (3.3), we have

$$\begin{aligned} \int_{\mathbb{R}^2} |f(z)|^2 dx &= \frac{1}{4\pi^2} \int_{D(S_0,\omega)} |F(t_1, t_2)|^2 \\ &\quad \times e^{-2y_1 t_1} e^{-2y_2 t_2} (2t_1)^2 dt_1 dt_2 \\ &\le \frac{M_\varepsilon}{4\pi^2} \int_{D(S_0,\omega)} |F(t_1, t_2)|^2 (2t_1) dt_1 dt_2 < \infty. \end{aligned}$$

Hence, the functions $f(z) \in H(T_{G[D(S_0,\omega)]})$ belong to the analytic Hardy class H^2 on T_{G_ε}. In this example, the functions $f(z) \in H(T_{G[D(S_0,\omega)]})$ are conversely characterized by the conditions that $f(z)$ belong to the analytic Hardy class H^2 on T_{G_ε} for any $\varepsilon > 0$ and have finite norms (3.7). See Stein-Weiss [14, Chapter III, Section 3].

Indeed, by Theorem 3.1 in Stein-Weiss [14, pp. 101-102], if $f(z)$ belongs to the analytic Hardy class H^2 on T_{G_ε} for any $\varepsilon > 0$, then $f(z)$ is expressible in the form

$$f(z) = \int_{G_\varepsilon^*} e^{i(z,t)} g(t) dt$$

for a function $g(t)$ satisfying

$$\int_{G_\varepsilon^*} |g(t)|^2 dt < \infty.$$

Hence, from the finiteness of the integral (3.7), we see that $f(z)$ is expressible in the form (3.2) with (3.3). Hence, $f \in H(T_{G[D(S_0,\omega)]})$.

(2)
$$S_0 = \{(y_1, y_2); y_2 = y_1, y_1 > 0\},$$
$$\omega(y) = 1 \quad \text{and} \quad D(S_0, \omega) = \{(t_1, t_2); t_1, t_2 > 0\}.$$

Then,

$$W(t; S_0, \omega) = \frac{1}{\sqrt{2}(t_1 + t_2)}.$$

Furthermore,

$$G[D(S_0, \omega)] = \{(y_1, y_2); y_1, y_2 > 0\}$$

and

$$K(z, \overline{z}; T_{G[D(S_0,\omega)]}) = \frac{\sqrt{2}}{32\pi^2} \frac{y_1 + y_2}{y_1^2 y_2^2}.$$

Hence,

$$K(z, \overline{w}; T_{G[D(S_0,\omega)]}) = \frac{\sqrt{2}}{4\pi^2 i} \frac{(z_1 + z_2) - \overline{(w_1 + w_2)}}{(z_1 - \overline{w}_1)^2 (z_2 - \overline{w}_2)^2}$$
$$= \frac{1}{\sqrt{2}} \left\{ \frac{i}{2\pi(z_1 - \overline{w}_1)} \frac{-1}{\pi(z_2 - \overline{w}_2)^2} + \frac{-1}{\pi(z_1 - \overline{w}_1)^2} \frac{i}{2\pi(z_2 - \overline{w}_2)} \right\}.$$

The kernels $\frac{i}{2\pi(z_1 - \overline{w}_1)}$ and $\frac{-1}{\pi(z_1 - \overline{w}_1)^2}$ are the Szegö kernel and the Bergman kernel on the upper half plane of z_1, respectively. Hence, Examples (1) and (2) give the interrelationships among reproducing kernels. Hence, we can know the interrelationships among the associated reproducing kernel Hilbert spaces, by the general theory of Aronszajn[2]. The reproducing kernels in (1) and (2) have simple forms and they are the reproducing kernels of Szegö type in our sense. However, these reproducing kernels do not appear in the framework of Gindikin's Szegö kernels.

In the following examples, we will see the interrelationship among $S_0, D(S_0, \omega)$ and $G[D(S_0, \omega)]$.

(3)
$$S_0 = \{(y_1, y_2); y_2 = -y_1 + a, y_1 > 0\} \quad (a > 0) \quad \text{and} \quad \omega(y) = 1.$$

Then, we have

$$W(t; S_0, \omega) = \frac{e^{-2at_2}}{\sqrt{2}(t_1 - t_2)} \quad \text{for} \quad t_2 < t_1.$$

(i) $D(S_0, \omega) = \{(t_1, t_2); 0 < t_2 < t_1\}$.
Then,

$$G[D(S_0, \omega)] = \{(y_1, y_2); y_2 > -y_1 + a, y_1 > 0\}$$

and

$$K(z, \overline{z}; T_{G[D(S_0,\omega)]}) = \frac{\sqrt{2}}{32\pi^2} \frac{1}{y_1^2 (y_1 + y_2 - a)}.$$

Hence, we have

$$K(z,\overline{w};T_{G[D(S_0,\omega)]}) = \frac{\sqrt{2}}{4\pi^2 i(z_1-\overline{w}_1)^2\{(z_1+z_2)-\overline{(w_1+w_2)}-2ai\}}.$$

(ii) $D(S_0,\omega)=\{(t_1,t_2);t_2<0,t_1>0\}$.
Then,

$$G[D(S_0,\omega)] = \{(y_1,y_2);y_2<a,y_1>0\}$$

and

$$K(z,\overline{z};T_{G[D(S_0,\omega)]}) = \frac{\sqrt{2}}{32\pi^2}\frac{a+y_1-y_2}{y_1^2(a-y_2)^2}.$$

Hence, we have

$$K(z,\overline{w};T_{G[D(S_0,\omega)]}) = \frac{\sqrt{2}[2ai+\{(z_1-z_2)-\overline{(w_1-w_2)}\}]}{4\pi^2 i(z_1-\overline{w}_1)^2(2ai-(z_2-\overline{w}_2))^2}.$$

(iii) $D(S_0,\omega)=\{(t_1,t_2);t_2<t_1<0\}$.
Then,

$$G[D(S_0,\omega)] = \{(y_1,y_2);y_2<-y_1+a,y_2<a\}$$

and

$$K(z,\overline{z};T_{G[D(S_0,\omega)]}) = \frac{\sqrt{2}}{32\pi^2}\frac{1}{(a-y_1-y_2)(a-y_2)^2}.$$

Hence, we have

$$K(z,\overline{w};T_{G[D(S_0,\omega)]}) = \frac{\sqrt{2}}{4\pi^2 i}\frac{1}{\{2ai-(z_1+z_2)+\overline{(w_1+w_2)}\}\{2ai-(z_2-\overline{w}_2)\}^2}.$$

Next, we will examine a typical case of $(\hat{S}_0)^\circ \neq \phi$, and we will determine the maximal domain $D^m(S_0,\omega)$ and $G[D^m(S_0,\omega)]$.

(4) $$S_0=\{(0,y_2);y_2>0\}\cup\{(y_1,0);y_1>0\} \quad \text{and} \quad \omega(y)=1.$$

Then,

$$W(t;S_0,\omega)=\frac{1}{2}(\frac{1}{t_1}+\frac{1}{t_2}) \quad \text{for} \quad t_1,t_2>0.$$

Hence,

$$D^m(S_0,\omega)=\{(t_1,t_2);t_1,t_2>0\},$$

and so,

$$G[D^m(S_0,\omega)]=\{(y_1,y_2);y_1,y_2>0\}.$$

Then,

$$\begin{aligned}
&K(z,\overline{z};T_{G[D^m(S_0,\omega)]})\\
&= \frac{1}{(2\pi)^2}\int_0^\infty\int_0^\infty e^{-2y_1t_1}e^{-2y_2t_2}\frac{2t_1t_2}{t_1+t_2}dt_1dt_2 \quad (t_1-t_2=y,t_1+t_2=x)\\
&= \frac{1}{16\pi^2}\iint_{\substack{-x<y<x\\ x>0}} e^{-(y_1+y_2)x}e^{-(y_1-y_2)y}\frac{x^2-y^2}{x}dxdy\\
&= \begin{cases} \frac{1}{16\pi^2(y_1-y_2)^2}\left\{\frac{1}{y_1}+\frac{1}{y_2}-\frac{2\log(y_1\div y_2)}{y_1-y_2}\right\}, & y_1\neq y_2\\ \frac{1}{48\pi^2y^3}, & y_1=y_2=y. \end{cases}
\end{aligned}$$

Hence, we have

$$K(z,\overline{w};T_{G[D^m(S_0,\omega)]}) = \frac{-1}{4\pi^2\{(z_1-z_2)-\overline{(w_1-w_2)}\}^2} \cdot \left\{\frac{2i}{z_1-\overline{w_1}}+\frac{2i}{z_2-\overline{w_2}}-4i\frac{\log\{(z_1-\overline{w_1})\div(z_2-\overline{w_2})\}}{(z_1-z_2)-\overline{(w_1-w_2)}}\right\}.$$

By taking a point mass as $\omega(y)d\mu(y)$, we can deal with the Cauchy-Szegö kernels for radial tube domains. For the Cauchy-Szegö kernels, see Stein-Weiss [14, Chapter III].

(5) For $S_0 = \{y^{(0)}\}$ and $n = 2$, we set $W(t;S_0,\omega) = e^{-2(y^{(0)},t)}$

and

$$D(S_0,\omega) = \{(t_1,t_2); t_1,t_2 > 0\}.$$

Then,

$$G[D(S_0,\omega)] = \{(y_1,y_2); y_1 > y_1^{(0)}, y_2 > y_2^{(0)}\}$$

and

$$K(z,\overline{z};T_{G[D(S_0,\omega)]}) = \frac{1}{4\pi^2}\frac{1}{2(y_1-y_1^{(0)})2(y_2-y_2^{(0)})}.$$

Hence, we have

$$K(z,\overline{w};T_{G[D(S_0,\omega)]}) = \frac{1}{16\pi^2(\frac{z_1-\overline{w_1}}{2i}-y_1^{(0)})(\frac{z_2-\overline{w_2}}{2i}-y_2^{(0)})}.$$

(6) In the case (3) (iii), Theorem 5.1 can be stated explicitly as follows :

The solutions $u(x,t)$ of the Schrödinger equation

$$\frac{\partial u(x,t)}{\partial t} - ib\Delta u(x,t) = 0 \qquad (b>0, n=2)$$

satisfying the condition

$$u(x,0) = g(x)$$

with

$$\int_{\xi_1>\xi_2>0} |g(\xi)|^2 \frac{\exp\frac{a\xi_2}{bt}}{\xi_2-\xi_1} d\xi < \infty$$

and

$$g(\xi) = 0 \quad \text{on} \quad \mathbb{R}^2 \backslash \{\xi_1 > \xi_2 > 0\}$$

can be extended analytically onto $T_{\{y_2<-y_1+a, y_2<a\}}$, and the analytic extensions $u(z,t)$ satisfy the identity

$$\|u(z,t)\exp\{\frac{-i}{4bt}(z_1^2+z_2^2)\}\|^2_{H(T_{\{y_2<-y_1+a,y_2<a\}})}$$
$$= \int_{-\infty}^{\infty}\int_{-\infty}^{\infty}\int_0^{\infty} |u(x_1+iy_1, x_2+i(-y_1+a),t)|^2$$
$$\cdot \exp\{\frac{x_1y_1-x_2y_1+x_2a}{bt}\}dx_1dx_2dy_1$$
$$= \frac{2bt}{\sqrt{2}}\int_{\xi_2>\xi_1>0}|g(\xi)|^2\frac{\exp\frac{a\xi_2}{bt}}{\xi_2-\xi_1}d\xi.$$

Acknowledgments

The travel expenses of the author for attending to the International Symposium were partially supported by the Gunma University Foundation for Science & Technology. The living expenses during the Symposium were supported by the Organizing Committee.

Research partially supported by Grant-in-Aid for Scientific Research 03804003.

REFERENCES

1. Aǐzenberg A.I., and Yuzhakov A.P., "Integral Representations and Residues in Multidimensional Complex Analysis," *Transl. of Math. Mono., Amer. Math. Soc.*, 1983.
2. Aronszain N., Theory of reproducing kernels, *Trans. Amer. Math. Soc.* **68** (1950), 337-404.
3. Fuks B. A., "Introduction to the Theory of Analytic Functions of Several Complex Variables," Fizmatgiz, Moscow 1962; English transl., *Transl. Math. Mono., Amer. Math. Soc., Providence, R. I.*, 1963.
4. Genchev T. G., Some theorems of Paley-Wiener's type, in "Complex Analysis and Applications," Sofia, 1985.
5. Genchev T. G., Paley-Wiener type theorems for functions in Bergman spaces over tube domains, *J. Anal. and Appl.* **118** (1986), 496-501.
6. Gindikin S. G., Integral formulas for Siegel domains of the second kind, *Soviet Math. Dokl.* **2** (1961), 1480-1483.
7. ———, Analytic functions in tubular regions, *Soviet Math. Dokl.* **3** (1962), 1178-1183.
8. Gradshteyn I. S. and Ryzhik I. M., "Tables of Integrals, Series and Products," Academic Press., New York, 1980.
9. Saitoh S., Hilbert spaces induced by Hilbert space valued functions, *Proc. Amer. Math. Soc.* **89** (1983), 74-78.
10. ———, The Laplace transform of L^p functions with weights, *Applicable Analysis* **22** (1986), 103-109.
11. ———, Generalizations of Paley-Wiener's theorem for entire functions of exponential type, *Proc. Amer. Math. Soc.* **99** (1987), 465-471.
12. ———, Fourier-Laplace transforms and the Bergman spaces , *Proc. Amer. Math. Soc.* **102** (1988), 985-992.
13. ———, "Theory of reproducing kernels and its applications," Pitman Res. Notes Series 189, Longman Science & Technical, 1988.
14. Stein E. M. and Weiss G., "Introduction to Fourier Analysis on Euclidean Spaces, " Princeton Univ. Press. Princeton. N. J., 1975.
15. Szmydt Z., "Fourier transformation and linear differential equation," D. Reidel Rublishing Company, Dordrecht-Holl and Boston, 1972.

NEW SADDLE POINT THEOREMS

Martin Schechter

Department of Mathematics
University of California
Irvine, California 92717

ABSTRACT

We show how saddle point theorems can be changed to accomodate a wider range of applications.

1. INTRODUCTION

The usual saddle point theorems are useful tools in solving non-linear problems. One simple form is as follows: Let M be a closed subspace of Hilbert space H, and let $N=M^{\perp}$. Assume that

$$0 < \dim N < \infty. \tag{1.1}$$

Let G be a continuously differentiable functional on H and assume

$$\inf_{M} G = m_0 > -\infty. \tag{1.2}$$

and

$$\limsup_{\substack{\|v\| \to \infty \\ v \in N}} G(v) = m < \infty. \tag{1.3}$$

The theorem states that if $m < m_0$, then there is a constant $c \geq m_0$ and a sequence $\{ u_k \} \subset H$ such that

$$G(u_k) \to c, \quad G'(u_k) \to 0. \tag{1.4}$$

This does not provide a solution of

Generalized Functions and Their Applications, Edited by
R.S. Pathak, Plenum Press, New York, 1993

$$G'(u) = 0, \; G(u) = c \tag{1.5}$$

but an additional hypothesis such as the Palais-Smale (PS) condition dose imply that a solution of (1.5) exists.

By changing the proof of (1.4) we have been able to show that the hypothesis $m < m_0$ can be dropped completely and that (1.2), (1.3) imply (1.4) even if $m_0 \le m$ and even if

$$\sup_{\substack{\|v\| = R \\ v \in N}} G(v) > m_0 \tag{1.6}$$

for every $R > 0$. In fact we have

Theorem 1. Assume (1.2), (1.3). Let ψ be a nonincreasing function in $(0,\infty)$ such that

$$\limsup_{R \to \infty} \int_R^{2R} \psi(r)\, dr = \infty . \tag{1.7}$$

Let

$$m_1 = \sup_N G . \tag{1.8}$$

Then there are a constant c and a sequence $\{u_k\} \subset H$ such that

$$m_0 \le c \le m_1 \tag{1.9}$$

and

$$G(u_k) \to c, \; G'(u_k) \,/\, \psi(\| u_k \|) \to 0. \tag{1.10}$$

We note that if there is an $R > 0$ such that the left hand side of (1.6) is less than m_0, then we can weaken (1.7) to be

$$\int_1^\infty \psi\,(r)\, dr = \infty . \tag{1.11}$$

However, in applications it is extremely difficult to verify the existence of such an R when $m > m_0$.

Theorem 1 does not guarantee that a solution obtained from it is not zero. However , we can obtain such a guarantee if we adjust the hypotheses slightly. For this purpose we assume that there is a finite dimensional subspace N_0 of M such that

$$G(v) \le m, \; v \in N \tag{1.12}$$

$$G(v) \le m_1 < \infty \;, v \in N_1 = N \oplus N_0 \tag{1.13}$$

$$G(w) \geq \rho > m, \; w \in M, \; \|w\| = \delta. \tag{1.14}$$

It does not matter if $m_1 > \rho$. We have

Theorem 2. Under hypotheses (1.12) - (1.14) for each nonincreasing function ψ in $(0,\infty)$ satisfying

$$\limsup_{R\to\infty} \int_R^{2R-\delta} \psi(r)\,dr = \infty$$

there are a constant c and a sequence $\{u_k\} \subset H$ such that

$$\rho \leq c \leq m_1$$

and (1.10) holds.

Theorem 2 guarantees that any solution of (1.5) cannot be zero since $G(0) \leq m < \rho \leq c$.

2. SOME APPLICATIONS

We now apply Theorems 1 and 2 to the study of the Dirichlet problem

$$-\Delta u = f(x,u) \text{ in } \Omega, u = 0 \text{ on } \partial\Omega \tag{2.1}$$

where Ω is a smooth bounded domain in $\mathbb{R}^n$, $f(x,t)$ is a Caratheodory function on $\Omega \times \mathbb{R}$ satisfying

$$|f(x,t)| \leq C(|t| + 1), x \in \Omega, t \in \mathbb{R} \tag{2.2}$$

and

$$f(x,t)/t \longrightarrow b_{\pm}(x) \text{ a.e. as } t \longrightarrow \pm\infty. \tag{2.3}$$

Let $0 < \lambda_0 < \lambda_1 < \ldots < \lambda_\ell < \lambda_{\ell+1} < \ldots$ be the eigenvalues of the linear problem

$$-\Delta u = \lambda u \text{ in } \Omega, \; u = 0 \text{ on } \partial\Omega . \tag{2.4}$$

We assume that for some $\ell \geq 0$

$$\lambda_\ell t^2 - W_1(x) \leq 2F(x,t) \leq \lambda_{\ell+1} t^2 + W_2(x) \tag{2.5}$$

where

$$F(x,t) := \int_0^t f(x,s)\,ds \tag{2.6}$$

and the $W_j(x)$ are functions in $L^1(\Omega)$. Finally, we assume that no eigenfunction of (2.4) is a solution of

$$-\Delta u = b_+ u^+ - b_- u^- \text{ in } \Omega,\ u = 0 \text{ in } \partial\Omega \qquad (2.7)$$

where $u^+(x) = \max\{u(x), 0\}$, $u^-(x) = u^+(x) - u(x)$. We have

Theorem 3. Under the hypotheses given above, the Dirichlet problem (2.1) has a solution.

If $f(x,0) \equiv 0$, then $u \equiv 0$ is a solution of (2.1). In this case the solution given in Theorem 3 may be the trivial one. We can avoid this by adding the following hypotheses :

$$\limsup_{t \to 0} 2F(x,t)/t^2 \le \lambda' < \lambda_\ell \qquad (2.8)$$

and

$$\lambda_{\ell-1} t^2 \le 2F(x,t) \qquad (2.9)$$

We now have

Theorem 4. If we add (2.8) and (2.9) to the hypotheses of Theorem 3, then (2.1) has a solution $u \ne 0$.

Theorems 3 and 4 can be improved considerably. In particular, assumptions (2.3) and (2.5) can be weakend substantially.

REFERENCES

1. H.Amann and P. Hess, A multiplicity result for a class of elliptic boundary value problems, *Proc. Royal Soc. Edinburgh* 84-A, 145-151 (1979).
2. A.Ambrosetti and G. Prodi, On the inversion of some differentiable mappings with singularities between Banach spaces, *Ann. Mat . Pura Appl.* 93, 231-247. (1973)
3. A.Ambrosetti, P. H. Rabinowitz, Dual variational methods in critical point theory and applications *J. Functional Anal.* 14, 349-381(1973).
4. P.Bartolo, V. Benci, D. Fortunato, Abstract critical point theorems and applications to some nonlinear problems with "strong" reasonance at infinity , *Nonlinear Anal. Th. Meth. Appl.* 7, 981-1012 (1983).
5. H.Berestycki and D.G. de Figueiredo, Double resonance in semilinear elliptic problems, *Comm. P. D. E.* 6, 91-120 (1981)
6. H.Brezis and L. Nirenberg, Remarks on finding critical points, to appear.

7. M.Berger and E. Podolak, On the solution of a nonlinear Dirichlet problem, Indiana Univ. Math. J. 24, 837-846 (1975).
8. M.S.Berger and M. Schechter, On the solvability of semilinear gradient operator equations, *Adv. Math.* 25. 97-132 (1977).
9. N.P.Cac, On an elliptic boundary value problem at double resonance, *J. Math Anal. Appl.* 132, 473-483 (1988).
10. N.P.Cac, on the number of solutions of an elliptic boundary value problem with jumping nonlinearity, *Nonlinear Analysis, Theory, Methods and Applications* 13, 341-351 (1989).
11. N.P.Cac, On nontrivial solutions of a Dirichlet problem whose jumping nonlinearity crosses a multiple eigenvalue, J. Diff. Eq. 80, 379-404(1989).
12. N.P.Cac, On a boundary value problem with non-smooth jumping non-linearity to appear.
13. A.Castro, Hammerstein integral equations with indefinite kernel , *International J. Math. Mathem. Sci.* 1, 187-201(1978).
14. E.N.Dancer, Multiple solutions of asymptotically homogeneous problems, to appear.
15. D.G.De Figueiredo, Positive Solutions of Some Classes of Semilinear Elliptic Problems, Proc. Symp. Pure Math 45, 371-379 (1986).
16. D.G.De Figueiredo and J.P. Gossez, Conditions de non-resonance pour certains problems elliptiques semi-lineaires, *C. R. Acad. Sci. Paris*, 302, 543-545 (1986).
17. D.G.De Figueiredo, P.L. Lions, R.D. Nussbaum, A priori estimates and existence results for positive solutions of semilinear elliptic equations, *J. math. Pures Appl.* 61, 41-63 (1982).
18. T.Gallouet and O. Kavian, Resultats d'existence et de non-existence pour certains problemes demi lineaires a l'infini, *Ann. Fac. Sci. Toulouse Math.* 3, 201-246 (1981).
19. P.Hess, On a nonlinear elliptic boundary value problem of the Ambrosetti-Prodi type, *Boll. U.M.I.* 17-A 189-192 (1980).
20. J.L.Kazdan and F. W. Warner, Remarks on some qasilinear elliptic equations, *Comm. Pure Appl. Math.* 28, 567-597 (1975).
21. A.C.Lazer, "Introduction to multiplicity theory for boundary value problems with asymmetric nonlinearities, Parial Differential Equations", F. Cordoso et al. eds., Lecture Notes in math. No. 1324, Springer Verlag, N.Y. 137-165 (1988).
22. S.S.Lin, Some results for semilinear differential equations at resonance, *J. Math. Anal. Appl.* 93, 574-592(1983).
23. P.L.Lions , On the existence of positive solutions of semilinear elliptic equations, *SIAM Rev.* 24, 441-457 (1982).
24. E.A.Landesman and A.C. Lazer, Nonlinear perturbations of linear elliptic boundary value problems at resonance, *J. Math. Mech.*, 19, 609-623 (1970).
25. J.L.Lions and E. Magenes, "Non-Homogeneous Boundary Value Problems I", Springer Verlag, Berlin, 1972.

26. A.C.Lazer and P.J.McKenna, Multiplicity results for a class of semilinear elliptic and parabolic boundary value problems, J. Math. Anal. Appl.107, 371-395 (1985).
27. A.C.Lazer and P. J. McKenna, Critical point theory and boundary value problems with nonlinearities crossing multiple eigenvalues, *Comm. in P.D.E.*, I, 10, 107-150 (1985), II, 11, 1653-1676 (1986).
28. A.C.Lazer and P. J. McKenna, Multiplicity of solutions of nonlinear boundary value problems with nonlinearities crossing several higher eigenvalues, *J. Reine Angew Math.* 368, 184-200 (1986).
29. J.Mawhin and M. Willem, Critical points of convex perturbations of some indefinite quadratic forms and semilinear boundary value problems at resonance, *Ann. inst. H. Poincare, Analyse Non-Lineaire* 3, 431-453 (1986).
30. L.Nirenberg, Variational and topological methods in nonlinear problems, *Bull. Amer. Math. Soc.* 4, 267-302 (1981).
31. L.Nirenberg, "Variational methods in nonlinear problems", Lecture Notes in Math, Vol.1365, Springer-Verlag, 100-119, (1988).
32. E.Podolak, On the range of operator equations with an asymptotically nonlinear term, *Indiana Univ. Math. J.* 25, 1127-1137 (1976).
33. P.H.Rabinowitz, Variational methods for nonlinear eigenvalue problems, Eigenvalues of Nonlinear problems (G. Prodi, Ed.) C.I.M.E.Ed. Cremonese, Roma 1975, 141-195.
34. P.H.Rabinowitz, "Minimax methods in critical point theory with applications to differential equations", Conf. Board of Math. Sci. Reg. Conf. Ser. in Math., No 65, Amer. Math. Soc., 1986.
35. B.Ruf, On Nonlinear elliptic boundary value problems with jumping nonlinearities, *Ann. Mat. Pura Appl.* 128, 133-151 (1980).
36. M.Schechter, Nonlinear elliptic boundary value problems at strong resonance, *Amer. J. Math.* 112, 439-460 (1990).
37. M.Schechter, solution of nonlinear problems at resonance, *Indiana Univ. Math. J.* 39, 1061-1080 (1990).
38. M.Schechter, A bounded mountain pass lemma without the (PS) condition and applications, *Trans. Amer. Math. Soc.*
39. M.Schechter, Nonlinear elliptic boundary value problems at resonance, *Nonlinear Analysis, Theory, Methods & Applications*, 14, 889-903 (1990).
40. M.Schechter, A variation of the mountain pass lemma and applications, *J. London Math. Soc.*
41. M.Schechter, The Hampwile theorem for nonlinear eigenvalues, *Duke Mathematical Journal*, 59, 325-335 (1989).
42. S.Solimini, some remarks on the number of solutions of some nonlinear elliptic problems, Analyse Nonlineaire , *Ann. Inst. Henri Poincare* 2, 143-156 (1985).
43. E.Tarafdar, An approach to nonlinear elliptic boundary value problems, *J. Austral. Math. Soc.* 34, 316-335 (1983).

44. K.Thews, A reduction method for some nonlinear Dirichlet problems, *Nonlinear Analysis, Theory, Methods and Applications,* 3, 794-813 (1979).
45. K.Thews, Nontrivial solutions of elliptic equations at resonance, *Proc. R. Soc. Edinb.* 85 A, 119-129 (1980).

MULTIPLICATION AND QUASIASYMPTOTIC BEHAVIOUR IN GENERALIZED FOCK SPACES

John Schmeelk

Department of Mathematical Sciences
Virginia Commonwealth University
1015 West Main Street
Richmond, Virginia 23284-2014

1. INTRODUCTION

A principle application for our development will be modeling a physical system consisting of an "infinite" number of particles. A system of n-particles termed bosoms or fermions has previously been modeled by n-dimensional symmetric or antisymmetric functions [1]. When this system is enlarged to contain an idealized infinite number of particles, the column vector describing the system has the form,

$$\Phi \Leftrightarrow \begin{bmatrix} \phi_0 \\ \phi_1(x_1) \\ \vdots \\ \phi_q(x_1, x_2, \cdots, x_q) \\ \vdots \end{bmatrix}, \tag{1.1}$$

where ϕ_0 is a scalar and in the classical model, $\phi_q(x_1, \cdots, x_q) \in L_q\ (\Omega^q)$, $(1 \leq q < \infty)$ and Ω^q can either be a subset of $\mathbb{R}^q$ or $\mathbb{C}^q$. This system describing the set of all such vectors can be viewed as a direct sum of Hilbert spaces,

$$\mathcal{H} = \sum_{n=0}^{\infty} \oplus\ \mathcal{H}^n,$$

equipped with a norm which when squared equals

$$|||\Phi|||_{L_2} \triangleq |\phi_0|^2 + \sum_{q=1}^{\infty} \int_{\Omega^q}\cdots\int |\phi_q(x_1, \cdots x_q)|\ d^q x\ . \tag{1.2}$$

Generalized Functions and Their Applications, Edited by
R.S. Pathak, Plenum Press, New York, 1993

However the physics of the more advanced problem leads the researcher into distribution theory. This is already seen in the n-particle case. For example, the creation operator, $a^{+}(\xi)$, [1] represented as

$$\begin{bmatrix} 0 & \delta(y,\xi) & \cdot & \cdot \\ 0 & 0 & \sqrt{2}\delta(x,y_1)\delta(y_2,\xi) & \cdot \\ 0 & \cdot & 0 & \sqrt{2}\delta(x_1,y_1)\delta(x_2,y_2)\delta(y,\xi)\cdots \\ \vdots & \vdots & \cdot & \vdots \end{bmatrix}$$

contains the products of Dirac delta functionals. The mathematical foundation for this multiplication contained in each of the entries of the matrix are usually taken in a very formal manner. Moreover as $n \to \infty$, the notion of asymptotic behavior previously had very little mathematical foundation supporting the effort.

We will briefly examine a few basic results in the New Generalized Function Theory developed by Columbeau [4-9] and further investigated by Todorov [26] in applying a sequential approach to the New Generalized Function Theory. We will introduce the space, $\mathcal{H}\left(\mathbb{R}^k \times \mathcal{S}'(\mathbb{R}^k)\right)$, which will have mixed variables in the sense that the time variable, $t \in \mathbb{R}^k$ and the space variable, $x \in \mathcal{S}'(\mathbb{R}^k)$. Herein the space, $\mathcal{S}'(\mathbb{R}^k)$, denotes the classical tempered distributions. A more general theory can be developed and can be found in [19]. We will then focus on the asymptotic behaviour of the matrix as $n \to \infty$. The notion of asymptotic is replaced by quasiasymptotic behaviour when one examines distribution theory. The notion of quasiasymptotic behaviour was developed by several leading soviet mathematicians [11, 12, 27]. It was extensively developed in a recent work of Pilipovic, Stankovic and Takači [17]. This paper is the outgrowth of several papers [8-12] where multiplication and quasiasymptotic behaviour were individually developed on Fock functionals. In this present treatment we develop Fock functionals having both properties simultaneously.

2. SOME NOTIONS AND NOTATIONS

We begin with recalling some fundamental prerequisites on our sequences of constants, and sequences of functions. The requirements on the sequences lead us in a natural way into the approach in [14] and then towards our ultradifferential forms termed ultra annihilation and creation operators.

Throughout the paper we suppose that a monotonically increasing sequence of positive numbers, $\mu = (m_q)_{q \in \mathbb{N}_0}$, is given. We assume that conditions (M_1)–(M_3) from [14] are satisfied:

(M_1) $m_q^2 \leq m_{q-1}m_{q+1}$, $q = 1,2,\ldots$;
(M_2) there exist numbers A and H such that $m_{q+1} \leq A \cdot H^q m_q$, $q = 0,1,2, \ldots$;
(M_3) $\sum_{q=1}^{\infty} \frac{m_{q-1}}{m_q} \neq \infty$.

It is convenient to take $m_0 = 1$. One easily checks that, for instance, the sequence $m_q = q^{sq}$, $s > 1$, satisfies the upper three conditions.

Also, we suppose that a sequence, $\mathcal{M} = (M_p(\cdot))_{q\epsilon\mathbb{N}_0}$, of continuous functions on $\mathbb{R}^n$ is given. We require that the usual conditions, as (P), (M) and (N) hold (see [13] or [16]), as well as the inequalities,

$$1 = M_0(t) \leq M_1(t) \leq \ldots, \quad t \in \mathbb{R}^n. \tag{2.1}$$

Then, an infinitely differentiable function, ϕ, on $\mathbb{R}^n$ is in the space, $\mathcal{K}(M_p)$, if for every $p \in \mathbb{N}_0$ the following norms are finite

$$||\phi||_p := \sup\{|\phi^{(j)}(t)|M_p(t) \,\Big|\, t \in \mathbb{R}^n, |j| \leq p\}. \tag{2.2}$$

The family of norms, $(||\cdot||_p)_{p\epsilon\mathbb{N}_0}$, defines a locally convex topology on $\mathcal{K}(M_p)$, which in view of (P) turns this space into a complete one and in fact, into a Frèchet space. The space, $\mathcal{D}$, endowed with its usual topology is then dense in $\mathcal{K}(M_p)$. This implies that the space of linear continuous functionals on $\mathcal{K}(M_p)$, denoted by $\mathcal{K}'(M_p)$, is a normal space of distributions. Let us denote by $||\cdot||_{-p}$ following norm on $\mathcal{K}'(M_p)$:

$$||x||_{-p} := \sup \{ \, | \langle x, \phi \rangle | \, \big| \, ||\phi||_p \leq 1\}. \tag{2.3}$$

Observe that $||x||_0 \geq ||x||_{-1} \geq ||x||_{-2} \geq \ldots$, for any $x \in \mathcal{K}'(M_p)$

The sequences of positive numbers, $\mu=(m_q)_{q\epsilon\mathbb{N}_0}$, and the sequences of continuous functions, $\mathcal{M}=(M_p(\cdot))_{q\epsilon\mathbb{N}_0}$, will play an essential role in the definition of the generalized Fock space, $\Gamma^{\mu,\mathcal{M}}$, in Section 3.

3. GENERALIZED FOCK SPACES $\Gamma^{s,\mu,\mathcal{M}_0}$ AND $\left(\Gamma^{s,\mu,\mathcal{M}_0}\right)'$

Let the sequences $\mu = (m_q)_{q=0}^{\infty}$ and $\mathcal{M} = (M_p(\cdot))_{p\epsilon\mathbb{N}_0}$ be given with the properties given in Section 2. We then define

$$a_q\colon \underbrace{\mathcal{K}'(M_p) \times \ldots \times \mathcal{K}'(M_p)}_{q\text{ - copies}} \mapsto \mathbb{C} \tag{3.1}$$

to be a multilinear continuous functional, $q \in \mathbb{N}$ and by definition, we take $a_0 \in \mathbb{C}$.

Then the formal sum

$$\Phi = \sum_{q=0}^{\infty} a_q[\underbrace{\cdot, \cdot, \ldots, \cdot}_{q\text{ - spaces}}] \tag{3.2}$$

is in the space, $\Gamma^{s,\mu,\mathcal{M}}$, $s > 1$, if the norm

$$|||\Phi|||^{(p)}_{s,\mu,\mathcal{M}} := \sup\left\{\frac{||a_q||_p \cdot m_q}{s^q}\,\middle|\, q \in \mathbb{N}_0\right\}, \tag{3.3}$$

is finite for every $p \in \mathbb{N}_0$. Here

$$||a_q||_p = \sup\left\{|a_q[x,\ldots,x]|\,\middle|\, ||x||_{-p} \leq 1,\ x \in \mathcal{K}'(M_p)\right\}. \tag{3.4}$$

Recall the definition of $\|x\|_{-p}$ as given in expression (2.3).

Remark. Physicists prefer to represent the elements from our generalized Fock space, $\Gamma^{s,\mu,\mathcal{M}_0}$, as column vectors, for instance, $\Phi = \begin{bmatrix} a_0 \\ a_1 \\ \vdots \end{bmatrix}$. Since this is convenient also when working with annihilation or creation operators, we shall do likewise.

Let us note first that in view of (3.3) and (3.4), the canonical inclusion

$$\Gamma^{s,\mu,\mathcal{M}} \mapsto \Gamma^{s',\mu,\mathcal{M}} \tag{3.5}$$

is continuous provided that $s' > s > 1$. So in view of reference [1] we can now give the following definition.

Definition 3.1. A generalized Fock space, $\Gamma^{\mu,\mathcal{M}}$, is the inductive limit of the spaces, $\Gamma^{s,\mu,\mathcal{M}}$, i.e.,

$$\Gamma^{\mu,\mathcal{M}} = \underset{s\to\infty}{\text{ind}}\ \Gamma^{s,\mu,\mathcal{M}}.$$

For the notion of the inductive limit one can consult reference [28]. In reference [1] the state vectors satisfy constraints of the form given in expression (1.2). In keeping with the spirit of such a constraint we shall indicate that the elements from the inductive limit, $\Gamma^{\mu,\mathcal{M}_0}$, are L^r summable for any $r \in (1,\infty)$. For this result we cite a well known lemma.

Lemma 3.1. Conditions (M_1) and (M_3) imply that for an $s \in \mathbb{R}$, we have

$$\sum_{q=0}^{\infty} \frac{|s|^q}{m_q} < \infty. \tag{3.6}$$

See reference [19].

Assuming the so called kernel representation [20] each $\Phi \in \Gamma^{s,\mu,\mathcal{M}_0}$ can be represented as

$$\Phi \Leftrightarrow \begin{bmatrix} \phi_0 \\ \phi_1 \\ \vdots \\ \phi_q \\ \vdots \end{bmatrix}, \tag{3.7}$$

where $\phi_0 = \phi_0$ is a scalar and ϕ_q, $q \geq 1$ are each rapid descent test functions. Thus the dual of our space, $(\Gamma^{s,\mu,\mathcal{M}_0})'$, contains generalized Fock functionals, F, having representation, $F \Leftrightarrow (F_0, F_1, \cdots, F_q, \cdots)$, and satisfies constraints of the form,

$$|||F|||^{(p)}_{-[s,\mu,\mathcal{M}_0]} = \sum_{q=0}^{\infty} (s)^q (m_q)^{-1} \| F_q \|_{-m} < \infty.$$

4. THE SPACE, $\mathcal{H}(\mathbb{R}^k \times \mathcal{S}' ; \mathbb{R})$

Holomorphic functionals belonging to the space $\mathcal{H}(\mathbb{R}^k \times \mathcal{S}' ; \mathbb{R})$ are $C^\infty(\mathbb{R}^k)$ in the variables, $(t_1,\ldots,t_k) \in \mathbb{R}^k$, and $C^\infty(\mathcal{S}')$ in the variable, $x \in \mathcal{S}'$, as defined in reference [21]. We replace the space, $\mathcal{H}(\mathcal{M}_p)$, with the specific space, $\mathcal{S}(\mathbb{R}^n)$ and $\mathcal{H}'(\mathcal{M}_p)$ with the space $\mathcal{S}'(\mathbb{R}^n)$.

Example 4.1. Select a fixed rapid descent test function, $\phi \in \mathcal{S}(\mathbb{R}^k)$ and a fixed classical function, $g(t_1,\ldots,t_k) \in C^\infty(\mathbb{R}^k)$. Then define the functional

$$F : \mathbb{R}^k \times \mathcal{S}' \mapsto \mathbb{R}$$
$$F : (t_1,\ldots, t_k) \sim\!\rightarrow g(t_1,\ldots,t_k) \cdot \langle x, \phi\rangle .$$

This functional, F, is a member of $\mathcal{H}(\mathbb{R}^k \times \mathcal{S}' ; \mathbb{R})$.

Example 4.2. Select a fixed classical function,

$$g(t_1,\ldots,t_k) \in C^\infty(\mathbb{R}^k) \quad \text{and} \quad \text{a } \Phi \in \Gamma^{p,sB}.$$

Then define the functional,

$$G : \mathbb{R}^k \times \mathcal{S}' \mapsto \mathbb{R}$$
$$G : (t_1,\ldots, t_k, x) \sim\!\rightarrow g(t_1,\ldots,t_k) \cdot \Phi(x) .$$

Again the functional, G, is a member of $\mathcal{H}(\mathbb{R}^k \times \mathcal{S}' ; \mathbb{R})$.

5. THE SPACE OF NEW TEMPERED FOCK FUNCTIONALS

The space of new tempered Fock functionals, $\mathcal{H}(\mathbb{R}^k \times \mathcal{S}';\mathbb{R})^{\mathcal{S}+}$, is in the spirit of ultrapowers developed by Todorov [26]. The set, $\mathcal{S}_+ = \mathcal{S} \times \mathbb{R}_+$, is a Cartesian product of rapid descent test functions and positive real numbers. The set, $\mathcal{H}(\mathbb{R}^k \times \mathcal{S}';\mathbb{R})^{\mathcal{S}+}$, denotes uncountable or countable sequences having $\mathcal{S}_+$ as their common domain and the set, $\mathcal{H}(\mathbb{R}^k \times \mathcal{S}';\mathbb{R})$, as their common range. The context of the point selected from $\mathcal{H}(\mathbb{R}^k \times \mathcal{S}';\mathbb{R})$ will determine if we are considering an uncountable, countable sequence or single functional. For convenience the domain and range points for these sequences of functionals, $G \in \mathcal{H}(\mathbb{R}^k \times \mathcal{S}';\mathbb{R})^{\mathcal{S}+}$, will be reflected in the notation,

$$G_{\phi \otimes \epsilon}(t_1,\ldots,t_k, x) , \tag{5.1}$$

where for each $\phi \otimes \epsilon \in \mathcal{S}_+$ the expression denoted in line (5.1) represents a mixed holomorphic sequence of functionals belonging to $\mathcal{H}(\mathbb{R}^k \times \mathcal{S}';\mathbb{R})$. The following constructions are an adaption of the new generalized function theory [4-9] to our Fock functionals.

Definition 5.1. The regularization for any tempered distribution, $T \in \mathcal{S}'(\mathbb{R}^k)$, and functional, $\Phi \in \Gamma^{\mu,\mathcal{M}_0}$, is defined as

$$T \cdot \Phi_{\phi \otimes \epsilon}(t_1,\dots,t_k,x) \triangleq < T(\xi) \, , \frac{1}{\epsilon^n} \phi\left(\frac{\xi - t}{\epsilon}\right) > \cdot \Phi(x) \, .$$

It is immediate that the "sequence", $\{T \cdot \Phi_{\phi \otimes \epsilon}\}_{\phi \otimes \epsilon}$, is a member of $\mathcal{H}(\mathbb{R}^k \times \mathcal{S}';\mathbb{R})^{\mathcal{S}+}$.

Example 5.1. The regularization of the Dirac delta functional, $\delta(\xi) \in \mathcal{S}'(\mathbb{R}^k)$, and $\Phi \in \Gamma^{p,SB}$ is

$$\triangle \cdot \Phi_{\phi \otimes \epsilon}(t_1,\dots,t_k,x) \triangleq \frac{1}{\epsilon^n} \phi(-t/\epsilon) \cdot \Phi(x) \, .$$

Definition 5.2. $G \in \mathcal{H}(\mathbb{R}^k \times \mathcal{S}';\mathbb{R})^{\mathcal{S}+}$ is moderate if and only if for every compact subset, $K \subset \mathbb{R}^k$, and every mixed derivative operator,

$$D_t^\alpha D_h^n \triangleq \frac{\partial^{\alpha_1 + \dots + \alpha_k}}{\partial t_1^{\alpha_1} \dots \partial t_k^{\alpha_k}} \bullet D_h^n \, ,$$

there is a $q_0 \in N$ such that if $\phi(t)$ a$\in \mathcal{S}_{q_0}(\mathbb{R}^k)$ and $x \in \hat{\mathcal{S}}'_{q_0}$ then there exists a positive constant, C and $\epsilon_0 > 0$ such that

$$|D_t^\alpha D_h^n G_{\phi \otimes \epsilon}(t_1,\dots,t_k,x)| < C \, \epsilon^{-q_0}$$

for every $t \in K$ and $0 < \epsilon < \epsilon_0$.

The set of all moderate "sequence" in $\mathcal{H}(\mathbb{R}^k \times \mathcal{S}';\mathbb{R})^{\mathcal{S}+}$ will be denoted $\mathcal{H}_M(\mathbb{R}^k \times \mathcal{S}';\mathbb{R})^{\mathcal{S}+}$.

Theorem 5.1. The regularization of a tempered distribution, $T(\xi)$ and $\Phi(x) \in \Gamma^{p,sB}$ is moderate.

Proof. See reference [21].

Definition 5.2. A sequence, $G \in \mathcal{H}(\mathbb{R}^k \times \, \mathcal{S}'; \, \mathbb{R})^{\mathcal{S}+}$, is null if and only if for every compact subset, $K \subset \mathbb{R}^k$, and every increasing positive real valued function, $\alpha(q)$, where $\lim_{q \to \infty} \alpha(q) = \infty$, and every mixed derivative operator,

$$D_t^\alpha D_h^n = \frac{\partial^{\alpha_1 + \dots + \alpha_k}}{\partial t_1^{\alpha_1} \dots \partial t_k^{\alpha_k}} D_h^n \, ,$$

there is a $q_0 \in N$ such that if $\phi(t) \in \mathcal{S}_q(\mathbb{R}^k)$ and $x \in \hat{\mathcal{S}}'_q$ for $q \geq q_0$ there exists an $\epsilon_0 > 0$ and $C > 0$ such that

$$\left| D_t^\alpha D_h^n G_{\phi \otimes \epsilon}(t_1, \dots t_k \, , x) \right| < C \, \epsilon^{\alpha(q) - q_0}$$

for every $t \in K$ and $0 < \epsilon < \epsilon_0$.

The set of all null sequencing in $\mathcal{H}(\mathbb{R}^k \times \mathbb{S}';\mathbb{R})^{\mathbb{S}}+$ will be denoted $\mathcal{H}_0(\mathbb{R}^k \times \mathbb{S}';\mathbb{R})^{\mathbb{S}}+$. We observe that $\mathcal{H}_0(\mathbb{R}^k \times \mathbb{S}';\mathbb{R})^{\mathbb{S}}+ \subset \mathcal{H}_M(\mathbb{R}^k \times \mathbb{S}';\mathbb{R})^{\mathbb{S}}+$ and the pointwise product of a G ϵ $\mathcal{H}_0(\mathbb{R}^k \times \mathbb{S}';\mathbb{R})^{\mathbb{S}}+$ is well defined and moreover by the Leibnitz Formula for higher order mixed derivatives implies the pointwise product of W-$G_{\phi \otimes \epsilon}(t_1 \cdots t_k, x)$ and $G \cdot W_{\phi \otimes \epsilon}(t_1,\cdots,t_k,x)$ again is a member of $\mathcal{H}_0(\mathbb{R}^k \times \mathbb{S}';\mathbb{R})^{\mathbb{S}}+$. We can therefore conclude that $\mathcal{H}_0(\mathbb{R}^k \times \mathbb{S}';\mathbb{R})^{\mathbb{S}}+$ is an ideal in $\mathcal{H}_M(\mathbb{R}^k \times \mathbb{S}';\mathbb{R})^{\mathbb{S}}+$.

Definition 5.3. Two sequences, G, H, ϵ $\mathcal{H}_M(\mathbb{R}^k \times \mathbb{S}';\mathbb{R})^{\mathbb{S}}+$ are related if and only if G–H ϵ $\mathcal{H}_0(\mathbb{R}^k \times \mathbb{S}';\mathbb{R})^{\mathbb{S}}+$. It is immediate that definition 5.3 defines an equivalence relation on the set $\mathcal{H}_M(\mathbb{R}^k \times \mathbb{S}';\mathbb{R})^{\mathbb{S}}+$. Distributional multiplication is well defined if we implement the standard multiplication procedure on the equivalence classes given by the equivalence relation in definition 5.3.

6. QUASIASYMPTOTICS IN $\left(\Gamma^{s,\mu,\mathcal{M}_0}\right)'$

A generalized Fock functional fundamental to our development which plays an essential role in the quasiasymptotic behaviour is given by

$$F_\alpha(t) \Leftrightarrow \begin{bmatrix} F_0 \\ H(t_1)t_1^{\alpha-1}/\Gamma(\alpha) \\ H(t_1,t_2)t_1^{\alpha-1}\, t_2^{\alpha-1}/\Gamma(\alpha)^2 \\ \vdots \\ H(t_1,\cdots,t_q)\cdot(t_1\cdot \cdots \cdot t_q)^{\alpha-1}/\Gamma(\alpha)^q \\ \vdots \end{bmatrix}. \tag{6.1}$$

Herein $H(t_1, \cdots, t_q) = H(t_1) \otimes \cdots \otimes H(t_q)$ with

$$H(t_i) = \begin{cases} 1 & t_i \geq 0 \\ 0 & t_i < 0 \end{cases} \qquad (1 \leq i \leq q)$$

and $\Gamma(\alpha)$ is the Gamma function evaluated at α. The α in our paper will be such that $\alpha > 1$. However this can be relaxed as in reference [6] and by a recursive relationship involving generalized derivations the definition of α can be extended to include $\alpha \leq 1$.

Lemma 6.1. Each component, $H(t_1,\cdots,t_q)\cdot(t_1 \cdot\cdots\cdot t_q)^{\alpha-1}/\Gamma(\alpha)^q$ of the Fock functional defined in expression (5.1) for $q \geq 1$ and $\alpha > 1$ is a regular tempered distribution.

Proof. See Reference [22].

Proposition 6.1. $F_\alpha \ \epsilon \left(\Gamma^{s,\mu,\mathcal{M}_0}\right)'$.

Proof. Again select $m > \alpha$ and together with lemma 6.1 provides us with the computations,

$$\sum_{q=0}^{\infty} ||F_q||_{-m} \frac{s_q}{m_q}$$

$$\leq \sum_{q=0}^{\infty} \frac{1}{(\Gamma(\alpha))^q} (\tfrac{\pi}{2})^q \frac{(s)_q}{m_q}$$

$$= \sum_{q=0}^{\infty} \frac{k_q}{m_q} < \infty$$

where $k \triangleq \frac{1}{\Gamma(\alpha)} \cdot \frac{\pi}{2} \cdot s$. □

As already indicated a functional, $F \in \left(\Gamma^{s,\mu,\mathcal{M}_0}\right)'$ enjoys the representation,

$$F \Leftrightarrow (F_0, F_1, \cdots, F_q, \cdots)$$

where F_0 is a scalar and F_q , $q \geq 1$ are distributions all having finite order $\leq m$. Moreover, F satisfies

$$|||F|||^{(p)}_{-[s,\mu,\mathcal{M}_0]} = \sum_{q=0}^{\infty} (s)^q (m_q)^{-1} \| F_q\|_{-m} < \infty.$$

When $\Phi \in \Gamma^{s,\mu,\mathcal{M}_0}$ and $F \in \left(\Gamma^{s,\mu,\mathcal{M}_0}\right)'$ then we have

$$F(\Phi) \triangleq \langle\langle F, \Phi\rangle\rangle = \sum_{q=0}^{\infty} \langle F_q, \phi_q\rangle.$$

Definition 6.1. A Fock functional, $F \in \left(\Gamma^{s,\mu,\mathcal{M}_0}\right)'$, has quasiasymptotic behaviour, (q.a.b.), at infinity of order, α, if and only if there exists a limit,

$$\lim_{t\to\infty} t^{-\alpha} F(t\Phi) = W(\Phi)$$

where this limit is taken in the sense,

$$\lim_{t\to\infty} \sum_{q=0}^{\infty} \langle \frac{F_q(tx_1,\cdots,tx_q)}{t^{\alpha q}}, \phi_q(x_1,\cdots,x_q)\rangle = \sum_{q=0}^{\infty} \langle W_q(x_1,\cdots,x_q), \phi_q(x_1,\cdots,x_q)\rangle.$$

Example 6.1. The Generalized Fock Functional

$$F_\alpha(x) \Leftrightarrow \begin{bmatrix} F_0 \\ H(x_1)\, e^{i\alpha x_1}/\Gamma(\alpha) \\ \vdots \\ H(x_1,\ldots,x_q)\cdot e^{i\alpha[x_1+\ldots+x_q]}/(\Gamma(\alpha))^q \\ \vdots \end{bmatrix},$$

has a q.a.b. of order -1 and moreover is in this sense equivalent to the generalized Fock functional,

$$\delta(x) \Leftrightarrow \begin{bmatrix} F_0 \\ \frac{i}{a}\,\delta(x_1) \\ \vdots \\ \left(\frac{i}{a}\right)^q \delta(x_1,\cdots,x_q) \\ \vdots \end{bmatrix}.$$

We observe that each component of each generalized Fock functional is separable in the sense that

$$H(x_1,\ldots, x_q)\cdot e^{i\alpha[x_1+\ldots+x_q]}\cdot/(\Gamma(\alpha))^q = \frac{H(x_1)e^{iax_1}}{\Gamma(\alpha)}\cdot\ldots\cdot\frac{H(x_q)e^{iax_q}}{\Gamma(\alpha)}$$

and

$$\left(\tfrac{i}{a}\right)^q\delta(x_1,\cdots, x_q) = \tfrac{i}{a}\cdot\delta(x_1)\otimes\tfrac{i}{a}\cdot\delta(x_2)\otimes\ldots\tfrac{i}{a}\cdot\delta(x_q) \text{ for } q\geq 1.$$

Again the calculations can be found in reference [22].

7. ULTRA ANNHILATION AND CREATION OPERATORS

The notion of q-particles diminishing to q−1 particles is mathematically treated with an annihilation operator. Similarly the increasing of q particles to q+1 particles is treated with a creation operator. Several physical theories depend only on the vacuum state. Future states are then "reached" by applying several operators such as the creation and annihilation operators to it. Successive application of the operator to the state vector leads us in a natural way to our notion of ultra annihilation and creation operators. They are termed in the spirit of ultradifferentiable operators as they refer to ultradistributions [14]. The techniques in this section are consequences of similar techniques found in [14]. The controls on the ultra annihilation and creation operators are within the framework of the preassigned sequences satisfying conditions of the form, (M_1), (M_2) and (M_3), described in Section 2. The generalized Fock space in Section 3 equipped with an inductive limit topology [28] would allow development within the framework of ultradistributions and hyper-functions. However we will be content with the distributional space, $K'(M_p)$.

If $h\epsilon\mathcal{K}'(M_p)$, then we define the operator, $D_{h,\mathcal{A}}:\Gamma^{\mu,\mathcal{M}}\mapsto\Gamma^{\mu,\mathcal{M}}$, in the following way.

$$D_{h,\mathcal{A}}:\Phi = \begin{bmatrix} a_0\\ a_1\\ \vdots\\ a_q\\ \vdots\end{bmatrix}\mapsto\Psi_1 = \begin{bmatrix} b_0\\ b_1\\ \vdots\\ b_q\\ \vdots\end{bmatrix}, \tag{7.1}$$

where b_q is the multilinear functional given by

$$b_q\,[\underbrace{\cdot,\ldots,\cdot,\cdot}_{\text{q-spaces}}] = A_q(1)\;a_{q+1}[h,\underbrace{\cdot,\ldots,\cdot}_{\text{q-spaces}}], \tag{7.2}$$

provided that the sequence, $\mathcal{A} = \left(A_q(1)\right)_{q\,\epsilon\,\mathbb{N}_0}$, satisfies the estimate,

$$|A_q(1)| \leq L \cdot \frac{m_{q+1}}{m_q}, \text{ for some } L > 0. \tag{7.3}$$

Let us remark that if in the case, $A_q(1) = q + 1$, $q \in \mathbb{N}_0$, $\mathcal{K}'(M_p) = \mathcal{S}'$, and $h \in \mathcal{S}'$, then our operator, $D_{h,\mathcal{A}}$, acts similarly to an annihilation operator.

Lemma 7.1. If (7.3) is satisfied, then the operator, $D_{h,\mathcal{A}}$, from (7.1) is well defined.

Proof. If $\Phi = \begin{bmatrix} a_0 \\ a_1 \\ \vdots \\ a_q \\ \vdots \end{bmatrix}$ is in $\Gamma^{\mu,\mathcal{M}}$, then there exists an $s > 1$ such that $\Phi \in \Gamma^{s,\mu,\mathcal{M}}$.

Let $\Psi_1 = \begin{bmatrix} b_0 \\ b_1 \\ \vdots \\ b_q \\ \vdots \end{bmatrix} = \begin{bmatrix} a_0 \\ a_1 \\ \vdots \\ a_q \\ \vdots \end{bmatrix} =: D_{h,\mathcal{A}}\Phi$, then we can find $s'>s$ such that $\Psi_1 \in \Gamma^{s',\mu,\mathcal{M}}$.

In fact

$$|||\Psi_1|||^{(p)}_{s',\mu,\mathcal{M}} := \sup\left\{\frac{m_q}{(s')^q} \cdot |A_q(1)| \cdot ||a_{q+1}||_p \,\middle|\, q \in \mathbb{N}_0\right\}$$

$$\leq ||h||_{-p} \cdot s \cdot \sup\left\{\frac{m_{q+1} \cdot ||a_{q+1}||_p}{s^{q+1}} \,\middle|\, q \in \mathbb{N}_0\right\} \cdot \sup\left\{\left(\frac{s}{s'}\right)^q \cdot \frac{m_q}{m_{q+1}} \cdot L \cdot \frac{m_{q+1}}{m_q} \,\middle|\, q \in \mathbb{N}_0\right\}.$$

So we obtain

$$|||\Psi_1|||^{(p)}_{s',\mu,\mathcal{M}} \leq C_p \cdot (||h||_{-p} \cdot s) \cdot L\left\{\sup\left(\frac{s}{s'}\right)^q \,\middle|\, q \in \mathbb{N}_0\right\} < \infty,$$

i.e., $\Psi_1 \in \Gamma^{s',\mu,\mathcal{M}}$. □

More generally, if we apply $D_{h,\mathcal{A}}$ v-times, then $D^v_{h,\mathcal{A}}$ maps $\Phi \in \Gamma^{\mu,\mathcal{M}}$ from (1.2) into $\Psi_v = \sum_{q=0}^{\infty} b_q^{(v)} a_q \underbrace{[\cdot, \ldots, \cdot]}_{q\text{-spaces}}$, where

$$b_q^{(v)} = A_q(v)\, a_{q+v}[\underbrace{h, \ldots, h}_{v\text{-times}}, \underbrace{\cdot, \cdot, \ldots, \cdot}_{q\text{-spaces}}], \; q \in \mathbb{N}. \tag{7.4}$$

For convenience, the v distributions that are all equal to h were put in the first v-spaces, while the necessary q empty spaces were left at the end.

One easily checks that the coefficients, $A_q(v)$, are given by the elements of the sequence $\mathcal{A} = (A_q(1))_{q \in \mathbb{N}_0}$:

$$A_q(v) = A_q(1) \cdot A_{q+1}(1) \ldots \cdot A_{q+v-1}(1), \; (q,v) \in \mathbb{N}_0 \times \mathbb{N}. \tag{7.5}$$

Now (7.5) and the condition (7.3) imply

$$|A_q(v)| \leq |A_q(1)| \cdot |A_{q+1}(1)| \cdot \ldots \cdot |A_{q+v-1}(1)|$$
$$\leq \left(L \cdot \frac{m_{q+1}}{m_q}\right) \cdot \left(L \cdot \frac{m_{q+2}}{m_{q+1}}\right) \cdot \ldots \cdot \left(L \cdot \frac{m_{q+v}}{m_{q+v-1}}\right),$$

so

$$|A_q(v)| \leq L^v \cdot \frac{m_{q+v}}{m_q} \text{ for each } q \in \mathbb{N}_0,\ v \in \mathbb{N}. \tag{7.6}$$

It is clear that $D^v_{h,\mathcal{A}}$ is an interior operation in $\Gamma^{\mu,\mathcal{M}}$, if (7.3) holds. We state this formally in the next proposition and, in particular, we estimate the p-th norm of $D^v_{h,\mathcal{A}}\Phi$ for any $v \in \mathbb{N}_0$.

Proposition 7.1. Let the operator, $D_{h,\mathcal{A}}$, be given by (7.1) and (7.2), where (7.3) is satisfied. Then for each $v \in \mathbb{N}$, the operator, $D^v_{h,\mathcal{A}}$, maps $\Gamma^{\mu,\mathcal{M}}$ into itself. Moreover this mapping is continuous.

Proof. See reference [22].

Let us analyse the action of the ultradifferential operator, $P(D_{h,\mathcal{A}})=\sum_{v=0}^{\infty} w_v D^v_{h,\mathcal{A}}$, on elements from $\Gamma^{\mu,\mathcal{M}}$. By definition if $\Phi\epsilon\Gamma^{\mu,\mathcal{M}}$ and Φ has the representation given in expression 3.2, then formally we put

$$\Psi := P(D_{h,\mathcal{A}})\, \Phi = \sum_{v=0}^{\infty} w_v\, D^v_{h,\mathcal{A}}\Phi. \tag{7.7}$$

Let us introduce the following condition.

There exist constants C and K such that for each $v = 0, 1, 2, \ldots$

$$|w_v| \leq C \cdot K^v/m_v \text{ (compare with [14]).} \tag{7.8}$$

Then for Ψ from expression 7.7, it holds for each $s' > s$ that $||\Psi||^{(p)}_{s',\mu,\mathcal{M}} < \infty$.

Remark. It is convenient to write condition (7.8) in the following form.

There exists a constant $K > 0$ such that

$$w_v = \mathcal{O}(K^v/m_v) \text{ as } v \to \infty. \tag{7.9}$$

Next we consider the construction of the creation operator.

Let ϕ be an arbitrary element from $\mathcal{K}(M_p)$ and let a sequence $\mathcal{C} = \left(C_q(1)\right)_{q \in \mathbb{N}_0}$ be given. We suppose that a constant $L > 1$ exists such that for every $q \in \mathbb{N}_0$ it holds

$$|C_q(1)| \leq L \cdot \frac{m_q}{m_{q+1}}. \tag{7.10}$$

Then the operator, $Q_{\phi,\mathcal{C}}$: $\Gamma^{\mu,\mathcal{M}} \mapsto \Gamma^{\mu,\mathcal{M}}$, is well defined, if $Q_{\phi,\mathcal{C}}(\Phi) = \Psi_1$, where

$$\Phi = \sum_{v=0}^{\infty} a_q[x, \ldots, x] \in \Gamma^{\mu,\mathcal{M}} \quad \text{and} \tag{7.11}$$

$$\Psi_1 = \sum_{v=0}^{\infty} b_q^{(1)}[x, \ldots, x], \quad \text{with}$$

$$b_q^{(1)} = \langle \cdot, \phi \rangle \cdot C_q(1) \; a_{q-1} \underset{(q\text{-}1)\text{-spaces}}{(\cdot, \ldots, \cdot)}. \tag{7.12}$$

It is immediate as in Lemma 7.1 that $\Psi_1 \in \Gamma^{s',\mu,\mathcal{M}}$, if $\Phi \in \Gamma^{s,\mu,\mathcal{M}}$ for $s' > s$. The v-th iteration of $Q_{\phi,\mathcal{C}}$ is denoted by $Q^v_{\phi,\mathcal{C}}$, $v = 1, 2, \ldots$, and let us put

$$Q^v_{\phi,\mathcal{C}}(\Phi) = \Psi_v, \tag{7.13}$$

where $\Phi \in \Gamma^{s,\mu,\mathcal{M}}$ is from section (3). Again we can prove similar results as in the annihilation operator [22].

I wish to thank Dr. J. Colombeau, Dr. S. Pilipovic, Dr. A. Takači and Dr. T. Todorov for their revealing and inspiring discussions at the Generalized Function, Convergence Structures and their Applications Conference in Dubrovnik, Yugoslavia during June 1987. I especially wish to extend sincere gratitude to the Dr's Arpad and Djurdjica Takači for the marvelous 1989-1990 academic year spent with us in the U.S.A. when Dr. Arpad Takači was a Fulbright Scholar. Their discussions with me revealed many profound results concerning Abelian and Tauberian theorems.

REFERENCES

1. F.A. Berezin, "The Method of Second Quantization," Academic Press, New York and London, (1966).
2. N.N. Bogoliubov, T.T. Logunov, and I.T. Todorov, "Introduction to Axiomatic Quantum Field Theory," W. A. Benjamin, Inc., Massachusetts, (1975).
3. R. D. Carmichael, "Analytic Functions related to Distributions of Exponential Growth," SIAM J. Math. Ann., 10: 1041-1068 (1979).
4. J.F. Colombeau, "Some Aspects of Infinite-Dimensional Holomorphy in Mathematical Physics," Aspects of Math. and its Applic., J.A. Barroso, editor, Elsevier, 253-263 (1986).
5. J.F. Colombeau, "Diffferential Calculus and Holomorphy," North Holland Mathematical Studies, No. 64, N. Holland Pub., Co., N.Y., (1982).
6. J.F. Colombeau, "New Generalized Functions, and Multiplication of Distributions," N. Holland Math. Studies #84. N.Y., (1984).
7. J.F. Colombeau, "Elementary Introduction to New Generalized Functions," N. Holland Math. Studies, No. 113, N. Holland Pub. Co., N.Y., (1985).
8. J.F. Colombeau, "A Multiplication of Distributions," J. of Math. Ann. and Applications, 94, No. 1:96-115 (1983) .
9. J.F. Colombeau, Roux, Le, A.Y., "Generalized Functions and Products," Appearing in Eq. of Physics, preprint.

10. F. Constantinescu, "Distributions and their Applications in Physics," Pergammon Press, N.Y., (1980).
11. Yu. N. Drozzinov, and B.I. Zavialov, "The Quasiasymptotic Behaviour of Generalized Functions and Tauberian Theorems in the Complex Domain," Mat. Sb. Vol. 102, (1977), 372-390; English Translation, Math. USSR Sb. Vol 31:329-345 (1977).
12. Yu. N. Drozzinov, and B.I. Zavialov, "Tauberian Theorems for Generalized Functions with Supports in Cones," Mat. Sb. Vol. 108:78-90(1979); English Translation, Math. USSR Sb. Vol. 36:75-86 (1980).
13. I. M. Gelfand and G.E. Shilov, "Generalized Functions," Vol. II, Acad. Press, N.Y. (1968).
14. H. Komatsu, "Ultradistributions, I, Structural Theorems and a Characterization," J. of the Faculty of Science, Tokyo, Section 1A Mathematics Vol. 20:25-105 (1973).
15. T.P.G Liverman, "Generalized Functions and Direct Operational Methods," Prentice-hall, Englewood Cliff, New Jersey, (1964).
16. S. Pilipović and A Takači, "Space $H'(M_p)$ and Convolutors," Proc. of the Moscow Conference on Generalized Functions, Moscow, 415-427 (1981).
17. S. Pilipović, B. Stankovic, and A. Takači, "Asymptotic Behaviour and Stieltjes Transformation of Distributions," B.G. Teuner, Leipzig, (1990).
18. J. Schmeelk, "Infinite Dimensional Parametric Distributions," Appl. Analysis, Vol 24:291-319 (1987).
19. J. Schmeelk, and A. Takači, "Ultra Creation and Annihilation Operators," Portugaliae Mathematica, Vol 49 (1992). In press.
20. J. Schmeelk, "Infinite Dimensional Fock Spaces and Associated Creation and Annihilation Operators," J. of Math. Analysis and Appl., 134, No. 2:432-461 (1988).
21. J. Schmeelk, "Multiplication and Mixed Differentiation in Generalized Fock Spaces," J. of Math. Analysis and Appl., 165, No. 1:216-238 (1992).
22. J. Schmeelk, and A. Takači, "Quasiasymptotic Behaviour in a Generalized Fock Space," Mathematics Series, Review of Research, (1992). In press.
23. L. Schwartz, "Theorie des Distributions." Hermann, Paris, (1966).
24. L. Schwartz, "Impossibilte de la Multiplication des Distributions," Comptes Rendus de L' Academie des Science, 239:847-848 (1954).
25. A. Takači, "A Note on the Distributional Stieljes Transformation," Math. Proc. Comb. Phil Soc., 94:523-527 (1983).
26. T. Todorov, "Sequential Approach to Colombeaus Theory of Generalized Functions," International Center for Theor. Physics, IC, 126, Trieste, Italy (1987).
27. V.S. Vladimirov, Yu. N. Drozzinov, and B.I. Zavialov, "Tauberian Theorems for Generalized Functions," Kluwer Acad. Pub., Maine, (1988).
28. V.V. Žarinov, "Compact Families of Locally Convex Topological Vector Spaces," Fréchet-Schwartz and Dual Fréchet-Schwartz Spaces, Russian Math. Surveys, Vol. 34, No. 4:105-143)1979).
29. A. Zemanian, "Distribution Theory and Transform Analysis," McGraw-Hill Book Co., N.Y., (1965).

SOME REMARKS ON THE DISTRIBUTIONAL HANKEL TRANSFORM

O.P. Singh

Department of Applied Mathematics
Institute of Technology
Banaras Hindu University
Varanasi, India

1. INTRODUCTION

Classically, the Hankel transform h_μ is defined for $\mu \geq -\frac{1}{2}$ as :

$$[h_\mu \varphi(x)](y) = \int_0^\infty \varphi(x)(xy)^{1/2} J_\mu(xy)\, dx. \qquad (1.1)$$

Zemanian [1] extended it to a certain distribution space H'_μ - the dual space of

$$H_\mu = \{ \varphi : (0,\infty) \longrightarrow \mathbb{C} \mid \varphi \text{ is } C^\infty \text{ and satisfies } (1.2)\}.$$

$$\gamma^\mu_{m,k}(\varphi) = \sup_{0<x<\infty} | x^m (x^{-1} D_x)^k x^{-\mu-\frac{1}{2}} \varphi(x) | < \infty, \quad (1.2)$$

for each pair of non-negative integers m and k.

H_μ is a topological linear space — the topology being generated by the countable family of separating seminorms $\gamma^\mu_{m,k}$. Then he proved [1] :

Theorem 1.1. For $\mu \geq -\frac{1}{2}$, h_μ is an automorphism on H_μ and hence on H'_μ.

Note that $\varphi \longrightarrow x^{\nu-\mu}\varphi$ is an isomorphism from H_μ onto H_ν since $\gamma^\nu_{m,k}(x^{\nu-\mu}\varphi) = \gamma^\mu_{m,k}(\varphi)$, $\varphi \in H_\mu$. So it was natural to

Generalized Functions and Their Applications, Edited by
R.S. Pathak, Plenum Press, New York, 1993

ask "Is there an isomorphism h from H_μ onto H_ν s.t. the following diagram commutes"

$$
\begin{array}{ccc}
H_\mu & \xrightarrow{\;h_\mu\;} & H_\mu \\
{\scriptstyle x^{\nu-\mu}}\downarrow & & \downarrow{\scriptstyle h} \\
H_\nu & \xrightarrow{\;h_\nu\;} & H_\nu
\end{array}
$$

The aim of the present paper is to find such h and once we do so Theorem 1.1 gets extended for all real μ. Further we give a representation for the differential operator $(x^{-1}D_x)^a$ for $a \in \mathbb{R}$ and show that it is an automorphism on the testing function space $H_{-1/2}$ and hence on its dual space $H'_{-1/2}$.

We shall follow the terminology of Zemanian [1].

2. THE MAIN RESULT

Theorem 2.1 for $\nu - \mu \in \mathbb{N}$ and

$$h = y^{\nu+1/2}\,(-y^{-1}D_y)^{\nu-\mu}\,y^{-\mu-1/2} \tag{2.1}$$

the diagram commutes.

Proof. Let $\varphi \in H_\mu$. Then $\varphi \longrightarrow h\,\varphi$ is an isomorphism from H_μ onto H_ν. Since

$$\gamma^{\nu}_{m,k}(h\,\varphi) = \sup_{0<x<\infty} \left| x^m (x^{-1}D_x)^k (-x^{-1}D_x)^{\nu-\mu}\, x^{-\mu-1/2}\varphi(x) \right|$$

$$= \gamma^{\mu}_{m,\,k+\nu-\mu}(\varphi). \tag{2.2}$$

Hence h is continuous. Let $\varphi_1, \varphi_2 \in H_\mu$ be such that $h\varphi_1 = h\varphi_2$. Then

$$\gamma^{\mu}_{m,k+\nu-\mu}(\varphi_1) = \gamma^{\mu}_{m,k+\nu-\mu}(\varphi_2)$$

for each m,k = 0,1,2,... Hence, in particular,

$$\gamma^{\mu}_{m,\nu-\mu}(\varphi) = 0 \;, \text{ where } \varphi = \varphi_1 - \varphi_2 \in H_\mu \;.$$

Therefore,

$$\frac{d}{dx}\left[(x^{-1} D_x)^{n-1} (\varphi)\right] = 0, \qquad n = \nu - \mu.$$

But $\varphi(x)$ is of rapid descent as $x \longrightarrow \infty$, we get $(x^{-1}D_x)^{n-1}\varphi = 0$. Repeating the process (n-1) times, we conclude that

$\varphi = \varphi_1 - \varphi_2 = 0$. Hence h is injective.

For $\nu - \mu = 1$ and $\Psi \in H_\nu$, let

$$\varphi(x) = -x^{\mu+1/2} \int_0^x t^{-\nu+1/2}\, \Psi(t)\, dt, \qquad (2.3)$$

then $\varphi \in H_\mu$ and $h\varphi = \Psi$. Using induction on $\nu - \mu$ it can be shown that h is onto. Now open mapping theorem shows that h is an isomorphism.

Let $\varphi \in H_\mu$, then

$$h[h_\mu(\varphi(x))(y)] = y^{\nu+1/2}(-y^{-1}D_y)^{\nu-\mu} y^{-\mu-1/2} \int_0^\infty \varphi(x)\, (xy)^{1/2}\, J_\mu(xy)\,dx,$$

differentiating under the integral sign we get

$$= y^{\nu+1/2} \int_0^\infty \varphi(x) x^{2\nu-\mu+1/2} (xy)^{-\nu} J_\nu(xy)\, dx$$

(since $(xy)^{-\nu}J_\nu(xy)$ is a smooth bounded function on $0<xy<\infty$, differentiation under the integral sign is valid).

$$= \int_0^\infty x^{\nu-\mu}\, \varphi(x)\, (xy)^{1/2} J_\nu(xy)\, dx$$

$$= h_\nu(x^{\nu-\mu}\, \varphi(x))(y).$$

Thus proving the theorem.

We also have the following theorem.

Theorem 2.2 For $\nu - \mu$ and h as given in Theorem 2.1 we have

$$h_\nu \circ h = y^{\nu-\mu} \circ h_\mu \qquad (2.4)$$

Proof. Let $\varphi \in H_\mu$. Then

$$h_\nu \circ h(\varphi) = \int_0^\infty [x^{\nu+1/2}(-x^{-1}D_x)^{\nu-\mu}\, x^{-\mu-1/2}]\varphi(x)(xy)^{1/2}J_\nu(xy)\,dx$$

For $\nu - \mu = 1$, it is easily shown that (using integration by parts)

$$h_\nu \circ h(\varphi) = y^{\nu-\mu} \circ h_\nu(\varphi).$$

Using induction on $\nu - \mu$, we complete the proof.

Zemanian has extended the classical Hankel transform h_μ to the space H_μ for $\mu \geq -1/2$. But (2.4) can be used to extend h_μ for all real μ; see also Zemanian [2, Theorem 1].

Definition. For $\mu \in \mathbb{R}$, define the Hankel transform h_μ on the space H_μ by

$$[h_\mu(\varphi(x))](y) = y^{\mu-\nu} \, [h_\nu \circ h(\varphi)](y), \qquad (2.5)$$

where ν is chosen so that $\nu - \mu \in \mathbb{N}$ and $\nu \geq -1/2$.

Now we extend Theorem 1.1 [Zemanian, 1] to get

Theorem 2.3. h_μ is an automorphism on H_μ and hence on H'_μ for all real μ.

Remark. Consider the map $T^a : H_{-1/2} \longrightarrow H_{-1/2}$ defined by

$$T^a(\varphi) = (h_\nu \circ x^{\nu+1/2})^{-1} \circ y^{\nu-\mu} \circ (h_\mu \circ x^{\mu+1/2})(\varphi), \qquad (2.6)$$

where $a, \mu, \nu \in \mathbb{R}$ are such that $a = \nu - \mu$. Then T^a is clearly an automorphism and $T^a(\varphi) = (-x^{-1}D_x)^a(\varphi)$ for $a \in \mathbb{N}$ (follows easily from (2.4)). Hence (2.6) gives a natural representation for the differential operator $(-x^{-a}D_x)^a$, $a \in \mathbb{R}$ restricted to the space $H_{-1/2}$.

For distribution $f \in H'_{-1/2}$, define T^a by

$$< T^a f, \varphi> = < f , T^a \varphi> , \quad \varphi \in H_{-1/2}. \qquad (2.7)$$

Then T^a is an automorphism on the dual space $H'_{-1/2}$.

An Open Problem

The map

$$h = y^{\nu+1/2} \, T^{\nu-\mu} \, y^{-\mu-1/2} ,$$

where $T^{\nu-\mu}$ is given by (2.6) with x and y interchanged, is clearly an isomorphism from H_μ onto H_ν (and as we have seen, it makes the diagram commutative for $\nu - \mu \in \mathbb{N}$), but it remains an open problem whether the diagram (1.1) will commute for $\nu - \mu \notin \mathbb{N}$.

REFERENCES

[1] A.H. Zemanian, The distributional Hankel transformation, *SIAM J. Appl. Math.*, 14: 561-576(1966).

[2] A.H. Zemanian, Hankel transforms of arbitrary order, *Duke Math. J.* 34(4) : 761-770(1967).

APPLICATIONS OF DISTRIBUTIONS AND CURRENTS IN COMPLEX ANALYTIC GEOMETRY

M. Sitaramayya

School of Mathematics & CIS
University of Hyderabad
Hyderabad 500134, India

INTRODUCTION

Since the concept of distributions was introduced by L. Schwartz [21] and its generalization as current by George de Rham [7], extensive studies were made on their theory from analytic point of view (cf, [3], [4], [5], [8], [10], [14], [15], [24], [26]). On the other hand, these analytic objects found numerous geometric applications in transcendental algebraic geometry and complex analytic geometry (cf. [5], [6], [8], [11]) besides in several other areas in analysis itself. In this talk I restrict myself to the application of these objects to a special category of compact complex spaces (resp. manifolds) called Moishezon spaces. During last 35 years, since their introduction by Moishezon [16], these spaces were extensively studied (cf. [1], [2], [9], [12], [13], [18], [19 to 25]) and our understanding of their geometry was complete after the proofs of Grauert-Riemann Schneider conjecture in 1987 by Siu [25] and Demailly [2] and the complete picture was presented in my survey paper [22]. We understand the obstructions for a Moishezon space (resp. manifold) to be projective algebraic in terms of certain special currents which are related to the Lelong numbers of fractional nature. In Section 1 we recall very briefly the relevant facts on distributions and currents and in Section 2 we recall some geometric facts on the geometry of Moishezon spaces and finally in Section 3 we interpret the obstructions in terms of special currents.

Generalized Functions and Their Applications, Edited by
R.S. Pathak, Plenum Press, New York, 1993

1. DISTRIBUTIONS AND CURRENTS

Let $C_c^\infty(R^n)$ be the vector space of compactly supported smooth functions on $\mathbb{R}^n$ equipped with the C^∞ - topology.

Definition 1.1. A distribution on $\mathbb{R}^n$ is a continuous linear map $T : C_c^\infty(\mathbb{R}^n) \longrightarrow C$ and $D'(\mathbb{R}^n)$ denotes the vector space of all distributions on $\mathbb{R}^n$.

Let $A_c^q(\mathbb{R}^n)$ denote the space of C^∞ q-forms on $\mathbb{R}^n$ with compact support and we can make $A_c^q(\mathbb{R}^n)$ in to a complete topological vector space by inducing topology on $A_c^q(\mathbb{R}^n)$ from $C_c^\infty(\mathbb{R}^n)$ component wise.

Definition 1.2. A current of degree q on $\mathbb{R}^n$ is a continuous linear map $T : A_c^{n-q}(\mathbb{R}^n) \longrightarrow C$ and $D^q(\mathbb{R}^n)$ denotes the space of all currents of degree q on $\mathbb{R}^n$.

Example 1.3. If C is a piecewise smooth oriented (n-q) - chain in $\mathbb{R}^n$, then C defines a current $T_C \in D^q\,\mathbb{R}^n$ by $T_C(\phi) = \int_C \phi$ for $\phi \varepsilon A_c^{n-q}(\mathbb{R}^n)$. Moreover, support of T_C is exactly C.

Example 1.4. Let ψ be a q form on $\mathbb{R}^n$ with local L^1-coefficients on $\mathbb{R}^n$. Then ψ defines a current $T_\psi \varepsilon D^q(\mathbb{R}^n)$ by

$$T_\psi(\phi) = \int_{\mathbb{R}^n} \psi \wedge \phi \; , \; \phi \in A_c^{n-q}(\mathbb{R}^n) \; .$$

Example 1.5 The exterior derivative d on smooth forms on $\mathbb{R}^n$ induces an operator $d : D^q(\mathbb{R}^n) \longrightarrow D^{q+1}(\mathbb{R}^n)$ on currents defined by $(dT)(\phi) = (-1)^{q+1}\,T(d\phi)$ for $\phi \in A_c^{n-q-1}(\mathbb{R}^n)$. Then $d^2 = 0$ and this d on chains coincides with the boundary operator δ up to sign in Example 1.3 and with the usual d operator on forms in Example 1.4.

1.6 We can localize the distributions and currents to an open subset U of $\mathbb{R}^n$ and hence we can extend these to a manifold M to get D(M) and $D^*(M)$ the spaces of distributions and currents on the manifold M.

1.7 A current $T \in D^q(\mathbb{R}^n)$ may be considered as a differential form with distributions as coefficients and we can smoothen a current by smoothening its distribution coefficients.

1.8 On a manifold M we have defined above (cf. 1.6) the *complex* of currents $(D^*(M),d)$. In view of Example 1.4 we can view smooth forms as currents and hence we get a natural map $I_1 : H^*_{deRham}(M,R) \longrightarrow H^*(D^*(M),d)$ and similarly in view of Example 1.3, we can regard singular cycles as currents giving a natural map $I_2 : H^*(M, \text{sing}) \longrightarrow H^*(D^*(M), d)$. If, in addition, M is a complex manifold, then we have a complex $(D^{p,q}(M), \bar{\delta})$ of currents of type (p,q) and interpreting (p,q) type forms as currents on M we get a natural map I_3 : $H^{p,q}_{\bar{\delta}}(M) \longrightarrow H^*(D^{p,q}(M), \bar{\delta})$ where $H^{p,q}_{\bar{\delta}}(M)$ is the Dolbeault cohomology group and we have that all these three natural maps I_1, I_2, I_3 are surjective isomorphisms called *the smoothing of cohomology*.

2. MOISHEZON SPACES

Definition 2.1. Let X be a compact complex space (manifold) of complex dimension n. X is called Moishezon if X admits n algebraically independent meromorphic functions over $\mathbb{C}$. X is called algebraic if X can be embedded in a projective space as closed subspace holomorphically. We refer for (1) the categorical place of Moishezon spaces, (ii) the characterization problem for X, (iii) the position of X relative to a 1-convex space to the survey paper [22].

2.2. Let X be a compact complex space with singular set S. Let $\pi : \hat{X} \longrightarrow X$ be a desingularization of X which exists due to Hironaka. The general algebraicity problem is to put conditions on X so that $\hat{X}$ becomes projective algebraic. It should be noted that X Moishezon is a natural solution to this problem. This leads us to the natural problem of studying the obstructions in "Pushing down" the algebracity from $\hat{X}$ to the Moishezon X.

2.3 Grauert's Example [9]. A theorem of Chow-Kodaira [1] says that every 2-dimensional Moishezon manifold is algebraic. On the other hand for the singular case we have Gruert's Example [9] of a 2-dimensional compact complex space which is Moishezon but not algebraic : Let M be a compact Riemann surface of genus $g \geq 2$ and F be a positive holomorphic line bundle over M with $H^1(M,F) \neq \phi$.

Let $\alpha = (\alpha_{ij}) \in H^1(M,F)$. Construct a holomorphic fiber bundle G over M with fiber C and the structure group the affine group by defining $(p,z) \in U_i \times C$ is equivalent to $(p,z') \in U_j \times C$ if $z' = f_{ij}(p)\, z + \alpha_{ij}(p)$ where (f_{ij}) are the transition functions of F with respect to the covering $\{U_i\}$. By completing each fiber of G by adding the point at ∞ we get a compact complex manifold $\bar{G}$ (over M) of complex dimension 2. Then the section A of $\bar{G}$ through ∞ can be contracted to a point $x_0 \in X$ where $\pi : \bar{G} \longrightarrow X$ is a proper map with $x_0 = \pi(A)$ and X is a compact complex space of complex dimension 2. This X is Moishezon but not algebraic.

Remarks 2.4(a) In fact, for each $n \geq 2$ there exists a compact complex space of complex dimension n which is Moishezon but not algebraic.

(b) In Grauert's construction by "blowing down" the section A through ∞, a special kind of singularity x_0 was introduced which became an obstruction for X to be algebraic.

(c) More generally Grauert's Example is a special point modification of hypersurfaces namely : Let X be a compact complex space of complex dimension n+1. Let A be a hypersurface of X such that X is non-singular in a neighbourhood of A. Let [A] be positive. Then X is obtained from the projectivization of the normal bundle $N_x(A)$ of A in X as a point modification by blowing down its section through ∞ to a point.

(d) Let X be a compact complex space. Let $p_0 \in X$ be an isolated singularity of X. Let $\pi : \hat{X} \longrightarrow X$ be a resolution of the singularity. p_0 is called an *analytical rational singularity* if $R^1\pi_* \, S(\hat{X}) = 0$, otherwise p_0 is called a *non-rational singularity* (here $R^1\pi_* \, S(\hat{X})$ denotes the first direct image of the structure sheaf of $\hat{X}$).

(e) Let X be a 2 - dimensional compact complex space having only isolated *rational singularities* and let X be Moishezon. Then X is algebraic (cf. [23]). Thus rational singularities are extremely good singularities and this result can be viewed as a generalization of Chow-Kodaira. From this and Grauert's Example we get

(f) Theorem . For a 2 - dimensional compact complex space X which is Moishezon to be algebraic the obstructions are non--rational isolated singularities and Grauert's point modifications introduce such non-rational singularities.

2.5 Obstructions in non - singular case : Hironaka - Moishezon Example : Let X be the projective 3 - space. Let C be a curve in X with only singularity a double point x_0 with distinct tangents. In a small neighbourhood U of x_0 , C has two branches C′ and C′′. First blow up along C′ as $\sigma_1 : U_1 \longrightarrow U$ and then blow up along C′′ as $\sigma_2 : \bar{U}_1 \longrightarrow U_1$. Let $\lambda = \sigma_2 \circ \sigma_1 : \bar{U}_1 \longrightarrow U$. On the other hand blow up $X - \{x_0\}$ along $C - \{x_0\}$ as $\mu : V \longrightarrow X - \{x_0\}$. Then μ and λ agree in $U - \{x_0\}$ and patch up these to get a compact 3 - dimensional complex manifold $\sigma : \hat{X} \longrightarrow X$. This $\hat{X}$ is Moishezon manifold which is not algebraic as $\hat{X}$ contains a curve γ which is homologous to zero.

Definition 2.6 Let M be a compact complex manifold. Let $C = \sum n_i C_i$ be an integral linear combination of irreducible smooth curves in M. C is called a "bad curve" if C is homologous to zero.

Remark 2.7 (a). The Hironaka - Moishezon example introduces a "bad curve γ" by a process of suitable successive blow ups in projective 3 - space and these bad curves are thus obstructions for the algebraicity of Moishezon spaces.

(b) The geometric meaning of Moishezon's basic result [16] that a Moishezon manifold M admitting a Kaehler metric g is algebraic is that the Kaehler metric g is killing all these obstructions "bad curves γ" of M via Stokes theorem.

3. Current Interpretation of obstructions in non - singular case

3.1 The information on the obstructions namely bad curves γ in X is contained in a pair (ω,θ) of differential forms on X where ω is a positive definite real (1,1) - form and θ is a real 2 - form on X such that (a) $d(\omega-\theta) = 0$ (b) $\int_C \theta = 0$ for all curves C in X. Then the presence of such a pair (ω,θ) on X annihilates the presence of bad curves in view of Seshadri's ample criterion [17], thus making X algebraic.

Remark 3.2 (a) For the Moishezon's Kahler case simply take ω = Kahler form and $\theta = 0$.

(b) Thus the non - existence of such pairs (ω,θ) contains information about the obstructions for the algebraicity of a Moishezon 3 - fold.

3.3 In a different context Harvey [10] studied currents in his studies on Holomorphic chains. Let X be a compact complex manifold of dimension 3. Let A^r and $A^{p,q}$ denote the spaces of smooth forms of degree r and type (p,q) respectively on X and $D^r(X)$ and $D^{p,q}(X)$ be the corresponding topological duals called the spaces of currents of degree r and of type (p,q) respectively on X.

Definition 3.4 A (p,q) current T is called real if $T = \overline{T}$ i.e. $\overline{T(\phi)} = T(\overline{\phi})$ and a real current T is called positive if $i^{p(p-1)/2}\, T\, (\phi \wedge \overline{\phi}) \geq 0$. Note that for a real $T \in D^{p,p}(X)$, $dT=0$ if and only if $\delta T = \overline{\delta} T = 0$.

Example 3.5 (a) Let M be a compact complex manifold, let $Z \subset M$ be a codimension p - analytic subvariety with $Z^* = Z - Z_s$ as its regular manifold. Then as in example 1.3, Z defines a positive closed current T_Z by $T_Z : A_c^{n-p,n-p}(M) \longrightarrow C$ defined by $T_Z(\phi) = \int_{Z*}\phi$ and the cohomology class $[T_Z] \in H^*(D^*(M),d) \cong H^*_{deRham}(M)$ is the fundamental class of Z.

(b) A smooth (1,1) form $\omega = \frac{i}{2} \sum_{i,j} h_{ij} dz_i \wedge d\overline{z}_j$ is real if $\overline{h}_{ij} = h_{ji}$ and is positive if $(\overline{h}_{ij})$ is positive definite and closed exactly when the hermitian metric $ds^2 = \sum_{i,j} h_{ij} dz_i \wedge dz_j$ is Kahler. Hence the exterior powers ω^p of the Kahler form ω define closed positive (p,p) currents. Thus we have the three types of currents (a) smooth currents (b) closed positive currents (c) currents supported by analytic varieties (as in (b) and (a) of this Example).

3.6 Let $D_+^{p,p}(M)$ denote the cone of positive (p,p) - currents and $D_{+c}^{p,p}$ denote the cone of closed positive (p,p) - currents and $D_{+a}^{p,p}(M)$ denote the cone of positive currents of the form $\lim_k \sum_{j=1}^{n_k} \lambda_{ij} T_{C_{ij}}$ where $\lambda_{ij} \in R$, C_{ij} are irreducible curves in X and $T_{C_{ij}}$ is the current defined by C_{ij}. Suppose now X is a Moishezon manifold. Then we can answer the question when such pairs (ω,θ) defined in 3.1 exist, namely, (ω,θ) Moishezon pair exists on X iff $D^{1,1}_{+a}(X) \cap d\, D^3(X) = (0)$. Putting these together we get

Theorem 3.7. Let M be a Moishezon 3-fold. Suppose there exists no $T \in D^{1,1}_{+a}(M)$ and no effective curve C in M such that $T+\epsilon$ is homologous to zero. Then M is algebraic.

Remarks 3.8. (a) The obstructions for algebraicity of a Moishezon 3 - fold lies in such special currents T and curves C. In Hironaka - Moishezon example a special curve C homologous to zero existed. Thus it will be interesting to construct an example of a Moishezon 3 - fold M having a special positive current T of bidegree (1,1) on M which is a weak limit of currents defined by irreducible curves in M which is homologous to zero.

(b) As in Moishezon's theorem a Kahler metric g annihilates all such obstruction elements of $D^{1,1}_{+a}(M) \wedge d\, D^3(M)$.

3.9 Let $T \varepsilon D^{p,p}_{+c}(U)$. Take $U = \mathbb{C}^n$, p = origin in $\mathbb{C}^n$. Let $B[r] = \{z\varepsilon\mathbb{C}^n : \|z\| \leq r\}$, $X(r)$ = characteristic function of $B[r]$ and $\omega = \frac{i}{2} \sum dz_i \wedge d\bar{z}_i$. We define the function

$\Theta(T,p,r) = \frac{1}{r^{2n-2p}} T(X(r)\omega^{n-p})$. Then $\Theta(T,p,r)$ is an increasing function of r.

Definition 3.10 The Lelong number of T at p is

$\Theta(T,p) = \frac{1}{\pi^{n-p}} \lim_{r\to 0} \Theta(T,p,r)$. Thus for a $T \varepsilon D^{p,p}_{+c}(U)$ and U open and p is any point in U, we can define its Lelong number $\Theta(T,p)$. By definition it is non - negative and it is zero for T a smooth current and if $T = T_Z$ defined by an analytical subvariety Z then $\Theta(T_Z,p) = \text{mult}_p(Z)$ and for closed positive currents it lies between 0 and 1.

Remark 3.11 In a recent work Demailly analysed the Lelong numbers in greater detail [6] and also Siu [24] studied the Lelong numbers and using their results we may understand elements of $D^{1,1}_{+a}(M)$ which are obstructions for algebraicity of Moishezon 3 - folds (which are weak limits of special currents) in terms of their Lelong numbers between 0 and 1 and may be able to construct a suitable example for 3.8(a).

Acknowledgments

The author thanks Prof.R.S. Pathak of Banaras Hindu University for giving an opportunity to present his ideas. Also the author is highly grateful to the SAREC for its financial assistance for the visit of the author at ICTP Trieste where these ideas were worked out.

REFERENCES

1. W.L. Chow and K.Kodaira, On analytic surfaces with two meromorphic functions. *Proc. Nat. Acad. Sci, U.S.A.* 38: 319-325(1952).
2. J.P. Demailly, Une preuve simple de la conjecture de Grauert - Riemen Schneider, Lect. Notes in Math. 1295: 24-47(1987).
3. J.P. Demailly, Surles nombres de Lelong associ'es a L'image direct d'un Courant positif ferm'e. *Ann. Inst. Fourier (Grenoble)* 32 : 37-66(1982).
4. J.P. Demailly, Nombres de Lelong-generalises, theoremes d'integralite et d' analyticite', *Acta Math.* 159 : 153-169(1987).
5. J.P. Demailly, Regularization of closed positive currents and intersection theory, preprint Inst. Fourier 176(1991).
6. J.P. Demailly, A numerical criterion for very ample line bundles Intl. Symp. TIFR 1992 Jan, 45 pages.
7. G.de Rham, "Differentiable manifolds", Herman, Paris(1955).
8. H. Federer, "Geometric measure theory", Springer Verlag, New York, 153(1969).
9. H. Grauert, Uber modifikationen und exzeptionelle analytische mengen, *Math. Ann.* 146 : 331-368(1962).
10. R. Harvey, Holomorphic Chains and their applications, *Proc. Symp. Pure Math.* 30 : 309-382(1977).
11. R. Harvey and H.B. Lawsen Jr, An intrinsic characterization of Kahler Manifolds, *Inv. Math.* 74 : 169-198(1989).
12. H. Hironaka, An example of a non-Kahlerian complex analytic deformation of Kahlerian complex structures, *Ann. of Math.* 75 : 190-208(1962).
13. C.Le Brun, Anti-self-dual hermitian metrics on the Blown-up Hopf surface. Inst. for advanced studies, Princeton, preprint 1990.
14. P. Lelong, Plurisubharmonic Functions and Positive Differential Forms, Gordon-Breach, New York(1969).
15. P. Lelong, Elements extremaux sur le cone des convants positifs fermes, Lelong seminar(1971) Lect. Notes. Math. 332 : 112-131(1972).
16. B. Moishezon, On n-dimensional compact complex varieties with n algebraically independent meromorphic functions uspehi , *Amer. Math. Soc. transl.* 63 : 51-177(1967).
17. D. Mumford,"Abelian Varieties", Oxford Univ. Press(1971).
18. T. Peternell, Algebraic criteria for compact complex manifolds. *Math. Ann.* 275 : 653-672(1986).
19. K. Rama, Geometry of Moishezon Spaces, Ph.D. thesis, Univ. of Hyderabad, 1984, 126 pages.

20. K. Rama, A note on characterization of Moishezon spaces. Publ. Math. 34 : 1-3(1990).
21. L. Schwartz, "Theorie Des Distributions", Hermann, Paris, (1956).
22. M. Sitaramayya, Geometry of convex spaces, ICTP Publications, Tech. report No. 382, 76 pages(1986).
23. K. Rama, Geometry of Moishezon spaces and algebraicity. Proc. Allahabad Math. Soc. Biennial conf. p. 17-25(1990).
24. Y.T.Siu, Analyticity of sets associated to Lelong numbers and the extension of closed positive current, *Inv. Math.* 27 : 53-156(1974).
25. Y.T. Siu, some recent results in complex manifold theory related to vanishing theorems for the semi positive case. Lect. Notes in Math. 1111, 169-192(1984).
26. H. Skoda, Prolongement consants positifs fermes de masse finite, *Inv. Math.* 66 : 361-376(1982).

STRUCTURAL PROBLEMS OF THE ASYMPTOTIC BEHAVIOUR OF GENERALIZED FUNCTIONS

Bogoljub Stanković

Institute of Mathematics,
University of Novi Sad, Yugoslavia

1. INTRODUCTION

Development of the theory of generalized functions and their applications in solving different mathematical models have pointed at the need of the adequate theory of the asymptotic behaviour of distributions. Let us point at integral transforms of distributions, Abelian and Tauberian type theorems, asymptotic behaviour of solutions of partial differential equations,... These and many other problems have pushed on the elaboration of a theory of asymptotic behaviours of generalized functions.

The simplest method to define the asymptotic behaviour of a distribution is related with so-called semiregular distributions. If a distribution T equals to a locally integrable function f in (w, ∞), $w \in \mathbf{R}$, then the eventual asymptotics of f is going to be, by definition, the asymptotics of the distribution T. The use of such a definition one can find by Lighthill [1]. Schwartz in his famous book [2] gave a basis for the shift asymtotics or the $S-$asymptotics. This idea has been ellaborated in many papers (see [3]). The $S-$asymptotic expansion of a distribution has been introduced by Brichkov and Shirokov [4] as a new approach in investigations in the quantum field theory.

In the literature one can find different approaches to the asymptotic behaviour of distributions. Our aim is not to count up all of them. We shall mention only those which has the largest use in mathematics and in applications.

The greatest influence to the theory and applications of the asymptotics of distributions has been made by Vladimirov and his pupils Drozhinov and Zav'yalov elaborating the quasi-asymptotics of distributions (see [5]). Equivalence at infinity of a distribution has been defined by Lavoine and Misra [6] and had nice applications in the theory of integral transforms of distributions (for references see [3]).

In this paper we shall ellaborate, first of all, structural properties of the quasi-asymptotics, equivalence at infinity and $S-$asymptotics and the relation between them, obtained in the last years. We shall deal only with one-dimensional case and with the asymptotic behaviour at infinity, to stress ideas and problems.

Generalized Functions and Their Applications, Edited by
R.S. Pathak, Plenum Press, New York, 1993

2. NOTATION

$\mathbf{N}$ denotes the set of natural numbers; $\mathbf{N}_0 = \mathbf{N} \cup \{0\}$. We shall always denote by L or $L_{a,b}$ a slowly varying function [7], namely a positive and locally integrable function on (w, ∞), $w > 0$ and such that $L(ux)/L(x) \to 1$, $x \to \infty$, $u > 0$.

$\mathbf{D}'$ is the space of Schwartz's distributions, $\mathbf{S}'$ the space of tempered distributions; $\mathbf{D}'_+ = \{T \in \mathbf{D}', \text{ supp} T \subset [0, \infty)\}$. Distributions f_a, $a \in \mathbf{R}$, belonging to $\mathbf{S}'_+$, are defined by

$$f_a(t) = \begin{cases} H(t)t^{a-1}/\Gamma(a), & a > 0 \\ D^m f_{a+m}(t), & a \leq 0,\ a+m > 0,\ m \in \mathbf{N}, \end{cases} \tag{1}$$

where D is the sign of derivative in the sense of distributions and H is the Heaviside function. If $T \in \mathbf{S}'_+$, then $T^{(-m)} = f_m * T$ ($*$ is the sign of the convolution); $f_{-n} = D^n\delta$, $n \in \mathbf{N}_0$.

3. DEFINITIONS

Three definitions in their original forms, of asymptotic behaviour of distributions, which can be basis for our discussions, are:

Definition 1. [8] Distribution $f \in \mathbf{S}'_+$ has the quasi-asymptotics related to u^a, $a \in \mathbf{R}$, if there exists

$$\lim_{u \to \infty} f(ut)/u^a = g(t), \text{ where } g \neq 0. \tag{2}$$

We shall write in short: $f \overset{q}{\sim} u^a \cdot g$ in $\mathbf{S}'$.

Definition 2. [9] Distribution $T \in \mathbf{D}'$ has the S–asymptotics related to a locally integrable and positive function c on $(0, \infty)$ if there exists

$$\lim_{h \to \infty} T(x+h)/c(h) = U(x) \text{ in } \mathbf{D}',\ U \neq 0. \tag{3}$$

We shall write in short: $T \overset{s}{\sim} c \cdot U$ in $\mathbf{D}'$.

Definition 3. [6] Distribution T is equivalent at infinity with Ax^a, $A \neq 0$, $a \in \mathbf{R}$, if there exist $m \in \mathbf{N}_0$ $m + a > 0$, $w > 1$ and a continuous function F on $\mathbf{R}$ such that $T = D^m F$ on (w, ∞) and

$$F(x)/x^{a+m} \to AC_{a,m}, \quad x \to \infty. \tag{4}$$

If $m > 0$, then $a \notin (-\mathbf{N})$; $C_{a,0} = 1$ and $C_{a,m} = 1/(a+1)\cdots(a+m)$, $m > 1$.

These definitions have been improved to this day. We shall comment some of such improvements.

3.1. Quasi - Asymptotics

The next theorem (see Theorem 1 in [5], secition 3.2) shows possibilities to improve Definition 1.

Theorem 1. Suppose that e is a positive and measurable function defined on $(0, \infty)$. If for a $f \in \mathbf{S}'_+$

$$\lim_{u \to \infty} f(ut)/e(u) = g(t) \quad \text{in } \mathbf{S}', \quad g \neq 0, \tag{5}$$

then $e(u) = u^a L(u)$ and $g = C f_{a+1}$, where $a \in R$, L is a slowly varying function and f_{a+1} is given by (1) (see Theorem 1, section 3.2 in [5]).

Hence, we can measure the quasi-asymptotic behaviour at infinity of tempered distributions by regular varying functions, $e(u) = u^a L(u)$ and the limit is always of the same form, namely $g = C f_{a+1}$.

3.2. S–Asymptotics

In Definition 2 the function c and the limit U have the form $c(h) = \exp(ah) \cdot L(\exp h)$, $h \geq h_0$ and $U(x) = C \exp(ax)$ ([3], p. 85). Because of the relation:

$$(T * \phi)(h) = \langle T(x+h), \phi(-x) \rangle, \quad T \in \mathbf{D}', \quad \phi \in \mathbf{D},$$

limit (3) can be given as

$$\lim_{h \to \infty} (T * \phi)(h)/c(h) = C_\phi,$$

where C_ϕ is a constant which depends on ϕ and $C_\phi \neq 0$ for some $\phi \in \mathbf{D}$. This form gives possibility to define the S–asymptotics for other generalized functions for which a convolution is defined.

There exist equivalent definitions to definitions of the quasi-asymptotics and S–asymptotics. The most important result in this sense is the Tauberian theorem of Drozhinov and Zav'yalov ([8], section 7). The paper [10] is devoted to such assertions for the quasi-asymptotics and for the S–asymptotics, as well. We shall cite here a theorem which gives sufficient conditions for the existence of the S–asymptotics and which is a consequence of a Weiner Tauberian theorem proved in [11].

Theorem 2. Let $T \in \mathbf{D}'$ be such that the set

$$\{T(x+h)/(\exp(ah)L(\exp h)), \quad h > 0\} \tag{6}$$

is bounded in $\mathbf{D}'$ and let $\phi \in \mathbf{D}$ be such that the Fourier transform of $\phi(x)\exp(-ax)$ is different from zero on $\mathbf{R}$. If

$$\lim_{h \to \infty} (T * \phi)(h)/(\exp(ah)L(\exp h)) = A \int_{\mathbf{R}} \phi(t) \exp(-at) dt, \quad A \neq 0, \tag{7}$$

then $T \overset{s}{\sim} A \exp(ah) L(\exp h) \cdot \exp(ax)$ in $\mathbf{D}'$.

By this theorem it is sufficient for the existence of limit (3) the boundness of the set in (6) and the existence of the limit in (7) for only one $\phi \in \mathbf{D}$.

3.3. Equivalence at Infinity

In [12] Lavoine and Misra replaced in Definition 3, x^a by $x^a \ln x$ and in [13] Takači introduced $x^a L(x)$ instead of $x^a \ln x$, where L is any slowly varying function. But the restriction $a \notin (-\mathbf{N})$ has been remained. We propose how to enlarge the definition of the equivalence at infinity to the case $a = -n$, $n \in \mathbf{N}$.

First, we have to introduce an operator I :

$$I(t^a L(t))(x) = \begin{cases} \int_w^x t^a L(t)dt, & a \geq -1 \\ (-1)\int_x^\infty t^a L(t)dt, & a < -1 \end{cases}, \quad x \in (w,\infty), \; w > 0. \tag{8}$$

I_k will be the $k-$times intereted I, $I^0 f = f$. In [14] we proved:

Lemma 1. The operator I has the following properties:

1. $D^k I^k f = f, \quad k \in \mathbf{N}$.
2. If $a \notin (-\mathbf{N})$, then $I^k(t^a L(t))(x) = x^{a+k} L_{k,a}(x), \quad x > w; \quad L_{k,a}(x)/L(x) \to 1/(a+1)\ldots(a+k)$, $x \to \infty$.
3. $I(t^{-1}L(t))(x) \equiv \hat{L}(x)$, $x > w$ and $\hat{L}(x)/L(x) \to \infty$, $x \to \infty$.
4. For $k = i + j$, $j \in \mathbf{N}_0$, $i \in \mathbf{N}$, $I^k(t^{-i}L(t))(x) = x^j(\hat{L}_{i-1,-i})_{j,0}(x)$, $x > w$, where $(\hat{L}_{i-1,-i})_{j,0}(x) \to \hat{L}_{i-1,-i}/j!$, $x \to \infty$; $\quad (\hat{L}_{i-1,-i})_{0,0} = \hat{L}_{i-1,-i}$; $\quad \hat{L}_{0,-1} = \hat{L}$; $L_{i-1,-i}(x)/L(x) \to (-1)^{i-1}/(i-1)!$, $i \geq 2$.

Definition 3'. Let $T \in \mathbf{D}'$ and let L be a fixed slowly varying function such that $\hat{L}_{n-1,-n}$ does not converge to a constant different from zero, when $x \to \infty$. It is said that T is equivalent at infinity with $Ax^{-n}L(x)$ if there exist $m \in \mathbf{N}_0$, $m - n \geq 0$ and a continuous function F on (w,∞), $w \geq 1$, such that $T = D^m F$ on (w,∞) and

$$F(x) \sim Ax^{-n+m}/(-n+m)! \;\; \hat{L}_{n-1,-n}(x), x \to \infty.$$

The correctness of Definition 3' is covered by

Theorem 3. Let $T \in \mathbf{D}'$ be equivalent at infinity with two functions: $Ax^{-n}L(x)$ and $A'x^{-n'}L(x)$. Then $n = n'$ and $A = A'$.

Proof. By Definition 3' we have: $T = D^m F$ on (w,∞) and $T = D^{m'}F'$ on (w',∞). Hence, $D^{m'}F' - D^m F = 0$ on $(\hat{w},\infty)$, $\hat{w} = \max(w,w')$. Let us suppose that $m' = m + p$, $p \geq 0$, then

$$D^{m+p}(F' - I^p F)(x) = 0 \;\text{ on }\; (\hat{w},\infty).$$

It follows that:

$$F' = I^p F + P_{m'} \;\text{ on }\; (\hat{w},\infty), \tag{9}$$

where $P_{m'}$ is a polynomial of a degree less or equal to $m' - 1$. By Lemma 1 we have

$$I^p(F(t))(x) \sim A/(-n+m+p)! \;\; \hat{L}_{n-1,-n}(x)x^{-n+m+p}, \quad x \to \infty.$$

We shall return to (9). Because of the asymptotic behaviour of F', $I^p F$ and $\hat{L}_{n-1,-n}$, the polynomial $P_{m'}$ has to be zero. Consequently $n = n'$ and $A = A'$.

4. STRUCTURAL PROPERTIES OF THE DEFINED ASYMPTOTIC BEHAVIOURS OF DISTRIBUTIONS

4.1. Quasi - Asymptotics

We shall start with the quasi-asymptotics because this is the best elaborated asymptotic behaviour of distributions and can be used to judge the results for the other two defined asymptotics.

The next theorem asserts that $\mathbf{S}'_+$ "naturally" suits to the quasi-asymptotics (Theorem 3, section 3.2 in [5]).

Theorem 4. Suppose $f \in \mathbf{D}'_+$ and $e(u) = u^a L(u)$, $a \in \mathbf{R}$. If

$$\lim_{u \to \infty} f(ut)/e(u) = g(t) \quad \text{in } \mathbf{D}', \tag{10}$$

then $f \in \mathbf{S}'_+$ and $f \overset{q}{\sim} e \cdot g$ in $\mathbf{S}'$.

A natural question may be formulated, now: If $f \in \mathbf{D}'$ and if the limit in (10) exists, does $f \in \mathbf{S}'$ and does the limit in (10) exist in the sense of convergence in $\mathbf{S}'$, as well? The answer on this questions is in the following theorem [15].

Theorem 5. Let $f \in \mathbf{D}'$ and there exists the limit in (10), where e is positive and measurable on (w, ∞), $w > 0$. Then:

(i) $e(u) = u^a L(u), \quad u \in (w, \infty), \quad a \in \mathbf{R}$.

(ii) $f \in \mathbf{S}'$.

(iii) If $a \notin (-\mathbf{N})$ and if $a = -1$, $1/L(x)$ is bounded on (w, ∞), then the limit in (10) exists in $\mathbf{S}'$, as well.

If we combine a result from Lemma 6 in [16], then (iii) is valued without any restriction on a.

The form of a tempered distribution having the quasi-asymptotics has been given by the next theorem which is a consequence of Theorem 2, section 3.2 in [5].

Theorem 6. A distribution $f \in \mathbf{S}'_+$ has the quasi-asymptotics at infinity related to $e(u) = u^a L(u)$ ($f \overset{q}{\sim} e \cdot C f_{a+1}$, $C \neq 0$) if and only if $f = D^m F$, where $m \in \mathbf{N}_0$, $m+a \geq 0$ and F is a continuous function on $[0, \infty)$ such that $F(x)/x^m e(x) \to C/\Gamma(m+a+1)$, $x \to \infty$.

Remark. It is important to remark that the quasi-asymptotics is, in general, a global property. Only in case $a > -1$, or $a = -1$ and $\hat{L}(x) \to \infty$ it gives a local property. Knowing a distribution T only in a neighbourhood of infinity, in case $a < -1$ or $a = -1$ and $\hat{L}(x) < \infty$, $x \geq w$, we can not give a precise assertion about the quasi-asymptotic behaviour of T. However, something can be said in this case, as well (see Theorem 13).

4.2. S–Asymptotics

The S–asymptotics is a local property. This is made precise in

Theorem 7. Suppose that the distribution T_1 equals T_2 over (w, ∞), $w \in \mathbf{R}$. If $T_1 \overset{s}{\sim} c \cdot U$ in $\mathbf{D}'$, then $T_2 \overset{s}{\sim} c \cdot U$ in $\mathbf{D}'$, as well ([3], p. 81).

The structure of a distribution T having the S–asymptotics is given by [17]:

Theorem 8. If $T \in \mathbf{D}'$ has the S–asymptotics related to c, then for the interval $(-r, r)$ there exist functions F_i, $i = 0, \ldots, m$, continuous on $(-r, \infty)$ such that for every $i = 0, \ldots, m$, $F_i(x+h)/c(h)$ converges uniformly for $x \in (-r, r)$, when $h \to \infty$ and the restriction of the distribution T on $(-r, \infty)$ can be given in the form:

$$T = \sum_{i=0}^{m} D^i F_i.$$

We have seen that the space $\mathbf{S}'$ is "natural" for the quasi-asymptotics. The question is which space plaies the same role for the $S-$asymptotics? This is the space $\mathbf{K}_1'$ introduced by Hasumi [18]. We recall the basis facts about $\mathbf{K}_1'$.

Let $\mathbf{K}_1$ be the space of all $\mathbf{C}^\infty-$functions φ in $\mathbf{R}$ such that

$$v_k(\varphi) = \sup_{x\in\mathbf{R}, o\le a\le k} (e^{kx}+e^{-kx})|\varphi^{(a)}(x)| < \infty, \quad k \in \mathbf{N}.$$

The toplogy in $\mathbf{K}_1$ is defined by the semi-norms v_k, $k \in \mathbf{N}$; $\mathbf{K}_1$ is a Fréchet space. The dual $\mathbf{K}_1'$ of $\mathbf{K}_1$ is the space of distributions of exponential growth. The space $\mathbf{D}$ is dense in $\mathbf{K}_1$. A distribution $T \in \mathbf{D}'$ is in $\mathbf{K}_1'$ if and only if there exist $p \in \mathbf{N}$, $k \in \mathbf{N}_0$ and a bounded continuous function f in $\mathbf{R}$ such that $T = D^p(e^{kx}+e^{-kx})f$. If $T \in \mathbf{K}_1'$ and $\varphi \in \mathbf{K}_1$, then the convolution $T * \varphi \in \mathbf{C}^\infty$ and $(T*\varphi)(x) = \langle T(x+y), \varphi(-y)\rangle$.

In [19] Pilipović proved

Theorem 9. Let $T \in \mathbf{D}'$, $T \overset{s}{\sim} c'(h)\cdot e^{ax}$ in $\mathbf{D}'$, where $c'(h) = e^{ah}h^bL(h), L$ is monotonous. If $S \in \mathbf{D}'$, supp$S \subset (w,\infty)$, $-\infty < w$ and if $T = S$ in some neighbourhood of infinity, then S is from $\mathbf{K}_1'$ and $S \overset{s}{\sim} c'\cdot e^{ax}$ in $\mathbf{K}_1'$, as well.

We shall show that the supposition that the slowly varying function L has to be monotonous can be rejected.

Theorem 10. If $T \in \mathbf{D}'$, supp$T \subset (-w,\infty)$, $w > \infty$ and $T \overset{s}{\sim} c\cdot Ae^{ax}$ in $\mathbf{D}'$, where $c(h) = \exp(ah)L(\exp h)$, then $T \in \mathbf{K}_1'$ and $T \overset{s}{\sim} c\cdot Ae^{ax}$ in $\mathbf{K}_1'$, as well.

Proof. By Theorem 8, for the interval $(-w,w)$, there exist functions F_i, $i=0,\ldots,m$, continuous on $(-w,\infty)$, such that

$$T = \sum_{i=0}^m D^iF_i \ \text{ on } \ (-w,\infty) \quad \text{and} \quad |F_i(x+h)/c(h)| \le M_i,\ h\in\mathbf{R}_+,\ x\in(-w,w).$$

By the Lemma proved in [17], for every $\varepsilon > 0$, supp$F_i \subset (-w-\varepsilon,\infty)$, $i=0,\ldots,m$. For a slowly varying function L there exists a slowly varying function $L^* \in \mathbf{C}^\infty_{(c,\infty)}, c>0$, such that $L^*(x)/L(x) \to 1$, $x\to\infty$. For the construction of L^* see [11]. Therefore, we can suppose, without any restriction, that L appeared in c is from $\mathbf{C}^\infty_{(c,\infty)}, c>0$. Let us denote by $M_i(h) = F_i(h)/c(h)$, $h\in\mathbf{R}$; $M_i(h)$ is a continuous and bounded function on $\mathbf{R}$ and $F_i = cM_i$. T is now

$$T = \sum_{i=0}^m D^i(cM_i) \quad \text{on} \quad (-w,\infty).$$

Hence, $T \in \mathbf{K}_1'$. The next step is to prove that for $h_0 \in \mathbf{R}$ the set $\{T(x+h)/c(h),\ h \ge h_0\}$ is weakly bounded in $\mathbf{K}_1'$.

We need the following inequality:

$$c(x+h)/c(h) = e^{ax}L(e^xe^h)/L(e^h) \le Ae^{ax+\delta|x|}, \quad x,h\in\mathbf{R},$$

where $\delta > 0$ (see [7], p. 25). Let $\varphi \in \mathbf{K}_1$, then

$$\begin{aligned}
|\langle T(x+h)/c(h), \varphi(x)\rangle| &\le \sum_{i=0}^m \int_{\mathbf{R}} |F_i(x+h)/c(h)|\ |\varphi^{(i)}(x)|dx \\
&\le \sum_{i=0}^m \int_{\mathbf{R}} |M_i(x+h)c(x+h)/c(h)|\ |\varphi^{(i)}(x)|dx \\
&\le \sum_{i=0}^m \int_{\mathbf{R}} e^{ax+\delta|x|}|\varphi^{(i)}(x)|dx \ .
\end{aligned}$$

Since $\mathbf{D}$ is dense in $\mathbf{K}_1$ and since we proved that $\{T(x+h)/c(h),\ h \geq h_0\}$ is weakly bounded in $\mathbf{K}_1'$, to make an end of the proof, it remains only to use the Banach-Steinhaus theorem.

4.3. Equivalence at Infinity

We have seen that Definition 3 and all the improved versions of it give a local property, a property in a neighbourhood of infinity. Therefore we can give the form of a distribution being equivalent at infinity with $Ax^aL(x)$ only in a neighbourhood of infinity.

Theorem 11. Distribution $T \in \mathbf{D}'$ is equivalent at infinity with $Ae(x) = Ax^aL(x)$, $a \notin (-\mathbf{N})$, if and only if it equals to a tempered distribution f in a neighbourhood of infinity and if $f \overset{q}{\sim} e \cdot A\Gamma(a+1)f_{a+1}$.

Proof. Suppose that $T = f$ on (w, ∞), $w > 0$, where $f \in \mathbf{S}_+'$ and $f \overset{q}{\sim} e \cdot Af_{a+1}$ in $\mathbf{S}'$. Then, by Theorem 4, there exist $m \in \mathbf{N}_0$, $m + a > 0$ and continuous function F such that $T = D^mF$ and

$$\lim_{x\to\infty} F(x)/x^m e(x) = A/\Gamma(m+a+1).$$

Hence T is equivalent at infinity with Ae. Let us suppose now that T is equivalent at infinity with Ae. Then, there exist $m \in \mathbf{N}_0$, $a + m > 0$, $w > 1$ and $F \in \mathbf{C}_{(w,\infty)}$ such that $T = D^mF$ on (w, ∞) and

$$\lim_{x\to\infty} F(x)/x^{a+m}L(x) = AC_{a,m}.$$

We shall denote by $\hat{F}$ the following function $\hat{F}(x) = F(x)$, $x \in (w+1, \infty)$ and $\hat{F}(x) = F(m+1)$, $0 \leq x \leq 1 + w$; $\hat{F}(x) = 0$, $x < 0$. Then $f = D^m\hat{F}$ is the requered tempered distribution. This construction shows that f is not unique.

5. RELATION BETWEEN THE THREE DEFINITIONS OF THE ASYMPTOTIC BEHAVIOUR OF DISTRIBUTIONS

It is not easy to compare the tree cited definitions because the $S-$asymptotics and the equivalence at infinity are local properties, while the quasi-asymptotics is, in general, a global property. We have also to fix a common space to which has to belong the distributions, we compare, and to choose the class of functions which will measure the asymptotic behaviour of distributions. We can illustrate this need by the following example: Tempered distribution $H(t)\exp(-t)$ has the $S-$asymptotics related to $\exp(-h)$, but it has no quasi-asymptotics (see [14]).

The common space in which we shall compare the quasi-asymptotics with the $S-$asymptotics will be $\mathbf{S}_+'$ and the class of functions which measures this behaviour will be the class of regular varying functions, namely functions of the form $e(x) = x^aL(x)$; $a \in \mathbf{R}$, $x > 0$.

The first theorem which gives a relation between the quasi-asymptotics and the $S-$asymptotics has been proved in [8].

Theorem 12. Let $f \in \mathbf{S}_+'$, $a > -1$ and $\phi \in \mathbf{S}$. If The Fourier transform of ϕ

equals 1 in a neighbourhood of zero and if there exists

$$\lim_{x\to\infty} x^{-a}(f*\phi)(x) = C\int_{\mathbf{R}}\phi(t)dt, \quad C\neq 0,$$

then $f \overset{q}{\sim} t^a \cdot Cf_{a+1}$ in $\mathbf{S}'$.

This theorem has been improved using $x^a L(x)$ instead of x^a (Theorem 6, section 3.3 in [5]). For the case $a \le -1$ one can consult [20] in which in the case $a < -1$ the limit which defines the quasi-asymptotics can be zero, as well. This fact diminishes preciseness of the result. A complete relation between the quasi-asymptotics and $S-$asymptotics is given by the following theorem:

Theorem 13. [14] If $f \in \mathbf{S}'_+$ and has the $S-$asymptotics related to $h^a L(h)$, it has the quasi-asymptotics, as well, related to $h^b L'(h)$, where b and L' are:

If $a > -1$, then $b = a$ and $L = L'$.

If $a = -1$ and $\int_a^\infty L(t)/t dt < \infty$, then $b = -1,\ L' = 1$.

If $a = -1$ and $\hat{L}(x) = \int_a^x L(t)/t dt \to \infty,\ \ x \to \infty$, then $b = -1,\ L' = \hat{L}$.

If $a < -1$, then b and L' have one of the form: $b = -r-1,\ r \in \mathbf{N},\ L' = 1;\ \ b = a \notin (-\mathbf{N}),\ L' = L;\ \ b = a = -n,\ n \in \mathbf{N},\ L' = \hat{L}_{n-1,-n}$.

It is quite natural that from the existence of the $S-$asymptotics of f in case $a < -1$, we can not precise the quasi-asymptotics of f. In this case the quasi-asymptotics does not depend only on the behaviour of f in a neighbourhood of infinity. But, however, we know that f has the quasi-asymptotics in all cases.

A corollary of Theorem 13 is:

Corollary 1. If $f \in \mathbf{S}'_+$ and $f \overset{q}{\sim} h^b L'(h) \cdot Af_{b+1}$ in $\mathbf{S}'$, then a necessary and sufficient condition that f has the $S-$asymptotics related to $h^a L(h)$ is the existence of $m \in \mathbf{N}_0$ and continuous, bounded functions $F_i,\ i = 0,\ldots,m$, on $(w,\infty),\ w > 0$, such that $F_i(x+h) \to C_i,\ h \to \infty,\ x \in (w,\infty),\ C_0 \neq 0$ and

$$f = D^m \sum_{i=0}^{m} I^{m-i}(t^a L(t) F_i(t)) \ \text{ on } \ (w,\infty), \tag{11}$$

where a, b, L and L' are related as in Theorem 13.

One can find distribution which have the quasi-asymptotics and have no $S-$ asymptotics. Such an one is the regular distribution $H(t)\cdot \sin t \in \mathbf{S}'_+$ (see [3], p. 91), where H is the Heaviside function.

Theorem 14. [14] If $T \in \mathbf{D}'$ and $T \overset{s}{\sim} c\cdot U$ in $\mathbf{D}'$ with $c(h) = \exp(ah)\cdot L(\exp h),\ a \neq 0$, then T can not have the quasi-asymptotics.

A relation between the quasi-asymptotics and the equivalence at infinity has been given by Theorem 11. To compare the equivalence at infinity with the $S-$asymptotics we have to restrain the function which measures the asymptotic behaviour. This function will be $e(x) = x^a L(x),\ a \notin (-\mathbf{N})$.

Theorem 15. If $T \in \mathbf{D}'$ and $T \overset{s}{\sim} e \cdot U$ in $\mathbf{D}'$, where $U = C,\ C \neq 0$ and $e(u) = u^a L(u),\ a \notin (-\mathbf{N})$, then T is equivalent at infinity with $Cx^a L(x)$.

Corollary 2. If $T \in \mathbf{D}'$ and is equivalent at infinity with $Cx^a L(x)$ a necessary and sufficient condition that $T \overset{s}{\sim} e \cdot U$ in $\mathbf{D}'$, $U \neq 0$, is that T is of the form (11).

The regular distribution given by the function $F(x) = x^{-1/2} + e^x \cos e^x = (2x^{1/2} + \sin e^x)'$ is equivalent at infinity with $x^{-1/2}$, but it has no S–asymptotics (see [3], p. 95).

6. S–ASYMPTOTICS OF DIFFERENT CLASSES OF GENERALIZED FUNCTIONS

We have discussed the S–asymptotics of distributions. But the S–asymptotic behaviour can be defined for more general classes of generalized functions. We shall comment two of them.

6.1. S–Asymptotics of Ultradistributions of Beurling Type

For this part we shall use Komatsu's paper [21]. Let $\{M_p : p \in \mathbf{N}_0\}$ be a sequence of positive numbers which satisfy the following conditions:

$$\text{(M.1)} \qquad M_p^2 \leq M_{p-1}M_{p+1}, \quad p = 1, 2, \dots$$

$$\text{(M.3)'} \qquad \sum_{p=1}^{\infty} M_{p-1}/M_p < \infty.$$

In some assertions the sequence $\{M_p\}$ will satisfy

(M.2)' There are constants A and H such that

$$M_{p+1} \leq AH^p M_p, \quad p = 0, 1, \dots$$

or

(M.2) There are constants A and H such that

$$M_p \leq AH^p \min_{0 \leq q \leq p} M_q M_{p-q}, \quad p = 0, 1, \dots$$

We denote by

$$\mathbf{D}_{[-r,r]}^{M_p,h} = \{u \in \mathbf{C}^\infty, \operatorname{supp} u \subset [-r, r], \sum_{p \in \mathbf{N}} h^p \|u^{(p)}\|_\infty / M_p < \infty\}, \ h, r > 0.$$

$$\mathbf{D}^{(M_p)} = \operatorname*{ind\,lim}_{r \to \infty} \ \operatorname*{proj\,lim}_{h \to \infty} \mathbf{D}_{[-r,r]}^{M_p,h}$$

$\mathbf{D}'^{(M_p)}$ will be the strong dual of the space $\mathbf{D}^{(M_p)}$ and we shall call its ellements *ultradistributions of the Beurling type.*

Suppose that $U \in \mathbf{D}'^{(M_p)}$ and $u \in \mathbf{D}^{(M_p)}$, then the convolution $U * u$ is defined by $(U * u)(x) = \langle U(x + y), u(-y) \rangle$. If $\{M_p\}$ satisfies, in addition, (M.2)', then: a) the convolution is hypocontinuous bilinear mapping: $\mathbf{D}'^{(M_p)} \times \mathbf{D}^{(M_p)} \to \mathbf{E}$, or $\mathbf{E}'^{(M_p)} \times \mathbf{D}^{(M_p)} \to \mathbf{D}$. b) The differential operator $P(x, D) = \sum_{k=0}^m a_k(x) D^k$ with $a_k \in \mathbf{E}^{(M_p)}$, $k = 0, 1, \dots, m$, is continuous and maps $\mathbf{D}'^{(M_p)}$ into $\mathbf{D}'^{(M_p)}$.

If $\{M_p\}$ satisfies, in addition, (M.2), then the ultradifferential operator

$$P(D) = \sum_{k=0}^{\infty} a_k D^k, \quad |a_k| \leq CL^k / M_k \text{ for some } C \text{ and } L$$

is continuous and maps $\mathbf{D}'^{(M_p)}$ into $\mathbf{D}'^{(M_p)}$.

Now we can define the $S-$asymptotics of a ultradistribution.

Definition 4. The ultradistribution U has the $S-$asymptotics related to a positive and measurable fuction c if and only if there exists

$$\lim_{h\to\infty}\langle U(x+h)/c(h),\ u(x)\rangle = \langle V,u\rangle \ \text{ for every } \ u\in\mathbf{D}^{(M_p)}.$$

The $S-$asymptotics of ultradistributions has many properties which can be proved just in the same way as for distributions. So, by the properties of the convolution, we cited, we have:

Suppose that $\{M_p\}$ satisfies, in addition, (M.2)', then:

1. If $V\neq 0$, then $c(h)=\exp(ah)L(\exp h),\ h\geq h_0,\ a\in\mathbf{R}$.

2. The $S-$asymptotics is a local property.

3. If $U\in\mathbf{D}'^{(M_p)}$ and has the $S-$asymptotics related to c with the limit V, then the ultradistribution $P(x,D)U$ has the $S-$asymptotics, as well, related to c and with the limit $P(x,D)V$.

If $\{M_p\}$ satisfies, in addition, (M.2), then:

4. If $U\in\mathbf{D}'^{(M_p)}$ and has the $S-$asymptotics related to c with the limit V, then the ultradistribution $P(D)U$ has the $S-$asymptotics, as well related to c and with the limit $P(D)V$.

These properties can be useful if we discuss the asymptotic behaviour of solutions of equations $P(x,D)U=W$ or $P(D)U=W$ in the space $\mathbf{D}'^{(M_p)}$. Just in the same way we can treat *ultradistirbutions of the Roumieu type.*

6.2. S-Asymptotics of Regular Convolution Quotients

$S-$asymptotics can be defined for some classes of generalized functions which have been introduced in a quite different way then functionals. We shall follow Boehme's definition of regular convolution quotients [22].

Aproximate identtity is a sequence of functions $\{u_n\}\in\mathbf{L}^1(R)$ satisfying the following conditions:

i) $\int_{\mathbf{R}} u_n(x)dx=1,\quad n\in\mathbf{N}$;

ii) there is an $M>0$ such that $\int_{\mathbf{R}}|u_n(x)|dx<M,\quad n\in\mathbf{N}$;

iii) there exists a sequence $\{k_n\}\subset\mathbf{R}_+$ such that $k_n\to 0$, when $n\to\infty$ and $\operatorname{supp}u_n\subset[-k_n,k_n],\quad n\in\mathbf{N}$.

Δ will be the set of all approximate identities and $\Delta^\infty=\{\{u_n\}\in\Delta : u_n\in\mathbf{C}^\infty,\ n\in\mathbf{N}\}$. A defining sequence for a regular convolution quotient is a sequence of pairs $\{(f_n,u_n)\}$, where $\{f_n\}\subset\mathbf{L}_{\text{loc}}(\mathbf{R}),\ \{u_n\}\in\Delta$ and for all $m,n\in\mathbf{N}$ the following convolution products are equal:

iv) $f_n * u_m = f_m * u_n$ (the asterisk is the sign of the convolution).

Two defining sequences $\{(f_n,u_n)\}$ and $\{(g_n,v_n)\}$ are said to be equivalent if:

v) $f_n * v_m = g_m * u_n$ for $n, m \in \mathbf{N}$.

By f_n/u_n we shall denote the equivalence class containing the defining sequence $\{(f_n, u_n)\}$. A *regular convolution quotient* X is an equivalence class of defining sequences. The regular convolution quotients are a vector space when the usual multiplication by scalars and addition of fractions is used; we denote it by $\mathbf{B}(\mathbf{L}_{\text{loc}}, \Delta)$. The space $\mathbf{B}(\mathbf{L}_{\text{loc}}, \Delta)$ contains $\mathbf{D}'$ (space of Schwartz's distributions) under the isomorphism: $\mathbf{D}' \ni T \Leftrightarrow (T * v_n)/v_n \in \mathbf{B}(\mathbf{L}_{\text{loc}}, \Delta)$, where $\{v_n\} \in \Delta^\infty$. Moreover, $\mathbf{B}(\mathbf{L}_{\text{loc}}, \Delta)$ contains the class of all regular Mikusinski operators. Both of these containments are proper.

Definition 5. A regular convolution quotient X has the $S-$asymptotics at infinity, related to a positive and measurable function c, with the limit $U = F_n/u_n \in \mathbf{B}(\mathbf{L}_{\text{loc}}, \Delta)$ if there exists $\{(f_n, u_n)\}$ belonging to the class X such that

$$\lim_{h \to \infty} (f_n * v_n)(h)/c(h) = (F_n * v_n)(0), \quad n \in \mathbf{N}.$$

for every $(v_n) \in \Delta^\infty$. We shall write $X \overset{s}{\sim} c \cdot U$.

For the definition and results concerning the $S-$asymptotics of regular convolution quotients see [23]. We shall cite some of them.

The limit U in Definition 5 does not depend on the defining sequence $\{(f_n, u_n)\}$ in the equivalence class X.

$S-$asymptotics of regular convolution quotients, defined by Definition 5, generalizes the $S-$asymptotics of distributions in the following sense: If a distribution T has the $S-$asymptotics, $T \overset{s}{\sim} c \cdot S$ in $\mathbf{D}'$, then the regular convolution quotient $X = (T * u_n)/u_n$, which corresponds to T, has the $S-$asymptotics, as well and $X \overset{s}{\sim} c \cdot (S * u_n)/u_n$.

If $X \in \mathbf{B}(\mathbf{L}_{\text{loc}}, \Delta)$ and $X \overset{s}{\sim} c \cdot U$, then $D^n X \overset{s}{\sim} c \cdot D^n U$. D is the differentiation operator in $\mathbf{B}(\mathbf{L}_{\text{loc}}, \Delta)$.

Suppose that $X \in \mathbf{B}(\mathbf{L}_{\text{loc}}, \Delta)$ and $X \overset{s}{\sim} c \cdot U$, where $U = F_n/u_n$. If $F_n \neq 0$ for one $n_0 \in \mathbf{N}$, then $c(h) = \exp(ah)L(\exp h)$, $h \geq h_0$ and $F_n(x) = C_n \exp(ax)$, where $a \in \mathbf{R}$ and $C_{n_0} \neq 0$.

The $S-$asymptotics behaviour of a regular convolution quotient is a local property.

If $X \in \mathbf{B}(\mathbf{L}_{\text{loc}}, \Delta)$, then X has compact support if and only if $X \overset{s}{\sim} c \cdot 0$ for any c positive and measurable function.

Acknowledgments: This research was supported by Science Fund of Serbia, grant number 0401A, through Matematički institut.

REFERENCES

[1] M. J. Lighthill, "Fourier Analysis and Generalized Functions", Cambridge University Press, Cambridge (1960).

[2] L. Schwartz, "Théorie des distributions", Nouvelle edition, Hermann, Paris (1966).

[3] S. Pilipović, B. Stanković and A. Takači, "Asymptotic Behaviour and Stieltjes Transformation of Distributions", Teubner Verlagsgesellschaft, Leipzig (1990).

[4] Yu. A. Brichkov and Yu. M. Shirokov, On the asymptotic behaviour of the Fourier transform, *Theor. Math. Phis.*, T. 4, No 3 (1970), 301-309 (in Russian).

[5] V. S. Vladimirov, Yu. N. Drozhinov and B. I. Zav'yalov, "Tauberian Theorems for Generalized Functions", Kluwer Academic Publishers, Maine (1988).

[6] J. Lavoine and O. P. Misra, Théorèmes Abéliens pour la transformation de Stieltjes des distributions, *C. R. Acad. Sci.* Paris 279 (1974), 99-102.

[7] N. H. Bingham, C. M. Goldie and J. L. Teugels, "Regular Variation", Cambridge University Press, Cambridge (1989).

[8] Yu. N. Drozhinov and B. I. Zav'yalov, Quasiasymptotics of generalized functions and Tauberian theorems in the complex domain, *Math. Sb. T.* 102 (1977), 372-390 (in Russian). *English transl. Math. USSR Sb.* 36 (1977).

[9] S. Pilipović and B. Stanković, $S-$asymptotic of a distribution, *Pliska Studia Math. Bulgar.* 10 (1989), 147-156.

[10] S. Pilipović and B. Stanković, Structural theorems for the $S-$asymptotic and quasi-asymptotic of distributions, *Mathematica Panonica* (to appear).

[11] S. Plilpović and B. Stanković, Wiener Tauberian theorems for distributions, *J. London Math. Soc.* (to appear).

[12] J. Lavoine and O. P. Misra, Abelian theorems for the distributional Stieltjes transformation, *Math. Proc. Cambridge Philos. Soc.* 86 (1979), 287-293.

[13] A. Takači, On the equivalence of distributions at infinity, *Univ. u Novom Sadu, Zb. Rad. Prirod. Mat. Fak., Ser. Mat.* 15, 1 (1985), 175-187.

[14] B. Stanković, Relation between quasiasymptotic, equivalence at infinity and $S-$asymptotic, *Univ. u Novom Sadu, Zb. Rad. Prir. Mat. Fak., Ser. Mat.,* (to appear).

[15] S. Pilipović, On the Quasiasymptotic of Schwartz Distributions, *Math. Nachr.* 137 (1988), 19-25.

[16] B. I. Zav'yalov, Asymptotic properties of functions that are holomorphic in tubular cones, *Math. Sb. (N. S.)* 136 (178) (1988), No 1, 97-114 (in Russian). *English transl. Math. USSR Sb.* 64 (1989), No 1, 97-113.

[17] B.Stanković, A structural theorem for distributions having $S-$asymptotic, *Publ. Inst. Math. (Beograd) (N. S.)* T. 45 (1989), 35-40.

[18] M. Hasumi, Note on the $n-$dimensional tempered ultra-distributions, *Tohoku Math. J.* (2), 13 (1961), 94-104.

[19] S. Pilipović, On the $S-$asymptotic of tempered and $\mathbf{K}_1'-$distributions, Part IV. $S-$asymptotic and the ordinary asymptotic, *Univ u Novom Sadu, Zb. Rad. Prir. Mat. Fak., Ser. Mat.,* 18.2 (1988), 191-195.

[20] S. Pilipović, The translation asymptotic and the quasiasymptotic behaviour of distributions, *Acta Math. Hung.* 55 (3-4) (1990), 239-243.

[21] H. Komatsu, Ultradistributions I, *J. Fac. Sci. Univ. Tokyo Sect. IA Mat.* 20 (1973), 25-105.

[22] T. Boehme, Two theorems on the differentiation of regular convolution quotients, in "Generalized functions, convergence structures and their applications", Proc. Int. Conf., Dubrovnik, Plenum Press, New York (1988).

[23] B. Stanković, Asymptotics of regular convolution quotients, *Internat. J. Math. Math. Sci.* (to appear).

DISTRIBUTIONAL FORM INVARIANT LINEAR SYSTEMS

A.K. Tiwari

Department of Mathematics
Regional College of Engineering
Nagpur 440011, India

1. INTRODUCTION

Every physical system defines a relation between its input and output signals. Various physically significant properties of the system linearity, stability, time invariance, form invariance, etc. induce analytical description for that relation. Conversely particular classes of analytical relations characterize the physical properties. Zemanian [7] presented a theory for convolution representation of continuous linear time invariant systems.

The form invariant systems are those systems such that a linear scaling of their input gives rise to a linear scaling of their output. Braccini and Gambardella [1] considered the most general input-output relationship for a linear one dimensional system as

$$g(u) = \int_{-\infty}^{\infty} f(t)\ w\ (t,u)\ dt \qquad (1.1)$$

The notations and terminology of this work follow that of Tiwari [2,3] and Zemanian [5,6,7]. $\mathcal{E}(I)$, $I = (0,\infty)$, is the space of all real or complex valued smooth functions defined on I. $\mathcal{E}'(I)$ is topological dual of $\mathcal{E}(I)$. $\mathcal{D}(I)$ is the space of all real or complex valued smooth functions with support in I. $\mathcal{D}'(I)$ is the topological dual of $\mathcal{D}(I)$.

A function $\varphi(t)$ defined on $-\infty < t < \infty$ is said to be a

Generalized Functions and Their Applications, Edited by
R.S. Pathak, Plenum Press, New York, 1993

member of $\mathcal{L}_{c,d,\alpha}$ if φ is smooth and

$$\rho_k(\varphi) = \sup_{-\infty<t<\infty} \left| \lambda_{c,d}(t)\, D_t^k\, \varphi(t) \right|$$

$$\leq C_k\, A^k\, K^{k\alpha} \;;\; k = 0,1,2...$$

The constants A and C_k depend on the function φ and

$$\lambda_{c,d}(t) = \begin{cases} e^{ct}, & 0 \leq t < \infty \\ e^{dt}, & -\infty < t < 0. \end{cases}$$

Making the change of variables $x = e^{-t}$ in the definition of the space $\mathcal{L}_{c,d,\alpha}$ we get the space $\mathcal{M}_{c,d,\alpha}$ [3].

The spaces $\mathcal{L}_{c,d}$ and $\mathcal{M}_{c,d}$ are countable union spaces of $\mathcal{L}_{c,d,\alpha_i}$ and $\mathcal{M}_{c,d,\alpha_i}$ respectively.

In this paper we define distributional form invariance property. We prove that form invariance implies that the system has Mellin type convolution representation.

2. THE DISTRIBUTIONAL FORM INVARIENT LINEAR SYSTEMS

Generalizing the input-output relationship [1] for a linear one dimensional system we write the distributional input-output relationship as

$$g(u) = \langle f(t)\,,\, w(t,u) \rangle \tag{2.1}$$

We now generalize the definition given by Braccini and Gambardella and define distributional form invariance.

Let $g_a(u)$ be the output of the system (2.1) when the input is replaced by $f(at)$, $a > 0$. We say that the system is distributional form invariant if and only if

$$g_a(u) = \alpha\, g(\beta\, u)\;; \tag{2.2}$$

$\alpha = \alpha(a)$, $\beta = \beta(a)$ are real functions of a. From (2.1)

$$g_a(u) = \langle f(at),\, w(t,u) \rangle$$

and

$$\alpha\, g(\beta u) = \alpha \langle f(t),\, w(t,\beta u) \rangle.$$

Substituting these values in (2.2) we get

$$< f(at), w(t,u) > = \alpha < f(t), w(t,\beta u) >. \tag{2.3}$$

so that

$$\frac{1}{a} < f(t), w(\frac{t}{a}, u) > = \alpha < f(t), w(t,\beta u) > ,$$

The above equation is true if and only if

$$\frac{1}{a} w(\frac{t}{a}, u) = w(t, \beta u) . \tag{2.4}$$

This eqation is in agreement with the equation (3) of [1].

3. DISTRIBUTIONAL MELLIN TRANSFORM

We define the distributional Mellin transform F(s) of a distribution $f \varepsilon \mathcal{M}'_{c,d,\alpha}$ as

$$F(s) = < f(x) , x^{s-1} > . \tag{3.1}$$

Corresponding to the pair $f , g \varepsilon \mathcal{M}'_{c,d}$, the mellin type convolution product $f \vee g$ is defined by [5]

$$< f \vee g , \theta > = <f(x) < g(u), \theta(x,u) >> \tag{3.2}$$

Also we have from [5] the following :

If $f \varepsilon \mathcal{M}'_{c,d}$ and $g \varepsilon \mathcal{D}(I)$, then, in the sense of equality in $\mathcal{D}'(I)$, $f \vee g$ is a smooth function on I, given by

$$(f \vee g) (u) = <f(x), \frac{1}{x} g(\frac{u}{x}) > \tag{3.3}$$

To develop representation of form invariant system we need the following theorems.

Theorem 3.1. The scaling operator $S_a : \theta(x) \longrightarrow \theta(ax)$, a>o is a topological automorphism on the space $\mathcal{M}_{c,d,\alpha}$.

Theorem 3.2 : The mapping $\theta(x) \longrightarrow \phi(t)$, $\phi(t) = e^{-t}\theta(e^{-t})$, $x = e^{-t}$, is an isomorphism from $\mathcal{M}_{c,d,\alpha}$ on to $\mathcal{L}_{c,d,\alpha}$.

Proofs of above theorems are easy and hence omitted .

In the next theorem we prove that the operator $w\oplus$ commutes with the scaling oprator S_a.

Theorem 3.3. If $a > o$, $w \varepsilon \mathcal{D}'(I)$ and $f \varepsilon \mathcal{E}'(I)$, then

$$S_a(w\oplus f) = w\oplus (S_a f)$$

Proof. Since $S_a : \theta(x) \longrightarrow \theta(ax)$ is an automorphism on $\mathcal{D}(I)$ we have

$$<S_a(w\oplus f), \theta(ux) > = <w\oplus f, \frac{1}{a}\theta(\frac{ux}{a}) >$$

$$= <w(x) < f(u), \frac{1}{a}\theta(\frac{ux}{a}) >> \qquad (3.4)$$

Also,

$$<w\oplus (S_a f), \theta(tx) >$$

$$= <w(x) <S_a f(t), \theta(tx) >>$$

$$= <w(x) < \frac{1}{a} f(t), \frac{1}{a}\theta(t\frac{x}{a}) >> \qquad (3.5)$$

From (3.4) and (3.5) our theorem is proved.

Definition. A single valued, continuous linear operator $\mathcal{R}$ defined on $\mathcal{E}'(I)$ to $\mathcal{D}'(I)$ is form invariant if $\mathcal{R}$ commutes with scaling operator S_a.

We use the following result of Zemanian [6].

If $\mathcal{R}$ is a single valued continuous linear time invariant mapping of $\mathcal{E}$ into $\mathcal{D}'$ then there exists a unique $w\varepsilon\mathcal{D}'$ such that $\mathcal{R}= w\,*$. Noting the fact that the change of variable $x= e^{-t}$ changes the space $\mathcal{D}(\mathcal{R})$ into $\mathcal{D}(I)$ we obtain the following represention theorem.

Theorem 3.4 . Let the system be described by $g=\mathcal{R}f$. If $\mathcal{R}$ is a single valued continuous linear form invariant mapping of $\mathcal{E}'(I)$ into $\mathcal{D}'(I)$ then there exists a unique $w \varepsilon \mathcal{D}'(I)$ such that

$$R = w\oplus \text{ ie. } \mathcal{R}f = w\oplus f \qquad (3.6)$$

Thus every form invariant system has Mellin type convolution representation.

REFERENCES

1. C.Braccini and G. Gambardella, Form invariant Linear Filtering and Applications, *IEEE Trans. Acco. Speech Signal Proce. ASSP*-34; 1612-28 (1986)
2. A.K.Tiwari, A Distributional Generalized stieltjes transformation. *Indian J. Pure Appl. Math.* 11 (8): 1045-1054(1980)
3. A.K.Tiwari, on a Distributional Generalized Laplace Transform, *Indian. Pure App. Math.* 11 (12): 1609-1616 (1980).
4. A.H.Zemanian, "Distribution Theory and Transform Analysis", McGraw-Hill, New York (1965).
5. A.H.Zemanian, "Generalized Integral Transformations", Wiley Interscience, New York (1968).
6. A.H.Zemanian, A Frequency-Domain Characterisation for the Casuality of Active Linear Systems, *IEEE Trans. Cricuit Theory*, 263-265 (1968).
7. A.H.Zemanian. The Hilbert Port, *SIAM J. Appl. Math.*18 98-138(1970).

ON CERTAIN TOTAL BIORTHOGONAL SYSTEMS IN BANACH SPACES

M. Valdivia(*)

Facultad de Matematicas
Dr. Moliner 50
46100 Burjassot (Valencia)
Spain

SUMMARY

Let Y be an infinite dimensional closed subspace of the Banach space X such that the density character of Y is greater or equal than the density character of X^* for the weak-star topology. If the closed unit ball of Y^* is a weak-star Corson compact then there is a total biorthogonal system $(x_i, \mu_i)_{i \in I}$ for X such that $[x_i : i \in I] = Y$.

1. INTRODUCTION

All vecctor spaces to be used in this paper are assumed to be defined over the field $\mathbb{R}$ of real numbers. We denote by $\mathbb{N}$ the set of all positive integers.

Whenever no confusion occurs $\|.\|$ will represent the norm of a normed space X; B(X) is the closed unit ball of X. We write X^* for the Banach space conjugate of X. If A is a subset of X^*, A_σ denotes such a subset endowed with the topology induced by the week-star topology of X^*; $A_\perp$ is the subspace of X orthogonal to A. If M is a subset of X, then lin M is the linear span of M and $M^\perp$ the subspace of X^* orthogonal to M. The usual duality between X and X^* is represented by $< . , . >$, i.e.if $x \in X$ and $\mu \in X^*$, $<x,\mu>=\mu(x)$. For two given closed subspaces X_1 and X_2 of X, we say that X_1 is an orthogonal complement of X_2 in X_1+X_2 whenever $X_1 \cap X_2=\{0\}$

(*) Suported in part by D G I C Y T

Generalized Functions and Their Applications, Edited by
R.S. Pathak, Plenum Press, New York, 1993

and the norm of the projection of $X_1 + X_2$ onto X_2 along X_1 has value one.

If A is a set, $|A|$ stands for the cardinal number of A. We shall denote by $|\alpha|$ the cardinal of the ordinal number α; ω is the first infinite ordinal.

The density character of a topological space M is the smallest cardinal number λ such that there is a dense subset A of M with $|A| = \lambda$. We then write dens M = λ.

If $(x_i)_{i \in I}$ is a family of vectors in the Banach space X, $[x_i : i \in I]$ is the closure of lin $\{x_i : i \in I\}$. A biorthogonal system for X,

$$(x_i, \mu_i)_{i \in I}, \quad x_i \in X, \quad \mu_i \in X^*, \quad i \in I,$$

is said to be total whenever lin $\{\mu_i : i \in I\}$ is dense in X^*_σ. If $[x_i : i \in I]$ coincides with X, then such system is said to be complete. When a biorthogonal system $(x_i, \mu_i)_{i \in I}$ is both complete and total it is called a Markushevich basis.

For a given set I, $\Sigma^{(I)}$ denotes the subspace of the topological space $\mathbb{R}^I$ formed by all elements $(x_i : i \in I)$ such that the set $\{ i \in I : x_i \neq 0\}$ is countable. A compact topological space homeomorphic to a subspace of $\Sigma^{(I)}$ is said to be a Corson compact.

We say that a Banach space X is class Σ if there is a continuous injective mapping from X^*_σ into $\Sigma^{(I)}$ for some set I. It is observed in [2] that the proofs given in [3] remain valid if we replace "X is of class Σ" by "$B(X^*_\sigma)$ is a Corson compact", which enables the construction, by means of projective resolution of the identity operator, of Markushevich basis in a Banach space X with $B(X^*)_\sigma$ a Corson compact. In this paper, we shall use some of the methods described in [3] to obtain certain total biorthogonal systems in Banach spaces.

2. TOTAL BIORTHOGONAL SYSTEMS

Let X be a Banach space. Let E and F be two subspaces of X and X^*, respectively. F is said to be 1-norming for E if for a given x in E we have

$$\| x \| = \sup \{ |\langle x, \mu \rangle| : \mu \in B(X^*) \cap F\}.$$

The following two results, to be used later are easy to

show :

(a) Let Y be an infinite dimensional separable, closed subspace of the Banach space X. If X^*_σ is separable then there is a total biorthogonal system $(x_n, \mu_n)_{n \in \mathbb{N}}$ for X such that $[x_n : n \in \mathbb{N}] = Y$.

(b) Let X be a Banach space. If λ is an infinite cardinal number, λ < dens X, then there is an infinite dimensional closed subspace Y of X with dens Y = λ.

For the next propositions , X will be a Banach space and Y an infinite dimensional closed subspace of X. We assume there is a continuous injective mapping φ from $B(Y^*)_\sigma$ into $\Sigma^{(J)}$ for some set J, which takes the origin of Y^* to the point of $\Sigma^{(J)}$ whose coordinates are all zero. For j in J, we denote by f_j the projection of $\Sigma^{(J)}$ onto its jth coordinate; we define $e_j = f_j \circ \varphi$. Let T be the linear algebra over $\mathbb{R}$ of all continuous real valued functions defied in $B(Y^*)_\sigma$ which reprsent the origin by zero. We provide T with the norm

$$\| f \| := \sup \{|f(\mu)| : \mu \in B(Y^*)\}, \ f \in T.$$

Let S be the subalgebra of T generated by the restrictions to $B(Y^*)$ of the elements of Y. We know, by Stone -Weierstrass theorem, that S is dense in T, and thus, for each j in J, we may select a countable subset Y_j of Y such that the restrictions of its elements to $B(Y^*)$ generate a subalgebra of T with e_j in its closure.

Proposition 1. Let E and F be two subspaces of X and X^*_σ respectively, satisfying the following conditions :

(i) E is closed and contained in Y.

(ii) F is 1-norming for E.

(iii) If j is an element of J for which there is μ_j in $F \cap B(X^*)$ with $e_j(\mu_j |_Y) \neq 0$, then Y_j is contained in E.

We then have the following properties :

(a) $F_\perp$ is an orthogonal complement of E in E + $F_\perp$, and the latter subspace contains Y.

(b) Let G be the closure of F in X^*_σ. If j is an element of J for which there is v_j in $G \cap B(X^*)$ with $e_j(v_j |_Y) \neq 0$, then Y_j is contained in E.

Proof. We take two elements x and z of E and $F_\perp$, respectively. Given $\varepsilon > 0$, there is $v \in F \cap B\{X^*\}$ such that $<x,v> > \|x\| - \varepsilon$. Then

$$\|x\| < <x,v> + \varepsilon = < x + z , v > + \varepsilon \leq \|x+z\| + \varepsilon$$

and hence

$$\|x\| \leq \|x + z\| ,$$

thus we have $E \cap F_\perp = \{0\}$ and the projection of $E + F_\perp$ onto E along $F_\perp$ has norm one. It follows that $F_\perp$ is an orthogonal complement of E in $E + F_\perp$.

For a given μ in X^*, we write $\Lambda(\mu) := \mu|_Y \in Y^*$. Hahn Banach theorem guarantees that $\Lambda : X^* \longrightarrow Y^*$ is an onto mapping. Let B denote the closure of $F \cap B(X^*)$ in X^*_σ. Let M be the linear hull in Y^*_σ of $\Lambda(B)$. let ψ be the mapping from M to E^* which assigns to each element of M its restriction to E. For every non-zero element w of M, we find $\lambda > 0$ such that

$$\lambda w \in \Lambda(B) \subset B(Y^*).$$

Select an element j of J such that $<e_j, \lambda w> \neq 0$. Then there is μ_j in $F \cap B(X^*)$ with $<e_j, \Lambda(\mu_j)> \neq 0$, hence we conclude that Y_j is contained in E. We may now find an element y of Y_j with $< y, \lambda w > \neq 0$. Thus, $<y,w> \neq 0$ and ψ is injective. Suppose now that $\Lambda(B)$ is different from $M \cap B(Y^*)$. We choose an element w_o in this set not belonging to $\Lambda(B)$. Thus, in E^*_σ, $\psi(w_o)$ does not belong to the compact absolutely convex set $\psi(\Lambda(B))$. Hahn-Banach theorem yields x in E such that

$$<x, w_o> = 1, \sup \{|<x, \mu>| : \mu \in B \} = \delta < 1.$$

We find v in $F \cap B(X^*)$ so that $<x,v> > \|x\| - (1-\delta)$. Hence, we have

$$\delta \geq | <x, v> | > \|x\| - (1-\delta) \geq (x, w_0> - (1-\delta) = \delta,$$

a contradiction. Therefore, $\Lambda(B) = M \cap B(Y^*)$. Krein-Smulian theorem, [1, p.246], assures that M is closed in Y^*_σ. As a consequence of this $\Lambda(G)$ coincides with M.

Suppose now that the Banach space $E + F_\perp$ does not contain Y. There exists an element μ in x^* not vanishing in Y and having zero value for the points of $E + F_\perp$. Then, μ is in G and $\Lambda(\mu) \neq 0$. Since ψ is injective, we have that $\psi(\Lambda(\mu))$, which coincides with $\mu|_E$, is non-zero, again a contradiction. We have thus shown property (a).

Let j be an element of J for which there is v_j in $G \cap \Lambda B(X^*)$ with $e_j(v_j|_Y) \neq o$. Since $\Lambda(G \cap B(X^*)) = B(Y^*) = \Lambda(B)$, we may find μ_j in $F \cap B(X^*)$ such that $e_j(\mu_j|_Y) \neq 0$. Then, Y_j is contained in E and propoerty (b) is achieved.

q.e.d.

Proposition 2. Let A_0 and B_0 be two infinite subsets of Y and $B(X^*)$, respectively, and let λ be a cardinal number such that $|A_0| \leq \lambda$, $|B_0| \leq \lambda$. There exists then two subspaces E and F of X and X^*_σ, respectively, satisfying the folowing conditions :

(a) E is closed, $E \subset Y$, $E \supset A_0$, dens $E \leq \lambda$.

(b) F is 1-norming for E, $F \supset B_0$, dens $F \leq \lambda$.

(c) If j is an element of J for which there is μ_j in $F \cap B(X^*)$ with $e_j(\mu_j|_Y) \neq 0$, then Y_j is contained in E.

Proof. For each x in Y, we select an element v_x in X^* such that

$$\| v_x \| = 1 \text{ and } \| x \| = \langle x, v_x \rangle.$$

We proceed by recurrence and assume that, for a non-negative integer m, we have found

$$A_m \subset Y,\ B_m \subset B(X^*),\ |A_m| \leq \lambda,\ |B_m| \leq \lambda.$$

Denote by C_m and D_m the linear spans over the rationals of A_m and B_m, respectively, We set

$$B_{m+1} : (D_m \cap B(X^*)) \cup \{v_x : x \in C_m\}.$$

Let J_m be the subset of J defined by the following condition :

$j \in J_m$ if and only if there is an element μ in B_{m+1} with $e_j(\mu|_Y) \neq 0$. We then write

$$A_{m+1} := C_m \cup (\cup \{ Y_j : j \in J_m\}.$$

Let E be the closed linear hull of $\overset{\infty}{\underset{m=0}{\cup}} A_m$ in X. Let F be the linear span of $\overset{\infty}{\underset{m=0}{\cup}} B_m$ in X^*_σ. Clearly, (a) is satisfied, $F \supset B_0$ and dens $F \leq \lambda$. Take any element x of E and $\varepsilon > 0$. We get hold of a positive integer m and z in A_m such that $\|x-z\| < \frac{1}{2}\varepsilon$. We have

$$v_z \in B_{m+1} \subset F \cap B(X^*),\ \langle x, v_z \rangle = \langle z, v_z \rangle - \langle z-x, v_z \rangle$$

$$\geq \| z \| - \| z-x \| \geq \| x \| - 2 \| z-x \| > \| x \| - \varepsilon$$

and thus F is 1-norming for E. Condition (b) is then fulfilled. Finally, let j be an element of J such that there is μ_j in $F \cap B(X^*)$ with $e_j(\mu_j|_Y) \neq 0$. There are a positive integer m and μ in B_{m+1} such that $e_j(\mu_j|_Y) \neq 0$. Then j is in J_m, hence $Y_j \subset A_{m+1} \subset E$ and we have shown (c).

Proposition 3. If dens $Y \geq$ dens X^*_σ, then dens $Y =$ dens X^*_σ.

Proof. Assume $\lambda :=$ dens $X^*_\sigma <$ dens Y. We take an infinite subset A_0 of Y with $|A_0| \leq \lambda$, and a dense subset B_0 of X^*_σ such that $|B_0| = \lambda$. We apply the former proposition to obtain E and F with the properties there stated. Proposition 1 tells us that $E + F_\perp$ contains Y and, since dens $E \leq \lambda <$ dens Y, it follows $F_\perp \neq \{0\}$. On the other hand, $F \supset B_0$ and consequently $F_\perp = \{0\}$, an obvious contradiction. q.e.d.

Let μ denote the first ordinal number of dens Y. We write

$$\{y_\eta : 0 \leq \eta < \mu\} \text{ and } \{v_\eta : 0 \leq \eta < \mu\}$$

for two dense subsets of Y and X^*_σ, respectively. We represent by the family of all ordered pairs

$$((x_i, \mu_i)_{i\in I}, \alpha)$$

satisfying the following conditions :

1. $(x_i, \mu_i)_{i\in I}$ is a biorthogonal system for X such that $x_i \in Y$, $i \in I$.
2. α is an infinite ordinal, $\alpha \leq \mu$, such that $|\alpha| = |I|$, $y_\eta \in [x_i : i \in I]$, $0 \leq \eta < \alpha$, and if G_I is the closure of lin $\{\mu_i : i \in I\}$ in X^*_σ, $v_\eta \in G_I$, $0 \leq \eta < \alpha$.
3. G_I is 1-norming for $[x_i : i \in I]$.
4. If j is an element of J for which there is μ_j in $G_I \cap B(X^*)$ with $e_j(\mu_j|_Y) \neq 0$, then Y_j is contained in $[x_i :\in I]$.

Proportion 4. $\mathcal{H}$ is not empty.

Proof. We take two countably infinite subsets A_0 and B_0 of Y and X^*, respectively, such that lin A_0 is of infinite dimension and

$$y_\eta \in A_0,\ v_\eta \in B_0,\ 0 \leq \eta < \omega .$$

We apply Proposition 2 to obtain E and F with the properties there stated. Let G be the closure of F in X^*_σ. Let ψ be the canonical mapping from X onto $X/F_\perp$. We identify, in the usual way, $(X/F_\perp)^*$ with G. In light of Proposition 1, $F_\perp$ is an orthogonal complement of E in $E + F_\perp$ and thus $\psi(E)$ is a subspace of $X/F_\perp$ which is isomorphic to E. We then have the following situation : $\psi(E)$ is an infinite dimensional separable closed subspace of $X/F_\perp$ and G is separable. Making use of result (a) we obtain a total biorthogonal system $(y_n,\mu_n)_{n\in\mathbb{N}}$ for $X/F_\perp$ such that $[y_n : n \in \mathbb{N}] = \psi(E)$. Take now x_n in E with $\psi(x_n) = y_n$, n = 1,2... Then, $(x_n,\mu_n)_{n\in\mathbb{N}}$ is a biorthogonal system for X with $[x_n : n \in \mathbb{N}] = E$ and lin $\{\mu_n : n \in \mathbb{N}\}$ is dense in G. It is now immediate that $((x_n,\mu_n)_{n\in\mathbb{N}}, \omega)$ belongs to $\mathcal{H}$.

q.e.d.

For two given elements of $\mathcal{H}$ we put

$$((x_i,\mu_i)_{i\in I},\alpha) < ((y_j,v_j),\beta)$$

whenever $\{(x_i,\mu_i): i \in I\}$ is strictly contained in $\{y_j : j \in J\}$ and $\alpha < \beta$.

Proposition 5. $(\mathcal{H}, \leq)$ is an inductive ordered set.

Proof. Clearly, $(\mathcal{H},\leq)$ is an ordered set. We consider a linearly ordered non-empty subset $\mathcal{L}$ of $\mathcal{H}$. We write $\{(x_k,\mu_k) : k \in K\}$ and γ for the union of the sets $\{(x_i,\mu_i) : i \in I\}$ and the supremum of all ordinals α, respectively, when $((x_i,\mu_i)_{i\in I},\alpha)$ ranges over $\mathcal{L}$.

We then have that $(x_k,\mu_k)_{k\in K}$ is a biorthogonal system for X such that $x_k \in Y$, $k \in K$. Also, γ is an infinite ordinal, $\gamma \leq \mu$, $|\gamma| = |K|$,

$$y_\eta \in [x_k : k \in K],\ 0 \leq \eta < \gamma,$$

and if G_K is the closure of lin $\{\mu_k : k \in K\}$ in X^*_σ ,

$$v_\eta \in G_{K'}\ 0 \leq \eta < \gamma.$$

We write G_I for the closure of lin $\{\mu_i : i \in I\}$ in X^*_σ, when $((x_i,\mu_i)_{i\in I},\alpha) \in \mathcal{L}$. It is plain that G_K coincides with the closure in X^*_σ of

$$F_K := \bigcup \{ G_I : ((x_i,\mu_i)_{i\in I},\alpha) \in \mathcal{L} \}.$$

Given x in $[x_k : k \in K]$ and $\varepsilon > 0$, we may determine $((x_i,\mu_i)_{i\in I},\alpha)$ in $\mathcal{L}$ and z in $[x_i : i \in I]$ so that $\|x-z\| < \varepsilon/4$. We take now v in $G_I \cap B(X^*)$ such that $\langle z,v\rangle > \|z\| - \varepsilon/2$. Then

$$\langle x,v\rangle = \langle z,v\rangle - \langle z-x,v\rangle \geq \| z \| - \frac{1}{2}\varepsilon - \| z-x \|$$

$$= \|x+z-x\| - \frac{1}{2}\varepsilon - \|z-x\| \geq \|x\| - \frac{1}{2}\varepsilon - 2\|z-x\| > \|x\| - \varepsilon$$

and thus F_K, as well as G_K, is 1-norming for $[x_k : k \in K]$.

If j is an element of J for which there is μ_j in $F_k \cap B(X^*)$ with $e_j(\mu_j|_Y) \neq 0$, we may take $((x_i,\mu_i)_{i\in I},\alpha)$ in $\mathcal{L}$ such that μ_j belongs to G_I. Hence,

$$Y_j \subset [x_i : i\in I] \subset [x_k : k \in K].$$

By applying Proposition 1, we have that if j is an element of J for which there is v_j in $G_K \cap B(X^*)$ with $e_j(v_j|_Y) \neq 0$, then Y_j is contained in $[x_k : k \in K]$.

Thereby we conclude that $((x_k,\mu_k)_{k\in K},\gamma)$ is an upper bound of $(\mathcal{L},\leq)$ in $(\mathcal{H},\leq)$.

q.e.d.

Theorem. Let Y be an infinite dimensional closed subspace of a Banach space X such that $B(Y^*)_\sigma$ is a Corson compact. If dens $Y \geq$ dens X^*_σ, then there is a total biorthogonal system $(x_i,\mu_i)_{i\in I}$ for X such that $[x_i : i\in I] = Y$.

Proof. We proceed by transfinity induction over the density of Y. Clearly, if Y is separable it suffices to apply result a) to reach the conclusion. Suppose now that Y is not separable and that for each Banach space W with an infinite dimensional closed subspace V, with dens $V \geq$ dens W^*_σ, $B(V^*)_\sigma$ a Corson compact and dens $v <$ dens Y, there is a total biorthogonal system $(\psi_h,v_h)_{h\in H}$ for W such that $[\psi_h : h\in H] = V$. Let μ denote the first ordinal number of dens Y. We define φ and $(\mathcal{H},\leq)$ as before. The former proposition yields a maximal element $((x_i,\mu_i)_{i\in I}, \alpha)$ of $(\mathcal{H},\leq)$. Assume $\alpha < \mu$. Then, $|\alpha| = |I| < |\mu|$. In light of Proposition 1, $Y \cap \{\mu_i : i \in I\}_\perp$ is an orthogonal complement of $[x_i : i\in I]$ in Y. Using result (b), we may find a subspace H of $Y \cap \{\mu_i : i\in I\}_\perp$ such that dens $H = |\alpha|$. Let A be a dense subset of H with $|A| = |\alpha|$. We apply proposition 2 for the particular case of

$$A_0 := \{ x_i : i \in I \} \cup A \cup \{ y_\alpha \},$$

$$B_0 := \{ \mu_i : i \in I \} \cup \{ v_\alpha \}$$

and $\lambda := |\alpha|$. We determine two subspaces E and F of X and X^*_σ, respectively, with properties (a), (b) and (c) there mentioned. Let P be the closure of lin $\{\mu_i : i\in I\}$ in X^*_σ. By Proposition 1, we have that $Y \cap F_\perp$ is an orthogonal complement of E in Y and hence it is clear that $B(E^*)_\sigma$ is a Corson compact. Let Φ denote the canonical mapping from X onto $X/F_\perp$. If $L := \Phi(E)$, then L is isometric to E and thus $B(L^*)_\sigma$ is also a Corson compact. Let G be the closure of F in X^*_σ. We identify in the usual manner G with the topological dual of $X/F_\perp$. Let M be the subspace of $X/F_\perp$ orthogonal to P. Then, $M \cap L$ is a topological complement of $\Phi([x_i : i \in I])$ in L, whence $B((M \cap L)^*_\sigma$ is a Corson compact. The situation is then as follows : M is a Banach space whose topological dual is G/P ; $M \cap L$ is a closed subspace of M with $B((M \cap L)^*)_\sigma$ a Corson compact such that

$$|\alpha| = \text{dens}\,(M \cap L) \geq \text{dens}\,(G/P)\cdot$$

Hence, there is a total biorthogonal system $(t_j, w_j)_{j\in J_1}$ for M, J_1 disjoint from I, such that

$$[\, t_j : j \in J_1 \,] = M \cap L.$$

Let ψ be the canonical mapping from G onto G/P. Since $\{\mu_i : i\in I\}_\perp$ is an orthogonal complement of $[x_i : i \in I]$ in $[x_i : i \in I] + \{\mu_i : i \in I\}_\perp$, it follows that $P+[x_i : i\in I]^\perp = X^*$, and then $\psi < G \cap [\, x_i : i \in I \,]^\perp = G/P$. For each j in J_1 we choose x_j in $E \cap \{\mu_i : i\in I\}_\perp$ and μ_j in $G\cap[x_i : i\in I]^\perp$ such that $\Phi(x_j) = t_j$ and $\psi(\mu_j) = W_j$. Set K : $I \cup J_1$. Then, $(x_i, \mu_i)_{i\in K}$ is a biorthogonal system for X such that $[x_i : i\in K] = E$ and lin $\{\mu_i : i \in K\}$ is dense in G. It is now easy to verify that $((x_i, \mu_i)_{i\in I}, \alpha+1)$ belongs to $(\mathcal{H}, \leq)$ and is strictly greater than $((x_i, \mu_i)_{i\in I}, \alpha)$. We have thus attained a contradiction. Hence, $\alpha=\mu$, and we conclude that $(x_i, \mu_i)_{i\in I}$ has the required properties.

q.e.d.

REFERENCES

1. J.Horvath : Topological Vector Spaces and Distributions I. Addison-Wesley, Massachussets-London, 1966.
2. J.Orihuela and M. Valdivia : Resolution of identity and first Baire selection in Banach spaces. Preprint.
3. M.Valdivia : Resolution of the identity in certain Banach spaces, *Collect. Math.* 39, 127-140 (1988).

ANALYTIC REPRESENTATIONS OF TEMPERED DISTRIBUTIONS USING WAVELETS

Gilbert G. Walter*

Department of Mathematical Sciences
University of Wisconsin-Milwaukee
P.O. Box 413
Milwaukee, WI 53201 USA

INTRODUCTION

The properties of the analytic representations of distributions on the boundary of the unit disk are well understood. Trigonometric Fourier series constitute the principal tool used to construct and analyze them [5].

If we move to the real line, no analogous theory based on orthogonal series seems to exist. Rather Fourier transforms are often used instead. A seemingly natural orthogonal system to use on the real line would be the Hermite functions, which, however they have a number of shortcomings [6] Recently a new category of orthogonal systems has been introduced [3]. These systems are composed of "wavelets", orthonormal functions on $\mathbb{R}$ consisting of dilations and translations of a fixed function. Each tempered distribution can be represented by a series of such wavelets [8].

Their analytic representations satisfy the same defining relationship, in this case a "dilation equation". They have a further advantage that their coefficients are easy to calculate by using the "decomposition" and "reconstruction" algorithms.

In this work we shall first present a short introduction to wavelets on $\mathbb{R}$ and "analytic" and "harmonic" wavelets in a half plane. We shall show that a series of them can be used to define the analytic representation of some tempered distributions.

*Research supported in part by NSF grant #DMS-901526

Generalized Functions and Their Applications, Edited by
R.S. Pathak, Plenum Press, New York, 1993

BACKGROUND

Each $f \in L^2(\mathbb{R})$ has an analytic representation in the form

$$f^{\pm}(z) = \frac{1}{2\pi i}\int_{-\infty}^{\infty} \frac{f(x)}{x - z}\,dx, \qquad \mathcal{I}z \neq 0, \tag{1}$$

i.e. a pair of functions one of which is analytic in the upper and the other in the lower half plane (corresponding to + and - respectively). It has the property that

$$f^{+}(x + i0) - f^{-}(x - i0) = f(x) \tag{2}$$

where the implied limit on the left is in the sense of $L^2(\mathbb{R})$ [1]. The function given by

$$h(x,y) = f^{+}(x + iy) - f^{-}(x - iy)$$

is harmonic in the upper half plane with boundary values on the real axis given by $u(x,0) = f(x)$.

These formulae can be extended to various classes of generalized functions on the real axis. For tempered distributions with compact support, $f^{\pm}(z)$ is defined as

$$f^{\pm}(z) = \frac{1}{2\pi i} \langle f(x), \frac{\theta(x)}{x - z}\rangle \tag{3}$$

where $\theta(x)$ is equal to 1 on the support of f and belongs to S. If f is in a Sobolev space, then its Fourier transform $\hat{f}(w)$ is of polynomial growth and is locally an L^2 function. Hence its analytic representation may be defined as

$$\begin{aligned} f^{+}(z) &= \frac{1}{2\pi}\int_{0}^{\infty} \hat{f}(w)e^{iwz}dw, \qquad \mathcal{I}z > 0, \\ f^{-}(z) &= -\frac{1}{2\pi}\int_{0}^{\infty} \hat{f}(-w)e^{-iwz}dw, \quad \mathcal{I}z < 0 \end{aligned} \tag{4}$$

This is turn may be used to obtain the analytical representation for $f \in S'$ by multiplication by a polynomial.

The operator which maps a distribution into its analytic representation is linear, and commutes with certain other operators. These include the derivative operator, the dilation operator, and the translation operator.

The theory of orthonormal wavelet bases of $L^2(\mathbb{R})$ may be found in a number of places. A detailed readable introduction may be found in [2] while a more complete development in $\mathbb{R}^n$ is found in [3]. Here we merely cover a few of the concepts of the theory that we shall use.

The wavelet basis is composed of functions $\psi_{mn}(t)$ given by

$$\psi_{mn}(t) = 2^{m/2}\psi(2^m t - n), \qquad m,m \in \mathbb{Z}$$

where ψ is a fixed function in $L^2(\mathbb{R})$, the so-called "mother wavelet". Such an orthonormal basis consisting of dilations and translations is difficult to construct. This construction is usually based on another function $\phi(t)$, the "scaling function". This ϕ belongs to the space S_r of real valued functions on $\mathbb{R}$ which are r times differentiable and whose derivatives are continuous and rapidly decreasing. That is, ϕ satisfies

$$|\phi^{(k)}(t)| \le C_{pk}(1 + |t|)^{-p}, \qquad k = 0,1,\ldots,r; \qquad p \in \mathbb{Z}, \quad t \in \mathbb{R}. \tag{5}$$

In order to qualify as a scaling function, ϕ must be associated with a multiresolution analysis of $L^2(\mathbb{R})$, i.e. a nested sequence of closed subspaces $\{V_m\}_{m\in\mathbb{Z}}$ such that

(i) $\{\phi(t - n)\}$ is an orthonormal basis of V_0

(ii) $\cdots \subset V_{-1} \subset V_0 \subset V_1 \subset \cdots \subset L^2(\mathbb{R})$

(iii) $f \in V_m \Leftrightarrow f(2\cdot) \in V_{m+1}$ (6)

(iv) $\bigcap_m V_m = \{0\}, \quad \overline{\bigcup_m V_m} = L^2(\mathbb{R}).$

Clearly $\{\sqrt{2}\phi(2t - n)\}$ is an orthonormal basis for V_1 by (i) and (ii). Since $\phi \in V_1$, it must have an expansion

$$\phi(t) = \sum_k c_k\sqrt{2}\phi(2t - k), \qquad \{c_k\} \in \ell^2, \qquad t \in \mathbb{R}. \tag{7}$$

This is the "dilation equation" for ϕ. We shall assume that $c_k = O(|k|^{-m})$ for each $m \in \mathbb{N}$.

The multiresolution analysis can be extended from $L^2(\mathbb{R})$ to S'_r [8]. We define the subspace T_0 of S'_r to be composed of all functions on $\mathbb{R}$ given by

$$f(t) = \sum_n a_n\varphi(t-n) \qquad a_n = O(|n|^k) \text{ for some } k \in \mathbb{Z}.$$

Since $\varphi \in S_r$, this series converges uniformly on compact sets to a continuous function of polynomial growth. It is shown in [8] that

$$\cdots \subset T_{-1} \subset T_0 \subset T_1 \subset \cdots \subset T_m \subset \cdots S'_r$$

$$\bigcap_m T_m = \{0\}, \quad \overline{\bigcup_m T_m} = S'_r$$

and that each $f \in S'_r$ has a wavelet expansion

$$f(t) = \sum_n a_n \varphi(t-n) + \sum_{m=0}^{\infty} \sum_n b_{mn} \psi_{mn}(t)$$

convergent in the sense of S'_r. The first series is the projection of f in S'_r onto T_0. This may also be given by

$$(P_0 f)(t) = \langle f, q(\cdot,t)\rangle,$$

where q(x,t) is the reproducing kernel of V_0 given by

$$q(x,t) = \sum_n \varphi(x-n)\varphi(t-n).$$

Similarly we have

$$(P_m f)(t) = \langle f, q_m(\cdot,t)\rangle,$$

where $q_m(x,t) = 2^m q(2^m x, 2^m t)$.

Example 1. A simple example which however does not satisfy (3) is given by $\phi(t) = \chi_{[0,1]}(t)$, the characteristic function of the unit interval. Its translates $\{\phi(t-n)\}$ are orthonormal and V_0 is composed of piecewise constant functions.

Example 2. The *Franklin wavelets* correspond to a V_0 composed of continuous piecewise linear functions with breaks at the integers. Its scaling function ϕ, in terms of the Fourier transform is

$$\hat{\phi}(w) = \frac{\sin^2 w/2}{(w/2)^2} \left\{1 - \frac{2}{3} \sin^2 w/2\right\}^{-1/2}. \tag{8}$$

Example 3. The *Daubechies wavelets* are based on the dilation equation. We look for $\{c_k\}$ in (7) such that the orthogonality condition is satisfied. This is again given in terms of the Fourier transform

$$\hat{\phi}(w) = \frac{1}{\sqrt{2}} \sum c_k e^{-ikw/2} \hat{\phi}\left(\frac{w}{2}\right) = m_0\left(\frac{w}{2}\right)\hat{\phi}\left(\frac{w}{2}\right) \tag{9}$$

which translates the orthogonality condition into

$$\left|m_0\left(\frac{w}{2}\right)\right|^2 + \left|m_0\left(\frac{w}{2} + \pi\right)\right|^2 = 1. \tag{10}$$

Here m_0 is a 2π periodic function with m(0) = 1. If c_k is chosen by

$$c_0 = (1 + \sqrt{3})/2\sqrt{2},$$

$$c_1 = (3 + \sqrt{3})/2\sqrt{2},$$

and

$$c_0 + c_2 = c_1 + c_3 = 1/\sqrt{2},$$

(and all the other $c_k = 0$), then (3.6) will be satisfied [2, p. 946].

This does not directly give $\hat{\phi}(w)$ or $\phi(t)$ which rather must be found recursively by starting with, say, $\phi_0(t) = \chi_{[0,1]}(t)$ and then defining

$$\phi_n(t) = \sum_k c_k\sqrt{2}\phi_{n-1}(2t - k).$$

Once we have the scaling function $\phi(t)$, we may use it to construct the "mother wavelet" $\psi(t)$. It must be chosen such that $\{\psi(t-n)\}$ is an orthonormal basis of the space W_0, given by the orthogonal complement of V_0 in V_1. Then

$$V_1 = V_0 \oplus W_0.$$

If such a $\psi(t)$ can be found, then $2^{m/2}\psi(2^{m/2}t-n) = \psi_{nm}(t)$ is an orthonormal basis of W_m, the dilation space of W_0. Indeed from (6) it follows that

$$\bigoplus_{m\in Z} W_m = L^2(\mathbb{R})$$

and hence $\{\psi_{n,m}\}_{n,m\in Z}$ is an orthonormal basis of $L^2(\mathbb{R})$.

The method of finding $\psi(t)$ involving the dilation equation is straightforward; $\psi(t)$ is defined as

$$\psi(t) := \sqrt{2} \sum_k c_{1-k}(-1)^k\phi(2t - k). \tag{11}$$

Then it is merely a matter of checking that the orthogonality conditions are satisfied [2, p. 944].

SERIES OF ANALYTIC WAVELETS

We assume that $\varphi \in S_r$ and hence $\psi \in S_r$ as well. This enables us to find their analytic representations $\varphi^{\pm}(z)$ and $\psi^{\pm}(z)$ by the usual integral formula (1).

Since the operations of dilations and translations commute with the analytic representation, it follows that $\varphi^{\pm}(z)$ satisfies the same dilation equation (7) as φ,

$$\varphi^{\pm}(z) = \sum_k c_k \sqrt{2}\, \varphi^{\pm}(2z-k). \tag{12}$$

The same is true for $\psi^{\pm}(z)$ which by (2.11) must be

$$\psi^{\pm}(z) = \sum_k c_{1-k}(-1)^k \sqrt{2}\, \varphi^{\pm}(2z-k). \tag{13}$$

This enables us to define the "analytic wavelets" $\psi^{\pm}_{mn}(z)$ obtained by translation and dilation of $\psi^{\pm}$ (or equivalently as the analytic representation of ψ_{mn}),

$$\psi^{\pm}_{mn}(z) = 2^{m/2}\psi^{\pm}(2^m z-n), \qquad m, n \in \mathbb{Z}. \tag{14}$$

The reproducing kernel $q_m(x,t)$ of V_m has its analytic representation given by

$$q^{\pm}_m(z,t) = \sum_{n=-\infty}^{\infty} 2^m \varphi^{\pm}(2^m z-n)\varphi^{\pm}(2^m t-n) \tag{15}$$

Since $\{q_m(x,t)\}$ is a quasi-positive delta-sequence [8] on $(-\infty,\infty)$, it follows that

$$q^{\pm}_m(z,t) \to \frac{1}{2\pi i}\frac{1}{t-z} \qquad \text{as } m \to \infty$$

uniformly on bounded sets in $\mathbb{R}\times\mathbb{C}$. (For quasi-positive delta sequences we have, for $f \in L'(\mathbb{R})$,

$$\int_{-\infty}^{\infty} q_m(x,t)f(x)dt \to f(t)$$

uniformly an interior subinterval of intervals of continuity of f. While $\frac{1}{x-z} \notin L'(\mathbb{R})$, it is easy to show the tails of the integral converge to zero, so the statement follows.)

For $f \in S'_r$ we have the wavelet expansion

$$f = \sum_{n=-\infty}^{\infty} a_n \varphi(\cdot-n) + \sum_{m=0}^{\infty}\sum_{n=-\infty}^{\infty} b_{mn}\,\psi_{mn} \tag{16}$$

convergent in the sense of S'_r [8]. The analytic representation represents a continuous map from S'_r to a corresponding space of analytic functions, i.e. if $f_m \to f$ in S'_r then $f^+_m(z) \to f^-(z)$ uniformly on bounded subsets of the upper half plane. This of course requires that the proper analytic representation be chosen. One such choice is as follows: Any $f \in S'_r$ (or S') may be given as a finite order derivative of a continuous function of polynomial growth, say F(x)

$$f = D^k F.$$

An analytic representation of F is given by

$$F^+(z) = \frac{1}{2\pi i}\int_{-\infty}^{\infty} \frac{F(x)}{x-z}\left(\frac{z^2+1}{x^2+1}\right)^P dx$$

where p is sufficiently large to assure that $F(x)(x^2+1)^{-P} \in L^2(\mathbb{R})$. Then $f^+(z)$ is defined by

$$f^+(z) = D^k_z F^+(z) \tag{17}$$

The approximation to f from the space T_m is the function

$$f_m(x) = \langle f, q_m(\cdot, x)\rangle.$$

Thus if the analytic representative is taken as above,

$$f_m^+(z) \to f^+(z).$$

We may also express f_m as

$$f_m = f_0 + f_m - f_0 = f_0 + \sum_{k=0}^{m-1} \sum_{n=-\infty}^{\infty} b_{kn} \psi_{kn}.$$

Thus if the inner series converges we have

$$f_m^+(z) - f_0^+(z) = \sum_{k=0}^{m-1} \sum_{n=-\infty}^{\infty} b_{kn} \psi_{kn}^+(z) + g_m(z) \tag{18}$$

where $g_m(z)$ is an entire function.

Lemma 1. Let $b_n = O(|n|^{r-\epsilon})$; $\epsilon > 0$; then

$$\sum_{n=-\infty}^{\infty} b_n \psi^+(z-n)$$

converges uniformly on compact subsets of the open upper half plane.

The proof is based on the fact ([3], p. 68) that

$$\int_{-\infty}^{\infty} x^p \psi(x)dx = 0, \qquad p = 0,1, \cdots, r.$$

Then

$$z^p \psi^+(z) = \frac{1}{2\pi i}\int_{-\infty}^{\infty} \frac{z^p - x^p}{x-z} \psi(x)dx + \frac{1}{2\pi i}\int_{-\infty}^{\infty} \frac{x^p \psi(x)}{x-z} dx = \frac{1}{2\pi i}\int_{-\infty}^{\infty} \frac{x^p \psi(x)}{x-z} dx \tag{19}$$

holds for $p \le r + 1$. Hence $|z^{r+1} \psi^+(z)|$ is uniformly bounded for $\mathrm{Im}\, z \ge \epsilon > 0$, and the conclusion follows.

Thus if the coefficients in (18) satisfy the hypothesis, the series converges.

Lemma 2. Let $f \in S_p'$, $\varphi \in S_r$, $p < r$; then the wavelet coefficients of f satisfy

$$|b_{mn}| = |\langle f, \psi_{mn}\rangle| \le C\,(|n| + 2^m)^{p+\frac{1}{2}} \tag{20}$$

We use the fact that each $f \in S_p'$ is of the form

$$f = D^p[(x^2+1)^j \mu]$$

for some integer j and finite measure μ on $\mathbb{R}$ ([9] p. 109). Also $\psi(x)$ satisfies the inequality

$$|\psi^{(k)}(x)| \leq \frac{C_j}{(1+|x|)^j} \qquad k = 0,1, \cdots, r; \qquad j \geq 0,$$

since $\psi \in S_r$. Hence after integrating by parts p times we obtain

$$|b_{mn}| \leq \int_{-\infty}^{\infty} (1+x^2)^j |\psi_{mn}^{(P)}(x)| d|\mu|$$

$$\leq \int_{-\infty}^{\infty} (1+|x|)^{2j} \frac{C_{2j} \; 2^{m/2 + pm}}{(1+|2^m x-n|)^{2j}} d|\mu|$$

$$\leq (1+|n|2^{-m})^{2j} C_{2j} \; 2^{m/2 + pm} \int_{-\infty}^{\infty} d|\mu|. \tag{21}$$

The inequality (20) follows by taking $2j = p + \frac{1}{2}$.

We now combine the two lemmas to obtain the convergence we need.

Lemma 3. Let $f \in S_P'$, $\varphi \in S_r$, $p < r$, $b_{mn} = \langle f,\psi_{mn}\rangle$; then the series

$$\sum_{m=0}^{\infty} \sum_{n=-\infty}^{\infty} b_{mn} \; \psi_{mn}^{+}(z)$$

converges uniformly on the half plane Im $z \geq 1$.

That the inner series converges is immediate, but we have to be a little more careful with the outer series. By (19) we find that

$$|(2^m z-n)^{r+1} \; \psi_{mn}^{+}(z)| \leq \frac{1}{2\pi} \int \frac{|x^{r+1} \; \psi(x)| 2^{m/2} \, dx}{|x-2^m z+n|}$$

$$\leq \frac{1}{2\pi} 2^{-m/2} \int |x^{r+1}\psi(x)| dx \tag{22}$$

for Im $z \geq 1$. Hence the inner series satisfies

$$\sum_n |b_{mn} \; \psi_{mn}^{+}(z)| \leq C' \sum_n \frac{(|n|+2^m)^{P+\frac{1}{2}}}{|2^m z-n|^{r+1}} \, 2^{-m/2}$$

$$\leq C' \sum_n \left\{ \frac{(|n|+2^m)^2}{(n-2^m \text{Re } z)^2+2^{2m}} \right\}^{\frac{P}{2}+\frac{1}{4}} \quad 2^{-m/2} \left[(n-2^m \text{Re } z)^2+2^{2m} \right]^{\frac{P}{2}-\frac{r}{2}-\frac{1}{4}}$$

since $|2^m z-n|^2 = (n-2^m \text{Re } z)^2 + (2^m \text{Im } z)^2$.

In order to get a bound on the expression in parentheses we observe that

$$\frac{(a+1)^2}{(a-b)^2+1} \le (b+1)^2 + 1 \qquad \text{for } a,b > 0.$$

By taking $a = |n|2^{-m}$ and $b = \pm \operatorname{Re} z$ we obtain a bound which depends only on z. The series is thus dominated by the remaining factor

$$2^{-m/2} \sum_{n=-\infty}^{\infty} [(n-2^m \operatorname{Re} z)^2 + 2^{2m}]^{p/2 - r/2 - 1/4}$$

$$\le 2^{-m/2} C'' \sum_{n=-\infty}^{\infty} (n^2+2^{2m})^{-3/4} \le C''' \, 2^{-m/2}.$$

Thus the terms in the outer series are dominated by a geometric series and converge uniformly on the half plane.

We now return to the analytic representation. By taking the limit in (18) as $m \to \infty$, we find

$$f^+(z) = f_0^+(z) + \sum_{k=0}^{\infty} \sum_{n=\infty}^{\infty} b_{kn} \psi_{kn}^+(z) + g_\infty(z). \tag{23}$$

where $g_\infty(z) = \lim_{m\to\infty} g_m(z)$ is also an entire function. We can drop this g_∞ in (23) since an analytic representation plus an entire function is again an analytic representation.

Theorem. Let $f \in S_p'$, $\varphi \in S_r$, $p < r$, and let $b_{mn} = \langle f,\psi_{mn}\rangle$, $m = 0,1,2, \cdots$, $n = 0, \pm1, \pm2, \cdots$, be the wavelet coefficients of f. Then an analytic representation of f is given by

$$f^{\pm}(z) = f_0^{\pm}(z) + \sum_{m=0}^{\infty} \sum_{n=-\infty}^{\infty} b_{mn} \psi_{mn}^{\pm}(z) \tag{24}$$

where the series converges for $\operatorname{Im} z > 1(+)$ or $\operatorname{Im} z < -1$ (-), where $f_0^{\pm}(z)$ is an analytic representation of f_0, the projection of f on T_0, and $\psi_{mn}^+(z)$ is given by (3.3).

We have only worked out the convergence for f^+ but that for f^- is parallel. Also we have only shown the convergence for $|\operatorname{Im} z| \ge 1$, but we may continue the sum of the series analytically to the upper (lower) half plane.

Corollary 1. Let the conditions of the theorem be satisfied; then

$$f^{\pm}(z) = \sum_{m=-\infty}^{\infty} \sum_{n=-\infty}^{\infty} b_{mn} \psi_{mn}^{\pm}(z) \tag{25}$$

where the series converges uniformly on compact subsets of the half plane $\operatorname{Im} z > 1$ (or $\operatorname{Im} z < -1$).

The proof follows by extending (24) to

$$f^{\pm}(z) = f^{\pm}_{-M}(z) + \sum_{m=-M}^{\infty} \sum_{n=-\infty}^{\infty} b_{mn} \psi^{\pm}_{mn}(z)$$

which involves changing only a finite number of times. Since $f_{-M} \to 0$ as $M \to \infty$, it follows that $f^{\pm}_{-M}(z) \to 0$ uniformly in compact subsets of the appropriate half plane.

Corollary 2. Let the conditions of the theorem be satisfied; let

$$w(x,y) = \sum_{m=-\infty}^{\infty} \sum_{n=-\infty}^{\infty} b_{mn}[\psi^{+}_{mn}(x+iy) - \psi^{-}_{mn}(x-iy)];$$

then $w(x,y)$ is a harmonic function in the upper half plane whose boundary value satisfies

$$\lim_{y\to 0^{+}} w(x,y) = f(x).$$

in S'_r.

This is merely a restatement of Corollary 1. If $f(x)$ is an ordinary function in $L^2(\mathbb{R})$, then this limit may be taken in the sense of L^2 or even pointwise where f is continuous [7]. Similar results should be possible for distributions but have not as yet been demonstrated.

Example. Let $f(t) = \sum_n d_n \delta(t-n)$ where $d_n \in \ell^2$. Then $f \in S'_r$ for any $r \geq 0$ and its projection onto T_m is

$$f_m(t) = \sum_n d_n q_m(t,n).$$

The analytic representation of f gives us f^+ as

$$f^+(z) = \frac{-1}{2\pi i} \sum_n d_n (z - n)^{-1}.$$

Since $q^+(z,n) = \sum_j \phi^+(z - j)\phi(n - j)$, it follows that in this case,

$$f^+_0(z) = \sum_n d_n \sum_j \phi^+(z - j)\phi(n + j)$$

$$= \sum_n \phi^+(z - j)\left\{\sum_n d_n \phi(n - j)\right\},$$

and similarly for $f^+_m(z)$ by a change of scale. The coefficients b_{mn} are given by the convergent series

$$b_{mn} = \sum_k d_k\, 2^{m/2}\, \psi(2^m k-n).$$

REFERENCES

[1] H. Bremermann, "Distribution, Complex Variables, and Fourier Transforms," Addison-Wesley, Reading, MA (1965).

[2] I. Daubechies, Orthonormal bases of compactly supported wavelets, Comm. Pure Appl. Math. **41** 909-996 (1988).

[3] Y. Meyer, "Ondelettes et Opérateurs I," Hermann, Paris (1990).

[4] E. M. Stein and G. Weiss, "Fourier Analysis on Euclidean Spaces," Princeton Univ. Press, Princeton, NJ (1971).

[5] G. G. Walter, Local boundary behavior of harmonic functions, Ill. J. of Math. **16** 491-501 (1980).

[6] G. G. Walter, Hermite series as boundary values, Trans. AMS **218** 155-171 (1976).

[7] G. G. Walter, Analytic representations with wavelet expansions, to appear in Journal of Complex Variables (1992).

[8] G. G. Walter, Wavelets and generalized functions, in "Wavelets - A Tutorial in Theory and Applications" C. Chui, ed., Academic Press, New York (1992).

[9] A. H. Zemanian, "Distribution theory and Transform Analysis," Dover, New York (1965).

A LOWER ESTIMATE FOR THE NORM OF A PSEUDODIFFERENTIAL OPERATOR MODULO COMPACT OPERATORS(*)

S. Zaidman

Department of Mathematics
University of Montréal
Montréal, P. Q., Canada

INTRODUCTION

The present research is related to our previous paper [1]. In that paper a special class of "symbols" - complex-valued measurable functions defined on $\mathbb{R}^n \times \mathbb{R}^n/\{0\}$, $a(x, \xi) = a(\infty. \xi) + (2\pi)^{-n/2} \int_{\mathbb{R}^n} e^{i<x, \lambda>} \hat{a}'(\lambda, \xi) d\lambda$, and satisfying convenient inequalities is given, and operators $\mathcal{A}(x, D)$, mapping $L^2(\mathbb{R}^n)$ into itself are defined by the formula

$$\mathcal{A}(x, D)u = \mathfrak{F}^{-1}[a(\infty, \xi)\, \hat{u}(\xi) + (2\pi)^{-n/2} \int_{\mathbb{R}^n} \hat{a}'(\xi-\eta, \eta)\, \hat{u}(\eta) d\eta] \tag{0.1}$$

$\forall u \in L^2(\mathbb{R}^n)$, where $\hat{u}$ and $\mathfrak{F}^{-1}$ are direct and inverse Fourier-Plancherel transform.

We remember the inequality (1.10) in [1], which is a corollary of Theorem 4 there. It states that

$$\max_{x \in \mathbb{R}^n} \overline{\lim_{|\xi| \to \infty}} |a(x, \xi)| \leq \inf_{T \in \tau_c} \| \mathcal{A}(x, D) + T \|_{\mathcal{B}(L^2, L^2)} \tag{0.2}$$

where τ_c is the set of all linear compact operators, $L^2(\mathbb{R}^n) \to L^2(\mathbb{R}^n)$.

In the present paper we shall establish the inequality (0.2) for operators $\mathcal{A}(x, D)$ corresponding to a different class of symbols and using a different method.

(*) This research was effectuated while in sabbatical leave at the Banaras Hindu University, Varanasi, India

Generalized Functions and Their Applications, Edited by
R.S. Pathak, Plenum Press, New York, 1993

1. We remember definitions and properties of operators $P_{(x_o, \xi_o)}$ and $Q_{(x_o, \xi_o)}$ associated to a point $(x_o, \xi_o) \in \mathbb{R}^n \times \mathbb{R}^n/\{0\}$ (as given in [2]):

$$(P_{(x_o, \xi_o)}u)(x) = |\xi_o|^{n/4}\, u(|\xi_o|^{1/2}(x-x_o))e^{i<x, \xi_o>} \tag{1.1}$$

$$(Q_{(x_o, \xi_o)}v)(x) = |\xi_o|^{-n/4}\, v(x_o + |\xi_o|^{-1/2}x)e^{-i<x_o + |\xi_o|^{-1/2}x, \xi_o>} \tag{1.2}$$

for $u \in \mathcal{S}(\mathbb{R}^n)$, $u \in C_b(\mathbb{R}^n)$ or $u \in L^2(\mathbb{R}^n)$, v is in the same spaces, (spaces of $C^\infty(\mathbb{R}^n)$ rapidly decreasing functions, space of continuous bounded functions in $\mathbb{R}^n$, space of Lebesgue's square integrable functions; they are all mapped in themselves by any of these two operators).

One sees that PQ = QP = Identity, and that the Fourier transformation $(P_{(x_o, \xi_o)}\, u)^\wedge(\xi) = |\xi_o|^{-n/4}e^{-i<x_o, \xi-\xi_o>}\, \hat{u}\left(\dfrac{\xi - \xi_o}{\sqrt{|\xi_o|}}\right)$ (for, say, $u \in \mathcal{S}(\mathbb{R}^n)$).

Next, let us consider symbols $a(x, \xi)$, $\mathbb{R}^n \times \mathbb{R}^n/\{0\} \to \mathbb{C}$ with following properties:

1° $a(x, \xi)$ is measurable and $a(\infty, \xi) = \lim\limits_{|x| \to \infty} a(x, \xi)$ is bounded on $\mathbb{R}^n/\{0\}$.

2° $a'(x, \xi) = a(x, \xi) - a(\infty, \xi)$ is expressed by a formula

$$a'(x, \xi) = (2\pi)^{-n/2} \int_{\mathbb{R}^n} e^{i<x, \lambda>}\, \hat{a}'(\lambda, \xi)d\lambda \tag{1.3}$$

where $\hat{a}'(\lambda, \xi)$ is measurable on $\mathbb{R}^n \times \mathbb{R}^n/\{0\}$, the partial function $\lambda \to \hat{a}'(\lambda, \xi)$ is measurable $\forall\, \xi \in \mathbb{R}^n/\{0\}$ and there is an estimate

$$|\hat{a}'(\lambda, \xi)| \le k(\lambda),\ (\lambda, \xi) \in \mathbb{R}^n \times \mathbb{R}^n/\{0\},\ \text{where } k(\lambda) \in L^1(\mathbb{R}^n) \tag{1.4}$$

To such symbols we associate the operator $\mathcal{A}(x, D)$ defined on $L^2(\mathbb{R}^n)$ by the above formula (0.1).

As seen in [2], an alternative representation formula for $\mathcal{A}(x, D)u$ holds, whenever $u \in \mathcal{S}(\mathbb{R}^n)$. Precisely, one has the equality

$$(\mathcal{A}(x, D)u)(x) = (2\pi)^{-n/2} \int_{\mathbb{R}^n} e^{i<x, \xi>}\, a(x, \xi)\, \hat{u}(\xi)d\xi,\ u \in \mathcal{S}(\mathbb{R}^n) \tag{1.5}$$

where the integral is absolutely convergent and maps $\mathcal{S}(\mathbb{R}^n)$ into $C_b(\mathbb{R}^n)$.

It follows immediately that

$$\mathcal{A}(x, D)(P_{(x_o, \xi_o)} u)(x)$$

$$= (2\pi)^{-n/2} \int_{\mathbb{R}^n} e^{i<x, \xi>} a(x, \xi)|\xi_o|^{-n/4} e^{-i<x_o, \xi - \xi_o>} \hat{u}\left(\frac{\xi - \xi_o}{\sqrt{|\xi_o|}}\right) d\xi, \; u \in \mathcal{S}, \tag{1.6}$$

and therefore

$$(Q_{x_o, \xi_o} \mathcal{A}(x, D)P_{(x_o, \xi_o)} u)(x)$$

$$= (2\pi)^{-n/2} \int_{\mathbb{R}^n} e^{i<x, \xi>} a(x_o + |\xi_o|^{-1/2}x, |\xi_o|^{1/2}\xi + \xi_o) \hat{u}(\xi)d\xi, \; \forall u \in \mathcal{S}, \tag{1.7}$$

which is a function in $C_b(\mathbb{R}^n)$.

Let us add another assumption on symbols, precisely

3° There exists a continuous function $\varphi(t)$, $[0, \infty[\to]0, \infty[$, such that $\sqrt{t}\, \varphi(t) \to 0$ as $t \to \infty$ and inequalities

$$|a(\infty, \xi) - a(\infty, \tau)| \le (1 + |\xi - \tau|)\varphi(|\xi|), \; \xi, \tau \in \mathbb{R}^n/\{0\} \tag{1.8}$$

$$|\hat{a}'(\lambda, \xi) - \hat{a}'(\lambda, \tau)| \le (1 + |\xi - \tau|)\varphi(|\xi|)\, k(\lambda), \; \xi, \tau \in \mathbb{R}^n/\{0\}, \; \lambda \in \mathbb{R}^n \tag{1.9}$$

hold, where

$$(1 + |\lambda|)k(\lambda) \in L^1(\mathbb{R}^n). \tag{1.10}$$

(Note that from 2° and 3° above one obtains an estimate

$$|a(x, \xi) - a(x, \tau)| \le c(1 + |\xi - \tau|)|)\varphi(|\xi|), \; x \in \mathbb{R}^n, \; \xi, \tau \in \mathbb{R}^n/\{0\}. \tag{1.11}$$

Then (Theorem 2 in [2]) we get the following result:

If $\lim_{p \to \infty} a(x_o, \xi_p) = c_o$ exists for some $x_o \in \mathbb{R}^n$ and some sequence $(\xi_p)_1^\infty$, $|\xi_p| \to \infty$, it follows that

$$\lim_{p \to \infty} (Q_{x_o, \xi_p} \mathcal{A}(x, D)P_{x_o, \xi_p} u)(x) = c_o\, u(x) \tag{1.12}$$

holds, $\forall u \in \mathcal{S}(\mathbb{R}^n)$.

Let us denote by $v_p(x)$ the function $((Q_{(x_o, \xi_p)}\mathcal{A}(x, D)P_{(x_o, \xi_p)} u)(x)$, $p = 1, 2, \dots$ where we take $u \in C_o^\infty(\mathbb{R}^n)$.

Using formula (1.7) above we obtain that

$$|v_p(x)| \le (2\pi)^{-n/2} \sup_{\mathbb{R}^n \times \mathbb{R}^n/\{0\}} |a(x, \xi)| \int_{\mathbb{R}^n} |\widehat{u}(\xi)| d\xi. \ \forall\, x \in \mathbb{R}^n.$$

Using now (1.12) we obtain that

$$\lim_{p \to \infty} |v_p(x)|^2 = |c_o|^2 |u(x)|^2$$

and applying the "Fatou's Lemma" we have the inequality

$$\int_{\mathbb{R}^n} |c_o|^2 |u(x)|^2 dx = |c_o|^2 \|u\|_{L^2}^2 \le \liminf_{p \to \infty} \int_{\mathbb{R}^n} |v_p(x)|^2 dx. \tag{1.13}$$

We now use the immediate relations

$$\|P_{x_o, \xi_p} u\|_{L^2} = \|Q_{x_o, \xi_p} u\|_{L^2} = \|u\|_{L^2}, \ u \in L^2(\mathbb{R}^n) \tag{1.14}$$

and it follows that

$$\int_{\mathbb{R}^n} |v_p(x)|^2 dx = \|\mathcal{A}(x, D)\, P_{x_o, \xi_p} u\|_{L^2}^2$$

and

$$|c_o|^2 \|u\|_{L^2}^2 \le \liminf_{p \to \infty} \|\mathcal{A}(x, D)\, P_{x_o, \xi_p} u\|_{L^2}^2. \tag{1.15}$$

Let us take now any linear compact operator T, $L^2 \to L^2$ and write the obvious estimates

$$\|\mathcal{A}(x, D)\, P_{x_o, \xi_p} u\|_{L^2}^2 \le (\|\mathcal{A}(x, D) + T\|_{\mathcal{B}(L^2)} \cdot \|u\|_{L^2}$$

$$+ \|T\, P_{x_o, \xi_p} u\|_{L^2})^2, \ p = 1, 2, \dots \tag{1.16}$$

Next, it is well-known (and easy to see) that, given $u \in C_o^\infty(\mathbb{R}^n)$, the sequence $u_p(x) = (P_{x_o, \xi_p} u)(x)$, $p = 1, 2, \dots$ is weakly convergent to 0 in the space $L^2(\mathbb{R}^n)$.

Therefore we get that $\lim_{p \to \infty} \|T\, u_p\|_{L^2} = 0$ and then (1.16) implies that $\liminf_{p \to \infty} \|\mathcal{A}(x, D)u_p\|_{L^2}^2 \le \|\mathcal{A}(x, D) + T\|_{\mathcal{B}(L^2)}^2 \cdot \|u\|_{L^2}^2$ and then $|c_o|^2 \|u\|_{L^2}^2 \le$

$\| \mathcal{A}(x, D)+T\|^2_{\mathcal{L}(L^2)} \|u\|^2_{L^2}$ and for $u \not\equiv 0$ we get

$$|c_o| \leq \| \mathcal{A}(x, D) + T\|_{\mathcal{L}(L^2)}, \quad T \in \tau_c \tag{1.17}$$

Let us define now

$$B(x) = \limsup_{|\xi| \to \infty} |a(x, \xi)|, \quad x \in \mathbb{R}^n.$$

If we fix any $x_o \in \mathbb{R}^n$ we can find a sequence $(\xi_p)_1^\infty \subset \mathbb{R}^n/\{0\}$, $|\xi_p| \to \infty$ such that

$$|a(x_o, \xi_p)| \to B(x_o), \quad p \to \infty.$$

Extracting a subsequence $(\xi_{p_j})_1^\infty$ we will have $\lim_{j \to \infty} a(x_o, \xi_{p_j}) = c_o$ where $|c_o| = B(x_o)$. Using (1.17) we get $|c_o| = B(x_o) \leq \| \mathcal{A}(x, D) + T\|_{\mathcal{L}(L^2)}, \forall T \in \tau_c$ and this gives (0.2) without difficulty.

REFERENCES

[1] S. Zaidman: On some simple estimates for pseudodifferential operators, J.M.A.A., vol. 39, No. 1, 1972, p. 202-207.

[2] S. Zaidman: Some inequalities for Kohn-Nirenberg pseudo-differential operators, Ann. Sc. Math. Québec, Vol. II, No. 1, 1978, p. 143-165.

CONTRIBUTORS

BRACKX,F., University of Gent, Department of Mathematical Analysis,Galglann 2,B-9000 Gent, Belgium

CARMICHAEL,R.D.,Department of Mathematics and Computer Science, Wake Forest Univertsity, Winston-Salem, N.C.27109, U.S.A.

CHAUDHRY,M.A.,Department of Mathematical Sciences, King Fahd University of Petroleum and Minerals, Dhahran 31261, Saudi Arabia

CHOUQUET-BRUHAT, Y., Universite Paris VI, Mecanique Relativiste, Tour 66-3^{e} etage, 4 Place Jussieu, 75252 Paris Cedex 05,France.

CONSTANTINESCU, F., Fachbereich Mathematik, Johann Wolfgang Geothe Universitat Frankfurt, Robert-Mayer-str.10, 6000 Frankfurt am Main, West Germany.

DEEBA, E.Y., Department of Applied Mathematical Sciences, University of Houston-Dountown, Houston, Texas 77002, U.S.A.

DE GRAAF,J.,Eindhoven University of Technology, Department of Mathematics and Computing Science, P.O.Box 513, 5600 MB Eindhoven, The Netherlands.

DELANGHE,R., University of Gent, Department of Mathematical Analysis Galglann 2, B-900 Gent, Belgium.

FUJITA,K,.Faculty of Science and Technology, Sophia University, Chiyoda-Ku, Tokyo 102, Japan.

HASUMI, M.,Department of Mathematics, Ibaraki University, Mito, Ibaraki 310, Japan.

HOSKINS,R.F., Department of Applied Computing and Mathematics, Cranfield Institute of Technology Cranfield, England.

KANWAL,R.P., Pennsylvania state University, University Park, Pennsylvania, 16802, U.S.A.

KOH,E.L.,Department of Mathematics and Statistics,University of Regina,Regina, Canada S4S OA2.

KOMATSU, H.,Department of Mathematics, Faculty of Science, University of Tokyo, Tokyo 113, Japan.

MILLER,J.B.,Dapartment of Mathemetics, Monash University, Clayton, Victoria 3168, Australia.

MIKUSINSKI,P.,Department of Mathematics, University of Central Florida, Orlando, Fl 32816, U.S.A.

MORIMOTO,M.,Faculty of Science and Technology, Sophia University, Chiyoda-ku, Tokyo 102, Japan.

PANDEY,G.S., School of Studies in Mathematics, Vikram University, Ujjain 456010, India.

PANDEY,J.N., Depart m e nt of Mathematics and Statistics, Carleton University, Ottawa, Ont., Canada K1S 5B6.

PATHAK,R.S.,Department of Mathematics, Banaras Hindu University, Varanasi 221005, India.

PILIPOVIC′,S., Institute of Mathematics, University of Novi Sad, 21000 Novi Sad, Yugoslavia.

PINTO,J.SOUSA. Departmento de Matemática, Universidade de Aveiro, 3800 Aveiro, Portugal.

SAITOH,S., Department of Mathematics, Faculty of Engineering, Gunma University, Kiryu 376, Japan.

SCHECHTER,M., Department of Mathematics, University of California,Irvine, California 92717, U.S.A.

SCHMEELK,J., Department of Mathematical Sciences, Virginia Commonwealth University, 1015 West Main Street, Richmond, Virginia 23248-2014, U.S.A.

SINGH,O.P., Department of Applied Mathematics, Institute of Technology, Banaras Hindu University, Varanasi 221005, India.

SITARAMAYYA,M., School of Mathematics, Univertsity of Hyderabad, Hyderabad 500134, India.

SOMMEN,F., University of Gent, Department of Mathematical Analysis, Galglaan 2, B-900 Gent, Belgium.

STANKOVIC′,B., Institute of Mathematics, University of Novisad, 21000 Novi Sad, Yugoslavia.

TIWARI,A.K., Department of Mathematics, Regional College of Engineering, Nagpur 440011, India.

VALDIVIA,M., Facultad de Matematicas, Dr. Moliner, 50, 46100 Burjasot, Valencia, Spain.

VANACKER,N.,University of Gent, Dapartment of Mathematical Analysis Galglaan 2, B-900 Gent, Belgium.

WADA,R., Faculty of Integrated Arts and Sciences, Hiroshima University, Hiroshima City, Hiroshima 730, Japan.

WALTER,G.G., Department of Mathematical Sciences, University of Wisconsin-Milwaukee, P.O.Box 413, Milwaukee, WI 53201,U.S.A.

ZAIDMAN,S., Department of Mathematics, University of Montreal, Montreal, P.Q.,Canada.

ZAYED,A.I., Department of Mathematics, University of Central Florida, Orlando, FL 32816, U.S.A.

INDEX